LEGENDS IN THE ROCKS

Legends

Early Tertiary limestone dipping steeply seawards at Haumuri Bluff, on the Marlborough coast. (Photo: B. W. Collins)

in the Rocks

AN OUTLINE OF NEW ZEALAND GEOLOGY

Maxwell Gage

WHITCOULLS PUBLISHERS

CHRISTCHURCH SYDNEY LONDON

PHOTOGRAPH LOCATIONS

Front cover: Tahunanui, Nelson

Back cover: Mount Tasman from the west

Page 13: Griffiths Stream and Mount Findlay,
Wilberforce Valley, Canterbury

First published 1980
© 1980 Maxwell Gage

WHITCOULLS PUBLISHERS
Christchurch, New Zealand

ISBN 0 7233 0602 8

Printed in New Zealand by Whitcoulls Limited

For Molly Rose

CONTENTS

page

FOREWORD—'Geologists' Paradise'   15

1   INTRODUCTION—Some Basic Principles   18
Main Branches of the Subject   19
Earth Materials   19
Minerals, Rocks and Soils   20
Three Major Groups of Rock-types   22
Rock in its Natural Context   23
Weathering, Erosion and the Geological Cycle   25
The Earth's Restless Crust   30
Mountains and Orogenies   31
The Time-scale of Earth History   36
Fossils and the Geological Dating of Rocks   37
The Time-span of New Zealand Rocks   41

2   THE OUTLINE OF NEW ZEALAND—Present and Past   44
Surface Relief of the Earth   44
Defining New Zealand Geologically   45
How Long has New Zealand had its Present Shape?   49
What was its Shape in the Past?   49
How Old is New Zealand?   56
Our Place in the Pacific   56

3   OUR MOST ANCIENT ROCKS—'No nehe noa atu'   60
The Age of the Oldest New Zealand Rocks   61
Vanished Ancient Terrains   65
Paleozoic Complex of North-west Nelson   69
Haupiri Group   72
'Aorere'—A Stratigraphic Shambles   74
Golden Bay Group   76
Mount Arthur Group   77
Early Paleozoic Rocks Farther South   79
Greenland Group—Paleozoic or Older?   81
Do We Have Silurian Rocks in New Zealand?   84
The Devonian—Further Puzzles   84
End of the First Episode—Tuhua Orogeny   86

4   NEW ZEALAND IN MIDDLE AGE—Geologically Speaking   88
The 'Greywacke'   88
The New Zealand Geosyncline   90
Double-handling of the Sediment   95
Time-span of the New Zealand Geosyncline   96
Whence Came the Sediment?   98
Volcanoes Near and Beneath the Sea   99
The 'Ultramafics'—Dunite and Serpentine   100

The Gross Thickness—What Would it Mean? 102
Carboniferous and Permian Sedimentary Rocks 103
Stages and Series; Formations, Groups and Systems 106
Time-stratigraphic Classification of the New Zealand
Permian Strata 107
Triassic and Jurassic Rocks 108
'Portmanteau' Names—Hokonui, Torlesse, Murihiku and Others 109
Haast Schist—Metamorphic Spine of the South Island 113
Rangitata Orogeny—Final Phase of the New Zealand Geosyncline 113

5   **METAMORPHISM AND METASOMATISM—**
Rock in the Cooking Pot 116
The Environmental Approach 116
The Geothermal Gradient 119
Mineral and Rock Transformations 120
Progressive Metamorphism 120
Repeated Metamorphism and Rock Folding 123
The Mapping of Metamorphic Rocks in New Zealand 123
Contrasting Metamorphic Realms 124
Dates of Metamorphism 126
Regional Metamorphism on Parade 128

6   **GRANITE—Rock of Ages?** 132
Geologically Unfortunate Metaphors 132
Granites and Granites 132
Essentially South Island Rocks 136
The Tuhua Granites (Tuhua Intrusive Group) 137
Is 'Tuhua Group' still a Valid Unit? 141
Tasman Intrusives—The 'Boulder Bank' Rock 144
Granites in the Far South 145
Layered Granites, Granite-gneisses and Gneissic Granites 148
Pegmatite—Prime Source of Mica and Gemstones 152
Dark-coloured Granitic Rocks 156
Granite, Metasomatism and Mineralised Veins 160
Ages of New Zealand Granites 161
Postscript—Granite Study Areas 163

7   **THE LATEST CYCLE—Beginning and Ending with**
Mountains 164
Bold Mid-Cretaceous Landscapes 164
Mountains Worn Down 167
The Late Cretaceous Peneplain 169
Rubbish on the Peneplain—The Quartz Conglomerates 171
Major Fuel Source—the Quartzose Coal Measures 173
Going Under the Sea Again 177
Lignite and Bituminous Coal—Why the Difference? 183
Clear Seas, Shoals and Volcanoes 184
Our Chief Source of Lime 185
Re-emergence—the 'Kaikoura' Uplift Begins 186
Younger Limestone Horizons 188

The Younger Coals 189
Great Conglomerate Formations and the 'Kaikoura' Climax 189
'Kaikoura Orogeny'—How does it Stand after Sixty Years? 192
Classification and Naming of the Younger Strata 193
The Beginning of the Pleistocene Period 194
Locating the Boundary in New Zealand 194
Deposits Younger than the Kaikoura Orogeny 195
Cycle Completed—the 'Notocene' 196

8   NEW ZEALAND VOLCANOES—Ancient and Modern 198
Our Long Volcanic Record 198
What Constitutes a Volcano? 198
Types of Eruption Reviewed 199
How Does Magma Arise—and Rise? 201
Hot Springs, Geysers, 'Blowholes' and Silica Terraces 203
Volcanoes under Water 204
The Evidence for Ancient Volcanoes 204
Geosynclinal and Orogenic Volcanoes 206
Our Oldest Known Volcanoes 207
New Zealand Geosyncline Volcanic Rocks 207
Igneous Activity during the Rangitata Orogeny 209
Mid-Cretaceous Volcanoes 210
More Submarine Volcanoes in the South Island 212
Later Tertiary Land-based Volcanoes in the South Island 214
North Island Mid-Tertiary Eruptions 216
Pleistocene Volcanic Trends 219
The Auckland Basalt Eruptions 220
Ignimbrite—Product of Hot Sand Flows 221
Volcanic Mudflows—'Lahars' 224
Pumice and Tephra Eruptions 225
The Tongariro Volcanoes 226
Tarawera, 1886 229
White Island Volcano 233

9   GLACIATION—The Ice Ages in New Zealand 234
Land of Ice and Fire 234
What is an Ice Age? 234
The Great Glacial Controversy 235
Ice Ages in the Remote Past 236
How Many Ice Advances? 238
Swinging Sea Levels 240
Glaciations, Interglacials, Stadials, etc. 241
The 'Glacial Period' in New Zealand 242
The Ross Glaciation—Our Earliest 242
Ice Cap or Valley Glaciers? 243
The Number of Glaciations 244
The Warmer Episodes 244
Dividing up the New Zealand Pleistocene—and Some Problems 246
Terminology 251
Dating the Glacial Events 252

Some Valley Glaciation Sequences in the South Island   253
Glaciation in the North Island   259
Non-glacial Cold-climate Effects   259
When did the Pleistocene Ice Age End?   262
The 'Little Ice Age'   263

10   THE FRAMEWORK—Structural Outline and Crustal Setting   266
A Pattern with a Meaning?   266
Thickness of the Crust in the New Zealand Region   267
What Determined the Pattern?   268
Changing Ideas about New Zealand Structure   270
The New Zealand Orogen—Long-lived Crustal Mobility   274
Structural Divisions of New Zealand   278
Divisions Determined by Rangitata or Earlier Events   279
Divisions Determined by Post-Rangitata Events   280
Fault Lines on the Maps—What they Can Mean   284
Slices of New Zealand—the Great Strike-slip Faults   285
The Alpine Fault   287
Structure of the New Zealand Alps   292
Structural Plan of the South-west Pacific   294

11   SHAPING THE LANDSCAPE—New Zealand Geomorphology   296
The Genealogy of Landscapes   296
Pushed up or Carved out?   297
Different Ways of Describing Landforms   298
Landscape-modelling Processes   298
Crustal Uplift and Landscape Evolution   299
Predominantly Stream-modelled Landscapes   301
Influence of Rock Structure—the Bones Showing Through   301
Crustal Movements and the River Pattern   305
Rivers that Flow through Mountain Ranges   306
River Plains and Shingle Fans   314
Intervention of Cold Climate   317
Where the Glaciers have been   317
Cold Climate without Glaciers   322
Terraces   323
Landscapes on the Move   325
Lakes   327
The Form of the Coastline   330
Sandspits and Bars   331
Sandhill Landscapes   334
Straight and Crooked Coasts   335
Uplifted Shorelines   338
Volcanoes as Landforms   339
How 'Old' is the Landscape?   340

12   STIRRINGS OF THE CRUST—Earthquakes and Crustal Warping   342
Little by Little . . . ?   342

Language of Seismology   344
The Causes of Earthquakes   346
Earthquakes and Volcanoes—the True Relationship   348
Earthquakes and Plate Tectonics   348
Earthquake Hazard in New Zealand   349
Historic Earthquakes in New Zealand   351
Prehistoric Earthquakes   354
Recent Crustal Movements   356

13   **GEOLOGY SERVES NEW ZEALAND—**
The Economic Incentive   358
The Purpose of Geology   358
Early Geological Explorations in New Zealand   359
Older and Newer Geological Surveys—a Brief Historical Outline   360
The Search for Gold   362
How Gold Occurs   363
Geological Assistance in Gold-mining   364
The Search for Coal   365
Grades of Coal   367
Geological Settings of New Zealand Coals   368
The Search for Oil and Gas   370
Iron Ores   374
Other Base Metals   376
Radioactive Minerals   379
Non-metallic Earth Products   379
The Search for Water   382
Geology and Civil Engineering   384
Energy from the Interior   386
Geology and the Environment   387
Geophysical Exploration   390

Appendices
1 Some Basic Geologic Structures   392
2 Some Characteristic New Zealand Fossils   396
3 Notes on the Geology of the Outlying Islands   399
4 Sources of Information about New Zealand Geology   402
5 Glossary of Rock and Mineral Terms   405

Index   413

COLOUR PLATES

	page
Distribution of the Major Rock Groups	34
1. Gravitational downslope movement of rock debris released by erosion	38
2. Schist shaped and smoothed by glacier ice, Franz Josef Glacier	38
3. Recently aggraded bed of Waitangiaona River, Westland	39
4. Greville Sandstone, typical Permian rock in eastern Nelson	42
5. Fossil ripple-markings in Permian strata, Wairoa River, Nelson	46
6. Fault-crushed Torlesse sediments, mainly volcanic tuff, inland Marlborough	46
7. Waimakariri Gorge is cut into Torlesse Supergroup strata	47
8. Lake Wakatipu scenery, ice-modelled in schist	50
9. Jointing in Haast Schist, Waiho Valley, Westland	54
10. Separation Point Granite, source of sparkling golden sand on Nelson beaches	54
11. Concretions in Waipara River, Canterbury; some contained reptilian fossils	55
12. Typical aspect of marine Tertiary beds in New Zealand, Tarakohe, Nelson	55
13. Vertical strata of Longford Formation (Miocene) near Murchison	58
14. Recent scoria deposits, originally deposited at this steep angle	58
15. Deposits from Tertiary sea-floor eruptions, subsequently tilted	59
16. Lava flow on Ngauruhoe, 1954	62
17. Coastal benches formed during Pleistocene interglacials, Westland	130
18. Deposits of a Pleistocene ice-margin lake, Westland	134
19. Complex minor folds in limestone near Kaikoura	134
20. Pleistocene frost-broken debris layers on roadside near Porters Pass	134
21. A sequence of cold-climate gravel deposits, Kowai River, Canterbury	135
22. Reverse fault with a record of intermittent movement, Hanmer River	135
23. Superposed gorge of Opihi River, South Canterbury	138
24. Ice-sculptured granite landscape, Victoria Range, Nelson	139
25. Arête ridge, cirques, moraines, Mount Cook and Hooker Glacier	142
26. Slopes modelled by solifluction near Burkes Pass	146
27. Terraces marking halts in downcutting by Awatere River, Marlborough	150
28. Two types of terrace in the Poulter Valley, Canterbury	150
29. Slump and earthflow features, Motunau, Canterbury	154
30. Okarito Lagoon and gravel spit, South Westland	158
31. Lake Pearson, impounded by alluvial fans in a glaciated valley	158
32. Lyttelton Harbour, erosional hollow in Pliocene volcano	159

Hewn and hollowed stones, there will be legends
In the rocks and rocky hills . . .

Ruth Dallas, *Country Road*

FOREWORD

'Geologists' Paradise'

The main aim of this book is to satisfy the curiosity of many New Zealanders about the rocks and geological features of their country. It is not intended either as a formal class book for students or as a compendium of research results for professional earth scientists, though I hope portions of it will be found readable and useful by these groups as well. The chief target has been what are believed to be the requirements of the general reader.

The subtitle above was inspired by the remarks of Dr James Mackintosh Bell, Canadian-born geologist and one-time Director of the New Zealand Geological Survey, in his preface to what proved highly successful as the first popular outline of New Zealand geology. This was *Geology of New Zealand* by Patrick Marshall, which appeared in 1912. It is probably a compliment to the late Dr Marshall that there has been no further attempt since to produce a comprehensive account for a general readership. Perhaps the best and fullest summary in recent years is that contained in Volume One of *An Encyclopedia of New Zealand* (Government Printer, Wellington, 1966; pages 769–804).

At the time when Bell wrote his preface for Marshall's book he had been living in New Zealand only for about six years. His enchantment with our great diversity of rocks and scenery, different in so many ways from those of his native Canada, is given expression in the poetic language in which he commented on the range of geological features, rocks, minerals and landscape types that may be seen in every part of these islands.

Within the first few decades of European settlement several major earthquakes and one very destructive volcanic eruption had drawn everyone's attention rather forcibly to the more spectacular forms of geological activity. These events, combined with the abundance of other interesting if less dramatic geological features in New Zealand may explain how it came about that men born and reared in this country achieved prominence in the geological world before the end of last century. New Zealand-trained geologists and mining engineers have always been strong contenders for professional jobs overseas, and a number have achieved notable careers. They may well have enjoyed a natural advantage from the beginning.

As stated above, this book is not written with textbook requirements in mind, yet if it is to succeed in making New Zealand geology accessible and interesting to that large majority of people who have had no opportunity to study the subject at school or university, a few essential principles must first be looked at. The most important of these are presented in a very condensed form in the introductory chapter, while others are explained as they arise in subsequent chapters. The meanings of many common geological terms, rock names, etc. are given briefly in the Appendices. With this help, it is hoped that readers will be able to follow and understand the remainder of the book. Those who have already been introduced to the subject may not find it necessary to read Chapter One.

A textbook or treatise should attempt to deal evenly with all aspects of its subject, but such a balance is not expected of a book of this type, and I do not pretend that the coverage is either complete or even. More space has been allowed for topics I think will be of wider interest, while others are minimised. I have assumed that many readers will seek further information about areas of particular interest to them from the published maps and regional Geological Survey *Bulletins*, and especially from the series of twenty-seven sheets which make up the 1:250,000-scale *Geological Map of New Zealand* (see Appendix 4).

Some readers may be surprised to discover that this book deals only incidentally with fossils. This very fact may help to dispel any misapprehension that geology is concerned with nothing else. I am well aware of a popular fascination with fossils, and know that many specimens are gathered by casual collectors as well as by those for whom paleontology is a serious hobby or a profession. Fossils play two vital roles in historical geology. First, because of the evolutionary parade of changing life through the geological past, they provide one of the main sources of guidance as to the ages of sedimentary rocks that contain them. Then the kinds of organisms represented give clues about the environmental conditions under which the containing rocks were formed. However, to present a minimum, useful amount of information about the fossils found in New Zealand would have meant either a considerable expansion of the book or a severe curtailment of the geological story I wished to tell. Neither alternative was attractive, and in any case I am not particularly well fitted to undertake the task. It was a pleasure and a relief, therefore, to find that a book about New Zealand fossils, aimed at satisfying the interests of a similar audience, is being planned at the present time. Meanwhile, to give readers at least some impression of what some of the fossil groups mentioned herein actually look like, and to show which groups are important in different geological periods, I have arranged to have reproduced in Appendix 2 a diagrammatic 'parade of New Zealand fossils' from a popular article by Sir Charles Fleming, to whom my particular thanks are due for this favour.

Geology has recently been passing through an exciting phase, stimulated by the results of new research into the question of what underlies the oceans. When it is recalled that the rocks of the crust over four-fifths of the earth's surface are hidden from direct observation by ocean waters, the importance of this new development can be appreciated. One result has been that some parts at least of many textbooks and treatises published more than a few years ago are outdated. New editions and new books are appearing continually, but this does not mean that reasonably modern editions of the numerous, standard introductory texts and reference books are worthless. The greater part of 'descriptive' geology has changed little in substance as far as the layman is concerned, although the increasingly detailed knowledge yielded by new, more sophisticated study methods now needs to be expressed for the scientist in more precise language.

The information gathered for this book came from many sources, only a very small portion of it from the author's own original research. The greater part came from published maps and *Bulletins* of the New Zealand Geological Survey and from papers that have appeared in the *New Zealand Journal of Geology and Geophysics* and in the *Transactions* and *Journal* of the Royal Society of New Zealand. In addition I was fortunate in being able to draw on the goodwill of my geological colleagues as regards unpublished information and advice. It is a pleasure to acknowledge valuable comments and criticism from Professor J. D. Campbell, Sir Charles Fleming and Dr W. A. Watters, who read drafts of some chapters. I am much indebted to the Director of the Geological Survey, Dr R. P. Suggate, for allowing me to examine proofs of *Geology of New Zealand*, but it must be understood that this book is in no way a summary of that work.

At appropriate places in the following pages I have acknowledged with thanks the assistance of individuals and institutions who have made material available for reproduction in this book. I am grateful to Lee Leonard of Christchurch for her excellent preparation of many diagrams and maps; to Mr Lloyd Homer of the N. Z. Geological Survey, Lower Hutt, for his help in obtaining many photographs from Survey sources; and Mrs Avice Black of Napier for retyping part of the manuscript. It is a pleasure also to acknowledge the whole-hearted co-operation of the editorial staff of Whitcoulls Publishers at every stage.

In conclusion, though this book is essentially an attempt to digest the results of New Zealand geologists' work known to me at the time of writing, it is also an individual view of New Zealand geology, subject inevitably to some personal bias. With the help acknowleged above, I trust it will be found factually correct as far as it goes, but it would be too much to hope for universal agreement with all of the interpretations and comments expressed herein.

M.G.

Nelson,
March 1978

Chapter One

INTRODUCTION

Some Basic Principles

Geology is perhaps the least familiar of the basic sciences. Yet the greater part of it is a lot more straightforward than is popularly imagined. One can in fact discover and appreciate a surprising amount about the geology of any region with no more background knowledge than the few essential principles outlined in this chapter. With no more sophisticated equipment than a pocket knife, a good magnifying lens, a small bottle of dilute hydrochloric acid and an observant eye for detail, it is possible to distinguish many of the common rock and mineral varieties that make up the bulk of accessible parts of the earth's crust.

Having taught the subject both within the university and outside it over many years, I believe the aspects of the science most people find difficult to grasp at the outset are those which have to do with unfamiliar scales of magnitude, not only with regard to the mass and dimensions of earth features, but particularly with regard to *time*. To give a typical example of the time-scale difficulty, geologists and geographers are very interested in the physical changes going on at the present time on the earth's surface and below it, and many ingenious measuring procedures and imitating experiments have been devised to find out just how these processes are working and how rapidly. But when it comes to applying the results, looking at features in rocks of various ages which seem to indicate that similar processes were going on in the more or less remote geological past, a psychological difficulty arises. We have to try to picture familiar processes such as the wearing away of rock surfaces by erosion, or slow changes of level of the ground surface which even now are measurable in some parts of the world, but over periods of time which are thousands or millions of times longer than the duration of the field measurements or experiments. No surprise, then, if many people require a little help in becoming accustomed to thinking about the tempo of geological processes. Some effort is certainly needed to comprehend the different orders of magnitude of time involved, sometimes thousands, more often millions, even thousands of millions of years. Later in this introductory chapter ways will be suggested for getting over this difficulty and developing a true sense of the time-perspectives of geology.

It is also an unusual experience for most people to be asked to contemplate the true scale of magnitude of prominent, familiar, surface features like the Tararua Range or volcanic mountains like Ruapehu, or to think about the amount of energy involved when earthquake waves travel around the world. Some of these quantities, too, seem inconceivably vast—until they are considered relative to the mass, the bulk, and the energy resources of the earth as a whole. It then comes as a surprise to discover what truly puny things on the global scale the great mountain ranges and deep ocean trenches really are. I suspect, however, that the widely published photographs taken from artificial earth satellites have made many more people aware of this.

This matter of perspective is worth stressing because I think when one begins to study geology it is important *not* to get an exaggerated idea of the significance of some of the more dramatic geological events, like the eruption of many cubic kilometres of volcanic debris in a great blast that occurred at Taupo about 1850 years ago. Should a similar thing happen again, its effects upon all New Zealanders would be incalculable; it was a tremendous geological cataclysm for the central North Island. Yet on the scale of the earth as a whole it was quantitatively rather insignificant.

Main Branches of the Subject

Like all the other sciences, geology has become increasingly subdivided into more or less specialised sub-disciplines. The names of many of them will be strange to readers. Some no doubt will know that 'paleontology' means the study of fossils, but fewer will be familiar with 'stratigraphy', which deals with the order of succession of the layered rocks ('strata'—the singular is 'stratum') of the earth's outer crust. While these two divisions should, and usually do, go together, there are specialist paleontologists whose main interest is centred in describing, classifying and working out the history of particular groups of organisms found as fossils in rocks. 'Mineralogy' is self-explanatory, but 'petrology' has nothing much to do with petroleum; it concerns the origin of different kinds of rock. 'Petrography' also has to do with rocks, but the main emphasis is upon describing, classifying and naming the different kinds. 'Geomorphology', shared with geographers, studies the origins of the earth's surface features that provide the basis of scenery.

'Structural geology' deals with the internal structural features of rocks. It provides the terminology for describing different kinds of layering ('stratification') in sediments, the fractures and dislocations ('joints', 'faults' etc.) and folding that show where rocks have suffered various kinds of distortion by stresses within the crust since the rock was formed. 'Volcanology' is another word that explains itself; likewise 'sedimentology', perhaps. 'Geophysics' and 'geochemistry' declare themselves to be dedicated to applying the basic physical sciences to the study of rocks and geological phenomena generally, but they also have direct, practical applications in mineral exploration. Geophysics is very valuable for testing the suitability of sites for proposed civil engineering work. Besides the above, there are hosts of more specialised fields; more will be explained as we come to them.

Earth Materials

The raw materials of geology include not only the rocky substance that makes up the comparatively rigid body of the earth ('lithosphere') but also the world-encircling gaseous envelope ('atmosphere') and the

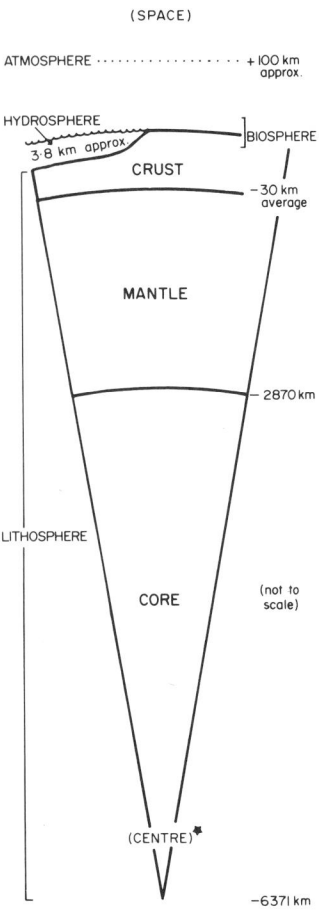

(SPACE)

ATMOSPHERE · · · · · · · · · · · · · · · · · +100 km
approx.

HYDROSPHERE
3·8 km approx.   BIOSPHERE

CRUST
—30 km
average

MANTLE

—2870 km

LITHOSPHERE

CORE   (not to
scale)

(CENTRE)

—6371 km

Fig. 1.1
Earth layers consist of concentric, nearly spherical shells within which a number of sub-shells are recognised. The upper layer of the mantle, beneath the relatively rigid crust is sometimes called the 'asthenosphere'. Note that the atmosphere is counted as an earth shell, its outer limit defined arbitrarily in terms of a minimum number of gas molecules in a given volume.

watery shell which now submerges about four-fifths of the lithosphere, and has, at different times in the geological past, been both more extensive and less extensive than now. The inner, less rigid 'mantle' layers beneath the rigid 'crust' and the central 'core' zones (which may be metallic) are all part of the total geological system. Geophysicists have established that concentric shells of different material exist within the earth, and they have been able to measure their thicknesses and other properties from the surface with remarkable precision. Although some of the boundaries between them are sharp and others transitional, it seems very probable that both material and energy have been exchanged between one shell and another throughout the earth's history.

What about exchanges between the earth and the rest of the solar system? This was a vital part of the earth's cosmic infancy, but for most geological questions we can afford to neglect the rest of the planets except our own satellite, which induces tidal effects. Otherwise, apart from so many processes depending upon the flow of energy from the sun, we can think of earth as essentially a self-contained system. It has been receiving its share of meteoritic space-junk, but this aspect has not been important in the later stages of earth history.

For the purposes of this book, at least, we will be able to neglect also the innermost earth shells, and be concerned mainly with the crust and outer layers of the mantle down to depths of around 100 kilometres.

Direct information about the chemical and physical make-up of accessible parts of the crust (which on average is about 33 kilometres thick) is augmented by indirect information about the upper sub-crustal layers. A mere eight of the known chemical elements make up ninety-nine percent of the mass of the crust, while silicon and oxygen together account for three-quarters of it. The metals useful to man are definitely *not* in the short list of greatest abundance, but for the most part are very sparsely distributed. Their concentration into economically workable amounts has resulted from some special geological happenings—which mainly accounts for the need to employ geologists to find mineral deposits and to help work them efficiently.

Minerals, Rocks and Soils

Only a few chemical elements occur naturally in the free, 'native' state. Besides oxygen, nitrogen and the inert gases of the atmosphere (helium, argon, etc.) accessible parts of the crust do contain free carbon, sulphur, iron, copper, gold and a few more, but except for carbon and gold these elements are rarely found in the free state.

Predominantly, the elements occur in rocks as more or less definite chemical compounds or as mixtures of compounds. Luckily, many of these compounds and mixtures have distinct forms and physical properties that enable them to be identified as belonging to particular mineral species. The composition of a mineral may be simple and virtually invariable as in the case of the crystalline mineral form of silica known as quartz; or the same chemical compound may occur as a different mineral species if formed under special physical conditions. For example, there is a high-temperature mineral version of silica called tridymite which has crystalline form and properties quite different from those of common quartz. Another example is calcium carbonate which occurs as the minerals calcite and aragonite, the latter less commonly in rocks, although it is the usual form of calcium carbonate in the shells of most marine animals. Minute amounts of 'impurities' or chemically absorbed additives can greatly alter the appearance of a mineral, especially its colour which, alas, is not often a safe guide in identifying minerals. Minute amounts of manganese or iron can convert colourless quartz into purple-tinted amethyst, or brown- and honey-coloured cairngorm.

At the other extreme, many minerals are composed of complex and variable mixtures, or what the physical chemist calls 'solid solutions' of compounds—silicates, oxides, hydroxides, etc.—themselves variable and subject to the insertion of stray substances into spaces within the crystalline structural framework. However, in order to meet the qualifications of mineral species, as demanded by mineralogists, such variations must be systematic and within defined limits.

Altogether, more than 2500 kinds of minerals have been accepted as distinct, valid mineral species. Many of them occur in two or more varieties distinguished by some minor difference, or by the way they occur, e.g. the iron oxide mineral hematite, the chief cause of red colorations in soils, which may occur in elegant, sparkling black crystals, or in dull red or brown massive forms.

The word 'mineral' tends to evoke thoughts of sparkling crystalline beauty or economic wealth, whereas on the whole minerals are rather mundane things, their function in the scheme of things being to make up the bulk of the material of earth. The high prices paid for those that occur handsomely as gemstones attest to their rarity, while the solemn fact about value to man is that only about two hundred species of mineral have economic importance.

Rocks of the crust and upper levels in the sub-crustal zone are known to be aggregates of grains or crystals of one or more mineral species. Crustal rocks may incorporate fragments of pre-existing rocks. The entire mantle shell may be composed of crystalline minerals, although there is some uncertainty about the physical character of materials under the enormous pressures of the deep earth zones. The total number of mineral species making up the bulk of crustal rocks is really quite small. To be reasonably familiar with the common rock-forming minerals, as they are called, one needs only to know about twenty-five mineral species. Most of these are mentioned in Appendix 5.

Quartz is the best known of all rock-forming minerals, and it is by far the most widely distributed in the rocks of land areas of the globe. Near land surfaces, where the rocks have been affected by decay, the 'clay minerals' (these are hydrous alumino-silicates and oxides of sodium, calcium, iron, etc., with often a multi-layered crystalline micro-structure) together with quartz form the mineral basis for most soils. Calcite is common too, especially as the cement between other mineral grains, and making up the greater part of limestone. Beneath the major ocean floors, however, quartz is only a rare constituent of the volcanic rock basement.

Three Major Groups of Rock-types

For well over a century the rock-types making up the accessible parts of the earth have been broadly classified into three major groups. It is a genetic classification, that is, it takes into account how the rocks are known or inferred to have originated. It still holds good, but the distinction between two of the groups is now less sharp, perhaps even non-existent when applied to rock that acquired its present character when situated many kilometres below the surface. The three groups are igneous, sedimentary and metamorphic rocks.

Igneous rocks formed from formerly molten material which solidified on cooling, either after being erupted by a volcano at the surface, or after being injected ('intruded') into pre-existing rock below the surface. Ninety-five percent of the upper 16km of the crust is estimated to consist of igneous rock. A widely known example of igneous rock is dense, dark-coloured basalt, familiar to Aucklanders as the 'blue-metal' used in road foundations. Basalt also makes up most of the extinct Banks Peninsula volcanoes and the worn-down stumps of ancient volcanoes that are a feature of east Otago landscapes.

Sedimentary rocks are accumulations of mineral grains, with or without particles of pre-existing rock or cementing material between the grains. They frequently contain the remains of traces of animals or plants. Familiar varieties of sedimentary rock include sandstone and mudstone (the suffix 'stone' generally implies that the rock is not a friable sediment, but is consolidated at least in some degree). The greywacke rock making up large parts of the main mountain ranges of both islands is a type of sandstone. Few people would not be able to recognise limestone, whether it be composed entirely of calcium cabonate as mineral grains, or as skeletal fragments of fossil organisms, or both. Coal too is a kind of sedimentary rock. Since the accumulation of sediment seldom progresses uniformly or continuously, but rather

Fig. 1.2
The 'strike' of layered rocks is the line of intersection between the layers and a horizontal plane. As the sea floods over this almost-level tidal rock platform at Tahunanui, Nelson, it becomes obvious that the strike of these steeply-dipping strata of Tertiary age swings round in a gentle curve.

intermittently and variably, sedimentary rocks tend to occur in layers; that is to say they are often 'stratified'.

Metamorphic rocks have been formed by the alteration of pre-existing rocks of almost any type by heat and pressure, essentially while in the solid state though in the presence perhaps of permeating, chemically-active fluids. An important difference between metamorphic rocks and the other two main groups is that the mode of origin of all metamorphics has to be inferred. Whereas we can actually see some igneous rocks in the process of forming at the surface—the cooling of lava to form solid volcanic rock—and we can observe how gravel and sand deposits can accumulate, a metamorphic origin can be determined only indirectly, by reasoning from features in the rock itself and inferences from what we can call the 'geological circumstances'.

Familiar kinds of metamorphic rock include slate, developed mainly by pressure from fine-grained sedimentary rock; marble, a product of the recrystallisation of limestone under pressure; and the hard, coarsely crystalline, sometimes layered gneiss (originally a German word; drop the 'g' and say 'nice') such as composes large parts of the Fiordland mountains and contributes to the special character of the West Coast near Charleston.

Another well known crystalline rock is granite. It will serve well to illustrate the link now recognised between the metamorphic rocks and igneous rocks formed at considerable depths below the surface. It has been shown that some rocks which from their mineral composition and texture should be called granite must have cooled and crystallised from a rock melt, very slowly, at depths of some kilometres while others equally entitled to be called granite can only have been transformed in place from pre-existing sedimentary rocks which had the same chemical composition. Granite formed in the first way is an igneous rock; granite formed in the second way is a metamorphic rock. No one seems to find it hard to accept this dual origin of granite nowadays, yet it was the subject of one of the most famous geological controversies. (See also Chapters Five and Six.)

Within the major rock groups thousands of varieties have been distinguished and more or less systematically defined and named. Many of the types, however, grade one into another, so the actual number of varieties is not finite but depends upon how finely distinctions are made. About twenty of the most commonly occurring types will be referred to repeatedly in this book; a brief note about each type mentioned will be found in Appendix 5.

Rock in its Natural Context

Attractive or unusual looking pieces of rock are often picked up on the beach or river bed just to be glanced at and thrown away, or perhaps to be taken for a collection or for lapidary work. I wonder how often it occurs to collectors that every piece was once attached to the solid rock somewhere, and was then still in continuity with the rest of the earth's crust? Just how long ago it was detached, from how far away, and what intervening adventures the piece may have had are likely to remain its own secret.

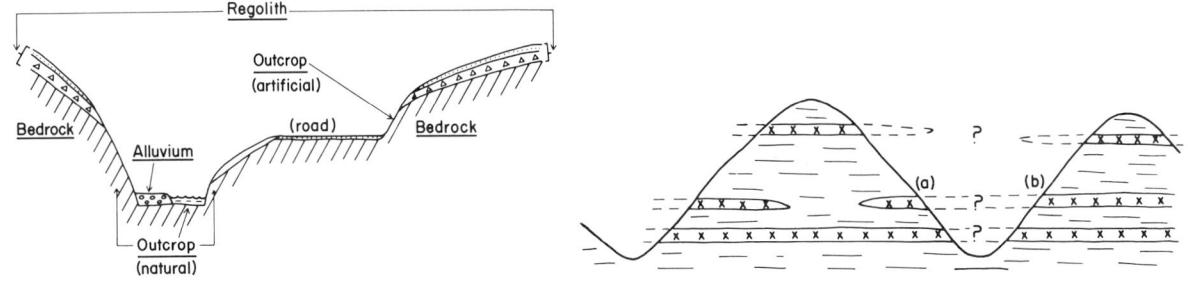

Fig. 1.3
So broad is the geological concept of 'rock' that the informal word 'bedrock' is needed to emphasise the distinction between the regolith (surface mantle of soil and weathered rock debris), or superficial alluvial deposits, and harder rock beneath. In other, non-geological contexts 'soil' includes unweathered soft rock (e.g., in parts of New Zealand, 'papa') as well as weakly compacted gravel and sand, all of which strictly count as 'rock' in geology. 'Outcrop' implies an absence of regolith.

Fig. 1.4 (right)
Strata have usually extended beyond the present outcrops where they are now exposed by erosion. Establishing that the layers now outcropping at (a) and at (b) are identical and were once joined up is an act of 'correlation'. This sketch could well represent a cross-section of two spurs and an intervening valley somewhere on the Christchurch side of the Port Hills, Canterbury, the prominent layers being basalt lava flows.

Those interested in the histories of ancient stone buildings sometimes like to trace the quarry or even identify a particular rock layer from which stone was extracted. It may not be feasible to determine how far the layer extends beyond what is exposed in the quarry face because it still lies below other rock layers and its edges elsewhere are hidden under rock debris and soil. This illustrates a few elementary geological concepts: rock-in-the-solid or 'bedrock'; overlying rock layers and soil, that is, the 'overburden' covering the bed or stratum we are interested in; the rock debris and soil which are the 'waste-mantle' or 'regolith'; and 'outcrop'. When bedrock shows through the waste mantle, for example as a ledge or bare rocky patch on a hillside, river bed or sea cliff, it is said to 'crop out' (old mining term), hence 'outcrops'. Continuous bedrock surfaces on high mountain crests provide extensive natural outcrops; exposures of solid rock in excavations and tunnels are artificial outcrops.

Looking at an exposed ledge of basalt lava rock on a hillside (Christchurch people see plenty of these on valley sides in the Port Hills and in cliffs near Sumner), it is easy to forget that these ledges are almost certainly *not* where the sheet of once-molten lava originally ended when it was poured down the flanks of the still-growing Lyttelton volcano some ten or fifteen million years ago. Like most of the innumerable lava sheets making up the Port Hills, its original extent has since been reduced and its continuity destroyed by the cutting of sea-cliffs and stream valleys after the volcano ceased to grow. The former continuation of any one lava layer may still exist, perhaps even be exposed, on an opposite valley wall, though to pick it out may be difficult or impossible. The point is that the original extent has been reduced by later erosion.

Before taking up the topics of geological erosion and weathering, a little more should be said about the meaning of soil in a geological context. The soil sciences are concerned in many different ways with the superficial zone in which, on land, we grow plants, found buildings, dispose of our wastes, and so on. From the geological point of view the soil is a surface accumulation of products from rock decay and disintegration mixed in the upper part with residues from plant decay.

Two further points might as well be disposed of here. Geologically, the term 'rock' includes not only hard rock; it is applied to soft sediments and loose uncompacted gravels, to hard sandstone and conglomerate, and to durable granite, all alike. And the word 'formation', popularly applied to striking or evocative shapes of natural rock exposures such as Frog Rock near Weka Pass, or the Devils Boot near Collingwood, has a different and important formal meaning in the language of stratigraphy. So the other use of the term is avoided by geologists.

Weathering, Erosion and the Geological Cycle

It would be hard to find a better beginning for this section than the following quotation from *Geology in the Service of Man* by W. G. Fearnsides and O. M. Bulman (Pelican Books, 1944, p. 51): '. . . the present state of the natural world . . . is to him [the geologist] the last still so far developed of a cosmic cinematographic film, many reels of which are forgotten or partially destroyed and others as yet unexposed. The high peaks of the Himalayas or the Alps are largely composed of rocks which were originally the sands and muds of an old sea floor. We think we know how, if not yet why, they rose to these prodigious heights; and we are certain that in time they will crumble to eroded vestiges no more awe-inspiring than Snowdonia or Lakeland, which in their days have had an alpine grandeur. And from their waste will be built up in another sea thick sheets of sediments destined to make yet another landmass, and perhaps again a mountain chain.' This passage, while not a complete description of the many-times-repeated sequence of events and processes embraced in the term 'Geological Cycle', nevertheless points up usefully the distinction between its two contrasting phases: the 'constructive' processes, these being the ones which form new rock and add materials locally at or near the surface of the crust, and the 'destructive' ones which modify, destroy or remove. The distinction is valid so long as it is remembered that the destruction cannot (by fundamental physical laws) be the end, only the transfer of the materials into some other form in some other geological environment. Nor is the distinction quite complete at any stage, since processes of both constructive and destructive kinds tend to be linked along various alternative paths of change (Fig. 1.5).

With the above reservations in mind, we will next take a brief look at processes which are essentially destructive. No rock exposed at the earth's surface entirely escapes being affected by changes, both physical and chemical, through reactions with some of the factors of the surface

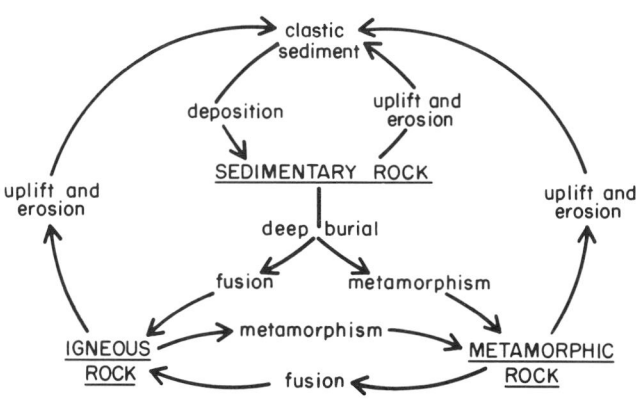

Fig. 1.5
The Geological Cycle represented as a 'flow-diagram' which generalises the sequences of events that can be experienced by materials of the crust following various possible paths.

environment, which is usually at least a little and often a lot different from the environment in which the rock was formed. It may be drier, wetter, cooler, hotter, the pressure may be less, etc. The kind and rate of change are governed by the composition of the rock, the climate, slope of the ground, vegetation and other factors. The physical or mechanical breakage of the rock combined with the chemical changes add up to 'weathering' in the geological sense. Thus:

rock disintegration + rock decay = weathering.

Most of the changes are progressive with time, but neither necessarily continuous nor at a constant rate.

If the products from rock weathering are able to accumulate where they are forming, the weathering processes are likely to become clogged and may slow down. Generally, therefore, if weathering is to keep going indefinitely the waste products must continually be removed by the transporting agencies (discussed later). Weathering combined with transportation of its wastes effects 'erosion' in the geological sense. Thus:

weathering + transportation = erosion.

Unfortunately the term 'erosion' has been used in various other ways, particularly in the sense of locally intense, accelerated wearing-away of land, whereas in the geological sense erosion is a normal process going on all the time at the earth's surface. This is confusing, but nothing can be done about it now.

Besides being one of the phases of a grand sequence of geological events, erosion plays a vital role in the shaping of landscape, and would in time destroy all topographic relief if the earth's crust were entirely immobile. It is also often the reason why useful mineral deposits have been concentrated naturally into economically workable amounts from widely-dispersed minute amounts in the parent rock.

The chief weathering agencies are:

the *atmosphere*, through chemical effects of oxygen, carbon dioxide,
    water, etc., and the physical effects of the wind;

Fig. 1.6
Weathering of limestone by solution on the surface or beneath the soil often produces ridges and grooves called 'lapies', as shown at this outcrop of the Thomas Formation at Castle Hill, inland Canterbury.
(Photo: R. Speight)

solar radiation, the energy-source for most physical processes;
water, through solution and other effects of rain and running water;
ice, through disintegrating effects of frost, and the wearing-away of
   rock by glaciers;
organisms, through the chemical effects of life activities, excretory
   products and decay products after death.
The relative importance of these agencies obviously differs enormously
in different terrestrial environments.

Wherever the daily range of temperature is large, rocks are
disintegrated by thermal expansion and contraction because of their
limited ability to conduct heat inwards from the surface. Large
temperature differences build up just below the surface, causing outer
layers to break off. Different minerals in the same rock can have
different expansion rates, and this can set up internal stresses during
both heating and cooling. Frost disintegrates rock because of stresses
due to the nine percent expansion, at the moment of freezing, of water
in cracks and inter-granular spaces. Obviously, this process needs
moisture to be present, and is most effective where the temperature
swings frequently across freezing-point. It is the main origin of rock
litter and scree in the mountains. Because some minerals, including
most of the clays, expand when wetted and contract on drying, there
are strong disintegrating effects where rocks containing clay are
exposed to repeated cycles of wetting and drying. Running water and
flowing glacier ice, if armed with abrasive rock particles, contribute
importantly to the wearing-away of rock, though the action may be
more localised. These disintegrating processes, and some other more
specialised ones, are the main source of sediment to be removed by the
transporting agencies.

Rocks decay at or near the surface through the action of oxygen,
carbon dioxide and other gases of the atmosphere and the soluble
products from life and decay processes in plants and animals.
Rainwater, soaking in, is the usual carrier. Each kind of rock has its
own pattern of decomposition and distinctive products, while the

Fig. 1.7 (left)
The destructive action of frost is plain to see in this shattered block of hard greywacke rock protruding from the grass-covered surface of an old glacial moraine near Lake Ohau, South Canterbury. Moisture penetrates into joints and other cracks in the rock, expands on freezing and forces the walls of the cracks apart. (Scale given by 10c coin. Photographer unknown)

Fig. 1.8
The 'Pinnacles' in the Harper River valley, Canterbury, are in part of an old landslide composed mainly of glacier deposits. Sculptured by the direct impact of rain-drops upon easily-eroded stony soils or stony clays, such features have the best chance of forming in sheltered places where rain tends to fall vertically. A pinnacle survives while it is protected by a pebble at the top, and is quickly lowered once the pebble has fallen. (Photographer unknown)

Fig. 1.9
The importance of rock-jointing in directing the work of the weathering agents and therefore in determining the shape of weathered fragments is well shown by the geometrical regularity of these fallen blocks of greywacke at Seventeen Mile Bluff, north of Greymouth.

Fig. 1.10
Depending upon situation and recent weather, upper layers are sometimes wet, sometimes dry. Permeable rock materials are permanently saturated with water for considerable distances below the water-table. Depth down to the water-table varies with the topography, generally being greater beneath hilltops, and intersecting the ground surface where there are flowing streams, lakes, springs and seepages, and at the seashore. Naturally, the water-table tends to be deeper in dry-climate regions and varies with weather cycles.

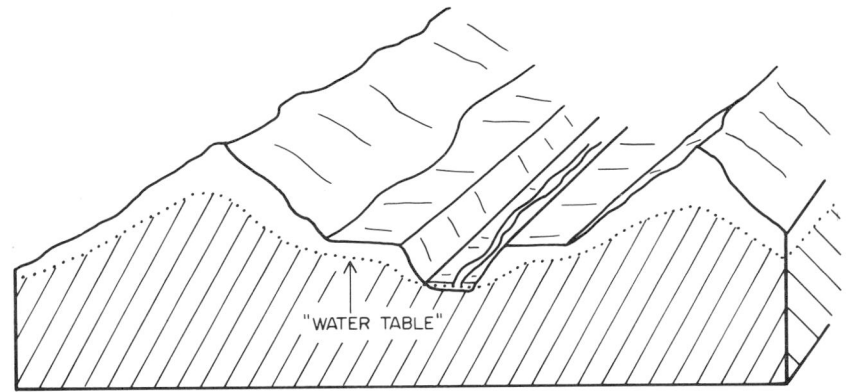

"WATER TABLE"

Fig. 1.11
Frost-shattering is the dominant kind of physical weathering in the high mountains. Here, on the eastern slopes of Browning Pass near the head of Wilberforce River, frost-riven debris streamed down to form an apron of scree which has since become sufficiently stable to support a weak cover of alpine vegetation. Alpine streams receive a major part of their sediment load from the products of frost action. (Photo: R. Speight)

nature of the decay processes is much affected by climate. Limestone, for example, in a moist climate is leached and dissolved to produce caves and 'pots' for the delight of the speleologist; in an arid, desert region the weathering will be chiefly through abrasive and disintegrating processes at the surface.

Among the most general and abundant products from decomposition of rock-forming minerals are quartz and clays, along with oxides and hydroxides of iron and silica in other forms. Soluble substances are conducted away by percolating water, while the solids await removal by other agencies. The surface zone of decomposition generally deepens with time, but the rate and the ultimate degree of decay are controlled by many factors, including particularly the depth to permanently water-saturated ground (the water-table).

Turning now to the agents of waste transportation—the other major partner in the erosion business—we usually find that gravity, acting either directly or indirectly, is the motivating force. Acting directly, gravity often is the first agent to start off the transportation of weathering products as soon as they are released from the parent rock, as rock-falls, landslides and other kinds of mass-movement. Then transport is taken over by streams and rivers, glaciers, the wind, ocean waves and currents. Organisms produce some odd transportation effects, including the floating of stones attached to seaweed far out from the land and the selective transport of favoured kinds of pebbles as gizzard-stones by birds, but these are not important quantitatively. There is much overlap and transition between different modes of sediment transportation. Just to give one example, it is not obvious where to draw the distinction between a very turbid, mud-laden stream and a very fluid mudflow.

It is not possible here to explain in detail the precise mechanisms of sediment transport, but one general comment should be made. Transport in some way always involves friction between individual sediment particles and the transporting medium, except in the case of direct fall due to gravity. Also, because in the vast majority of cases the sediment particles are more dense than the medium, some turbulent or eddying motion is necessary to compensate for the downward pull of gravity; otherwise transportation would soon come to an end.

Sediment is deposited wherever (and as soon as) the transporting agent no longer has sufficient spare energy to carry it along. The agent may later regain energy to raise sediment particles from rest and move them further, but if not, progressive accumulation may begin, and continue until a sedimentary deposit forms, which may in the future become a consolidated sedimentary rock. Sometimes one can tell easily from the kind of stratification and texture (especially from the range and variation of particle sizes, and from fossils, if any) whether a sedimentary rock layer was deposited from a river, a glacier, the ocean, etc. Not always, however, and there have been many controversies about the mode of formation of sedimentary rocks.

Once the products of weathering have reached the ocean or even a deep, long-enduring lake, their chances of continuing to accumulate and so building up a substantial thickness of sediment become much

Fig. 1.12
The sediments in this roadside outcrop near Rakaia Gorge, Canterbury, show features of deposition by water under varying conditions. Coarse gravel at the top must have been brought to the site and deposited by a strong river current capable of transporting pebbles of this size; the horizontally-bedded sands had been laid down earlier in the quieter waters of a lake which formerly existed here, probably while the Rakaia Glacier of the Ice Age was receding. The weakly-consolidated sands were then scoured by currents to form hollows which accommodated fine gravels and sand, brought in perhaps by a side stream. (Photo: B. W. Collins)

better. The long geological record, however, tells us that the accumulation is unlikely to remain undisturbed for ever. Many mountain ranges are made up largely of sedimentary rock which accumulated and consolidated beneath the sea, and has since been uplifted.

The Earth's Restless Crust

The oldest known rocks are crystalline and severely altered by metamorphism, but some sedimentary rocks of ages exceeding 1000 million years did escape alteration. Their structures and textures still reveal something about the conditions under which the sediment was produced and deposited in a very remote past time. The evidence strongly suggests that the destructive phase of the Geological Cycle operated vigorously even then. Since the same can be said about similar evidence from sedimentary rocks of all later ages as well, an important question arises. How is it that all relief on the earth's surface was not completely worn away long ago? Inevitably, other processes must have been at work throughout recorded geological time to compensate for erosion and to restore relief. Changes of level of the crust and the ocean surface, relative to one another, must continually have brought up tracts of new land so that erosion and sedimentation could keep the cycle going, for there seems never to have been a time, as far back as we can make out, when sedimentary rocks of all kinds were *not* being formed.

Besides the sudden displacements of the ground which accompany strong earthquakes, there are plenty of other signs that the crust is far from being static at the present time in many parts of the world. Even over a few years, small continuing displacements can be measured in California, Japan, the Baltic region and elsewhere. Geologically, and over longer periods of up to a few million years, there is no escaping the fact that vertical displacements of some kilometres and horizontal displacements of tens, perhaps hundreds of kilometres have occurred. It turns out not to be simple, however, to reconcile the crustal distortion measurable at the surface today with the displacements of a larger order that are indicated by geological evidence. At least we can say that good grounds exist for believing the larger, long-term displacements such as the 500-kilometre offsetting of the rocks on either side of the Alpine Fault of New Zealand to be the sum-total of many small increments over long spans of time.

To the foregoing must be added the directly constructional processes which compensate for erosion by building up the land surfaces of the crust. These include some types of volcanic activity and the laying down of large thicknesses of gravels etc. by rivers and streams ('aggradation'). Such additions to the surface, however, are not all pure gain, since the crust is slightly depressed by the additional loading ('isostatic adjustment').

Mountains and Orogenies

Within the space of this very condensed introduction to geology it is not possible to deal adequately with one of the most fascinating aspects of the subject, namely, that concerning large-scale crustal folding, the elevation of mountain belts ('orogeny') and the crustal and sub-crustal mechanisms proposed to explain not only the uplift but also the curious distribution of mountains, both above and below the sea. The various phenomena of crustal ups and downs are summed up in the word 'diastrophism'.

Vast areas of crustal surface, land or sea floor, have only slight relief, the high mountain chains being confined to restricted belts taking up only a small percentage of the total area. The New Zealand chain is part of a belt that almost completely encircles the Pacific Ocean. Another belt extends from South-east Asia westwards through central Asia, the Himalaya region, the Caucasus and the European Alps as far as the eastern shore of the Atlantic. The Urals are a separate feature, and the Appalachian chain of eastern North America is a relic of an earlier pattern. Lines of active volcanoes and belts of strong seismicity (i.e. of frequent, strong earthquakes) are associated with the geologically young mountain chains. Hawaii and the Azores are examples of lofty volcanic mountains built up from the deep ocean floor. Besides these, strong relief is provided locally by the long mid-oceanic lines of submarine volcanoes, the deep ocean trenches and the submerged canyons cutting across the outer continental shelves. It would be as well to point out, though, that relief maps of the ocean floors which appear in popular scientific magazines greatly exaggerate the heights of submarine features and the steepness of slopes.

In contrast with the localised zones of strong relief, often referred to as the earth's 'mobile belts', the greater part of the continental areas are flat and essentially stable except for broad warping and purely vertical movements. These are the 'continental platforms', also called 'cratons', and they contain the most ancient rocks known. One more term which it would be very difficult to avoid using repeatedly in subsequent chapters is 'tectonism'. This refers to the extreme distortion suffered by the rocks involved before and during the elevation of mountain chains. 'Tectonics' usually means the study of rock features that reveal this distortion.

Advances in the theory of mountain-building and tectonics over the past decade have been revolutionary in many aspects, largely from under-sea geological exploration. Yet neither the basic information from the direct study of accessible parts of the crust nor the inferred sequence of geological events which lead to the creation of new mountain chains has changed very much, even if the interpretation is now different. Within the limitations of this book, the following outline of the typical history of a mountain chain must suffice. Further information will be found in later chapters describing mountain-building episodes in the geological history of New Zealand.

The typical sequence of events begins with a 'geosynclinal' phase wherein narrow belts of the crust are depressed to accommodate great thicknesses of the sediment being eroded at the same time from adjoining land areas (foreland). It is no longer supposed that the load of sediment is the cause of the sinking. Filling and deepening of the 'geosyncline' may continue for tens of millions of years. Certain types of rock are regarded as distinctive of the geosyncline situation; these include the greywacke type of sandstone which with dark-coloured shale and siltstone makes up most of the New Zealand axial ranges, and rather special types of volcanic rock, chiefly varieties of basalt, erupted from volcanoes near and under the sea. The rate of accumulation of geosynclinal rocks must be rapid. Structural features within the sedimentary strata show that a great deal of slumping, and also of contemporary sea-floor erosion and redepositing of the material goes on before it finally becomes deeply buried, compacted, and the enclosed water expelled.

There is no clear separation between the geosynclinal phase and the succeeding 'tectonic' phase, in which the sediments of the geosyncline furrow, their basal parts already compacted into hard rock, are strongly compressed with much folding, dislocation of the strata, and metamorphism. Again, the boundary is vague between the tectonic phase and the upheaval of the whole belt above sea level ('orogenic phase'). Different stages can be reached at the same time in different parts of the geosyncline, and there is much overlap. Orogenic uplift is accompanied by erosion of the rising mass—vigorous at first because the rate of uplift is rapid—and mountain landscapes are sculptured from it. It is the real mountain-building phase. Deep within the upheaving mass, sediments have been altered to metamorphic schist and gneiss, and some converted into granite which may become molten and be injected into the roots of the mountains in large masses called 'batholiths' or 'plutons'.

When compression and upheaval eventually cease, erosion catches up and the mountainous relief fades. The ancient continental platforms of today cut across geosynclinal rock types and structures. It is accepted that erosion has totally destroyed the mountainous relief which would have accompanied the orogenic phase, producing the flat or gently undulating erosional surface called a 'peneplain'. This requires a long, quiet interval, after which the site of the previous crustal upfold or orogen may provide sediment for a new geosynclinal phase. The full duration of the cycle seems usually to be more than 100 million years.

The foregoing story is of a remarkably standardised sequence of events, the outline of which has emerged from the study of stratigraphy and structure in rocks of all ages from Precambrian to Tertiary in many different parts of the world. The material evidence is strong, and little doubt remains that such grand events really happened, over and over again. But how and why?

These questions are in a different category. Answers cannot come from direct observations. We need to know what has been going on, now and in the past, far below the accessible parts of the earth's crust.

Ever since man first conceived of the earth as a ball it has been suspected that its internal state is far from being cold and static. Earthquakes and volcanoes have to be accounted for and the old mythological explanations ceased to be satisfying to the bolder and more curious minds in Renaissance times. Geology progressed to become an integrated, systematic observational science from the eighteenth century onwards, the major events of earth history were pieced together, and the vital questions began to be perceived. Physical concepts of matter were still evolving and means of testing the more fanciful hypotheses were lacking, as was the modern spirit of candid, scientific criticism. A serious handicap for the earlier investigators of geological processes, one that cramped their imaginations and misled them, was an apparent shortage of time to accommodate all the happenings that seemed to be recorded in the rocks—if, indeed, the earth was really only a few thousand years old. Until well into the nineteenth century it was contrary to rigid religious beliefs to suggest otherwise, and a hundred years earlier, positively dangerous! The changes in the earth for which geology was producing incontrovertible evidence therefore had to be conceived of as being sudden and cataclysmic.

Influenced by the solid logic of James Hutton at the very end of the eighteenth century, and by the substantiating evidence gathered and publicised by Charles Lyell during the next few decades, thinking men began openly to doubt the truth of cataclysms as a general component of geological history, and were ready to welcome suggestions that the earth was very much older; millions, then hundreds and eventually thousands of millions of years old. Meanwhile geologists teamed up with physicists interested in how to measure the properties of the earth's inner materials from ground level. Their joint findings provided some real basis, in place of speculation, for theories to explain all the phenomena of geosynclines, tectonism and orogenies, and the uneven distribution of mountains, earthquakes and volcanoes.

The pace of discovery in these fields of inquiry has quickened enormously over the past twenty years, especially since the world's thirst for petroleum and other industrial requirements has provided vast sums for sub-oceanic exploration. The results of these

DISTRIBUTION OF THE MAJOR ROCK GROUPS

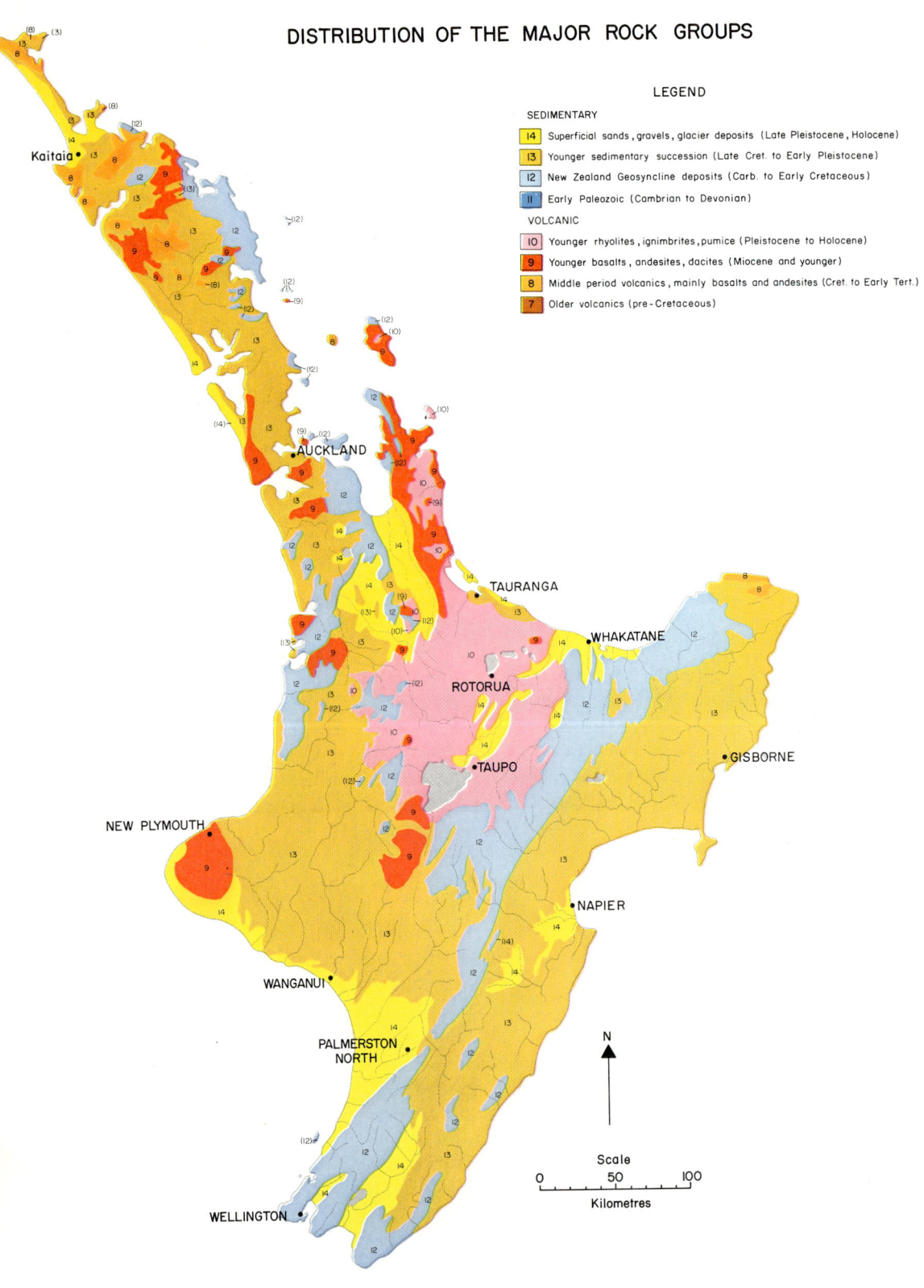

LEGEND

SEDIMENTARY

14 Superficial sands, gravels, glacier deposits (Late Pleistocene, Holocene)

13 Younger sedimentary succession (Late Cret. to Early Pleistocene)

12 New Zealand Geosyncline deposits (Carb. to Early Cretaceous)

11 Early Paleozoic (Cambrian to Devonian)

VOLCANIC

10 Younger rhyolites, ignimbrites, pumice (Pleistocene to Holocene)

9 Younger basalts, andesites, dacites (Miocene and younger)

8 Middle period volcanics, mainly basalts and andesites (Cret. to Early Tert.)

7 Older volcanics (pre-Cretaceous)

Kaitaia

AUCKLAND

TAURANGA

WHAKATANE

ROTORUA

TAUPO

GISBORNE

NEW PLYMOUTH

NAPIER

WANGANUI

PALMERSTON NORTH

WELLINGTON

N

Scale

0   50   100

Kilometres

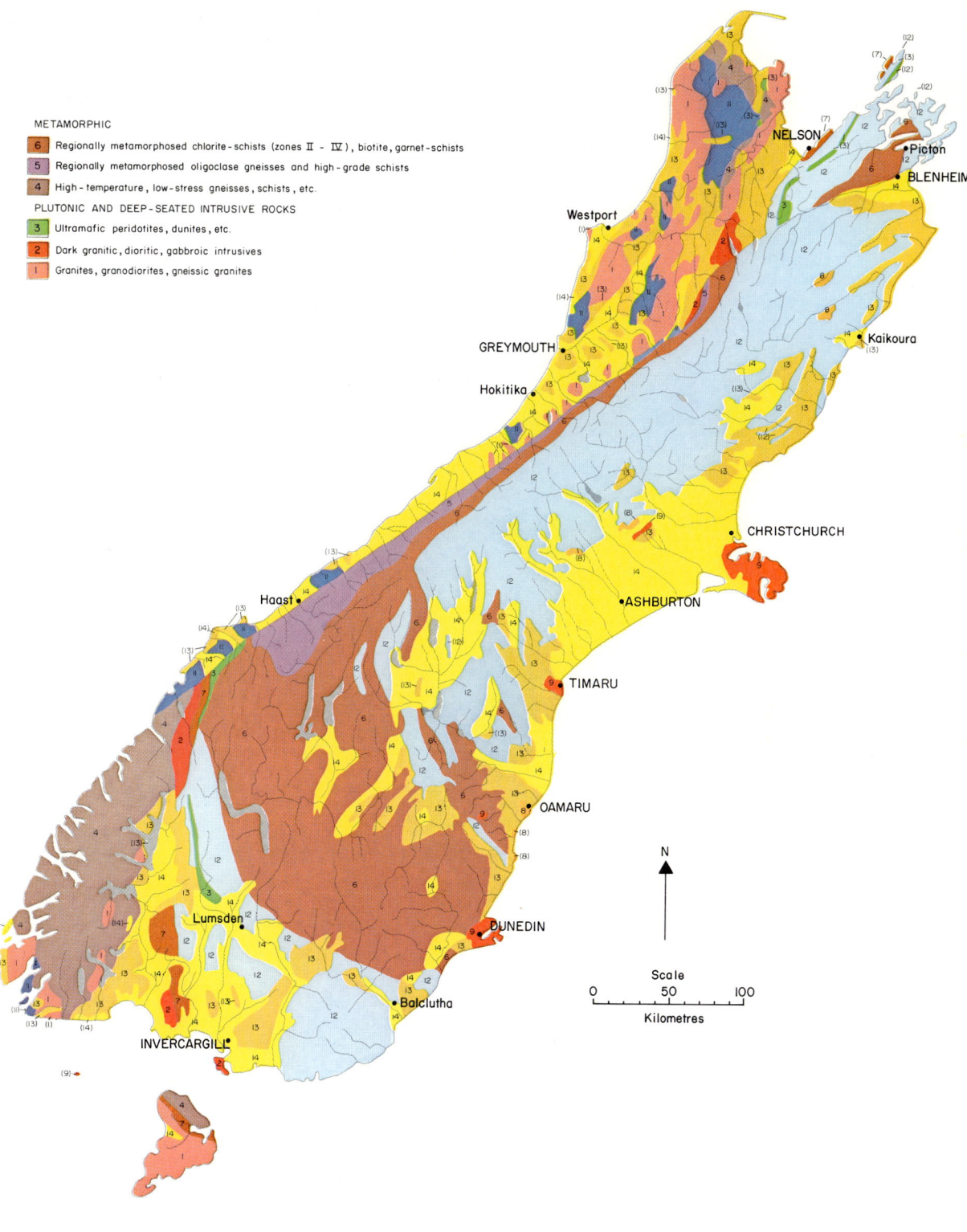

METAMORPHIC

6   Regionally metamorphosed chlorite-schists (zones II - IV), biotite, garnet-schists

5   Regionally metamorphosed oligoclase gneisses and high-grade schists

4   High-temperature, low-stress gneisses, schists, etc.

PLUTONIC AND DEEP-SEATED INTRUSIVE ROCKS

3   Ultramafic peridotites, dunites, etc.

2   Dark granitic, dioritic, gabbroic intrusives

I   Granites, granodiorites, gneissic granites

NELSON

Picton

BLENHEIM

Westport

Kaikoura

GREYMOUTH

Hokitika

CHRISTCHURCH

ASHBURTON

Haast

TIMARU

OAMARU

Lumsden

DUNEDIN

Balclutha

INVERCARGILL

N

Scale

0   50   100

Kilometres

explorations, mapping the surface features of the ocean bottoms, measuring their magnetic properties and drilling hundreds of metres beneath them from surface vessels, have made irrelevant about ninety percent of the argument of conventional twentieth-century tectonic theory, but have tended to confirm some of the better guesses of an inspired few. At the same time, a curious thing has been happening. Little by little rather than suddenly, the emphasis has shifted to questions of energy production and flow within the earth and sub-crustal processes, while the dramatic sequences of crustal geology—the drifting of continents, sinking of geosynclines, orogenies, formation of granite plutons, all now accepted as geohistorical facts—have dropped back into the role of inevitable, rather passive and superficial effects from the mighty heat-engine of the earth's interior.

Let it be remembered, though, that it was while trying to account for the past events inferred from the geology of the accessible crust that we first identified what *were* the basic questions, the ones that are now being answered not on land but under the sea. In fairness to the tectonics enthusiasts who framed the questions but could not answer them, let it be noted too that the rapid advances of the past few years make use of remote-measuring methods undreamed of half a century ago, and also of modern facilities for high-speed computing without which it would have been impossible to handle the flood of geophysical data now available for testing hypotheses.

The Time-scale of Earth History

The vast periods of time mentioned in the foregoing outline of mountain-building histories emphasise how important it is to acquire some sense of scale in all the dimensions of geology. To begin with, it is so easy to succumb to feelings of being overwhelmed by thoughts of millions of years, and of the enormous quantities of time needed to accommodate not just one but perhaps several complete cycles of mountain-building and destruction in the one region. But it begins to dawn upon us that geological time is by no means a commodity in short supply. We can draw on the several thousand million years since the most ancient known rocks came into existence.

An old but remarkably enduring geological doctrine sees the geological processes and changes which can be observed and measured today as being essentially the same as those which from earliest times have produced the features of rocks of all ages. To apply the doctrine, we have to envisage currently-active processes as having been operating over vast periods of past time. This is difficult because the time-quantities are so much beyond ordinary human experience of time, or the span of human history, as to be incomprehensible. Also we have to be able to visualise both short and long periods of geological time in a proper perspective. Anyone who has been reduced to a mind-boggled state by all this should try the experiment of translating the abstract quantities of time into some tangible measure, something more easily pictured, whether in small amounts or large amounts.

One way is to construct an imaginary scale in which 30 cm (1 ft) of linear distance represent a time interval of one million years. Applying this scale to some contrasting time-spans, we can visualise without much trouble a distance of 1.06 km (⅔ mile); for Auckland citizens this is roughly the distance up Queen Street to the Town Hall; in Christchurch from Moorhouse Avenue up Colombo Street to Cathedral Square. Those distances, on that scale, are equivalent to the age of the oldest known rocks on earth. The age of the oldest rocks to have yielded traces of abundant, varied organic life is equivalent to only 200 m (10 chains), say, the length of two short city blocks. The time since the first appearance of mammals is represented by a distance of 21 m (70 ft). Now come the shockers. Two million years ago (say, 60 cm back) the Southern Alps were just beginning to become a lofty mountain range, and the total span of man's presence on earth is only about half of that! The birth of Christ would be about 0.6 millimetres back. Yet the age of the oldest known rocks is represented by a distance of 1.06 kilometres.

The foregoing exercise may make it easier to appreciate how several successive cycles of major geological events, each cycle involving hundreds of millions of years, can be fitted into a regional history when the evidence from the rocks points that way. It also enables us to evaluate properly the newspaper report of the discovery of some fossil that lived a whole ten million years ago! Finally, perhaps it helps us to size up the true importance of man and his works against the background of the earth's (geological) history.

Fossils and the Geological Dating of Rocks

Modern techniques of isotope chemistry make it possible to measure very minute amounts of the radioactive breakdown products that are retained in some kinds of rocks. This opened the way to determining how much time has elapsed since the minerals yielding these decay products have existed under present conditions. It led to a variety of means of estimating minimum ages for types of rock that contain the right minerals for the task. We cannot go into details, but suffice to say that the methods of radiometric dating, despite many pitfalls and difficulties of interpretation, enable us with greater confidence than ever before to assess the true span of geological time, and to express its subdivisions in terms of multiples of the secular time unit, the year.

Geologists are likely to go on using their traditional 'scale' of time divisions that has been evolved over the past two centuries from innumerable studies of rock strata sequences and the fossils contained in them in all parts of the world. For one thing, so far only a relatively few kinds of rock can be made to declare their ages by the radiometric methods, whereas many sedimentary rocks contain fossils. Using the principle that in any sequence of strata which has not been inverted by folding or other distortion of the crust, each layer is at least a little younger than the one below, and a little older than the one above it, and assuming also that the parade of organic life through past geological ages, as represented by the fossilised relics of life from successive periods, has never gone back on itself, a very practical and reliable means was evolved for subdividing geological past time. Fossil-bearing rocks, wherever found, can be referred to a standard

1. *Under the constant pull of gravity, debris from this frost-weathered exposure of limestone near Castle Hill, Canterbury continually works its way downslope towards the stream in the valley below. This is one of the various forms of mass movement of waste providing a link between weathering and stream transportation.*

2. *The effectiveness of glacier ice in shaping and smoothing hard rock is handsomely shown here. When the photograph was taken, the schist rock had only recently appeared from beneath the receding Franz Josef Glacier. (Photo: R. Warburton)*

3. *As a result of accelerating erosion in its mountain catchment, the Waitangiaona River in Westland has had to carry excessive loads of waste, and has aggraded the floor of its lower valley by several metres, with far-reaching consequences.*

time-sequence, that is to say, 'dated' geologically. It works; it has served such stern masters as the oil industry, but it has its difficulties, partly because some creatures populated the earth for very long periods without changing, but mainly because many rocks do *not* contain recognisable fossils. The system is still being refined and tested continually by stratigraphers and paleontologists all around the world, and the international geological time scale is now fairly stable and consistent. In addition, most countries have their own local detailed classification of the strata of the region, more or less firmly correlated with the world-wide scale.

GEOLOGICAL DIVISIONS OF TIME * Table 1–1			
ERA	**Period**	**Began (years ago)**	**Duration (years)**
QUATERNARY	Holocene or Recent	10 000	10 000
	Pleistocene	2 000 000	2 000 000
TERTIARY	Pliocene	7 000 000	5 000 000
	Miocene	26 000 000	19 000 000
	Oligocene	38 000 000	12 000 000
	Eocene	54 000 000	16 000 000
	Paleocene	65 000 000	11 000 000
MESOZOIC	Cretaceous	136 000 000	71 000 000
	Jurassic	190 000 000	54 000 000
	Triassic	225 000 000	35 000 000
PALEOZOIC	Permian	280 000 000	55 000 000
	Carboniferous	345 000 000	65 000 000
	Devonian	395 000 000	50 000 000
	Silurian	430 000 000	35 000 000
	Ordovician	500 000 000	70 000 000
	Cambrian	570 000 000	70 000 000
PRECAMBRIAN		? 4 500 000 000	? 4 000 000 000

* There are other versions of the geological time scale. In America it is usual to replace Carboniferous by two units called Mississippian (older) and Pennsylvanian (younger); to give Tertiary the status of a period and reduce the periods Paleocene to Pliocene, inclusive, to the lower order of epoch in the hierarchy of time units; and to combine Tertiary and Quaternary in a Cenozoic Era. There are considerable variations in the dates attached to the beginning of the Pliocene, Pleistocene and Holocene. The Cenozoic Era version was the one adopted for the 1:1 000 000 Geological Map of New Zealand (1972).

This kind of scale of past time is, however, a purely relative one, an ordering of the rocks and the phases of geological history they record into a true time sequence, but it does not lead directly to an assessment of rock ages in years. Until the methods for assaying very precisely the amounts of radioactive decay products remaining in some rocks became available within the last few decades, no other approach to the direct measurement of ages of rocks gave sensible and consistent results, although some ingenious schemes were proposed. Now, even if the range of rock types to which the radiometric methods apply is still fairly limited, we can at least write in the 'absolute' or 'secular' ages, in multiples of years, alongside the international geological time scale, or geological column. (Table 1-1, p. 40.)

The Time-span of New Zealand Rocks

Within the last few years we have become certain that strata representing every one of the major geological time divisions since the beginning of the Paleozoic Era occur in New Zealand. Indeed, for the Permian and Cretaceous Periods and the Tertiary Era as a whole we seem to have an unusually full record. It is now less certain than a few years ago that Precambrian rocks exist here.

Tables 3-1, 4-1, 4-2, 7-1, 7-2 and 9-1 present the classification of our stratified rocks according to their ages, as now accepted and used by a majority of New Zealand geologists, and its relationship with the conventional international scale. Of course there are differences of opinion about details, and further refinements may be expected, especially as further groups of animals and plants occurring as fossils in New Zealand rocks come in for study. Major changes in the present classification and naming seem unlikely.

The classification presented in Table 1-1 is essentially a break-down of past time into units suggested by the time ranges of individual fossil species and fossil groups as measured by the sequences of strata in which they occur. Quite independently of this classification, there are also separate classifications of the sequences of strata themselves, as distinct from the time spans they represent. These classifications are based upon the different kinds of rock found in succession in different areas, regardless of age, or of time except in so far as is implied by position in the sequence, and they are described as 'lithostratigraphic' classifications. It is important that the divisions of lithostratigraphic classifications are not confused with the other system which refers time spans to rock sequences and fossil content, and is therefore called 'time-stratigraphic' or 'chronostratigraphic'.

Lithostratigraphic classification is most useful for local, practical purposes because it does not require a knowledge of fossils, because it is more stable, being usually unaffected by any upsets or new developments in paleontology, and because it is based upon properties of the rocks themselves. For working out geological histories, and comparing the timing of events in different regions, or with overseas countries, time-stratigraphic classification is required. These concepts will be explained more fully in Chapter Four (p. 107).

4. Permian sandstones of the Greville Formation, deposited early in the sedimentary history of the New Zealand Geosyncline in Permian times, exposed alongside the Lee River, Nelson (p. 104).

Chapter Two

THE OUTLINE OF NEW ZEALAND

Present and Past

In the preceding chapter we faced the problem of how to cope
mentally with unfamiliar magnitudes of time and to see geological past
time in true perspective. This chapter begins with an attempt to put the
magnitude of the earth's surface features into perspective as well.

Surface Relief of the Earth

The present shape of any land area on the map or as seen from a high
orbiting satellite is outlined by the intersection of the ocean surface
with a slight outwards bulge of the surface of the crust. Geologically, it
is only an instantaneous look at a picture which has been changing
continually, and will certainly change in the future. If the bulge is a
relatively smooth one, the coastline is short and simple in plan; if it is
wrinkled the coastline is long and indented with bays and sounds.

The relief of the bulge as a whole, its shape in detail as seen in profile
and the steepness of slopes, determines what its map-outline would
have been like in the past, even just a few thousand years ago when sea
level was different from now by as much as 100 metres. The less steep
the ground slope, the greater the change of outline and area when sea
level changed a given amount. Because the seas surrounding western
Europe are shallow over broad continental shelf areas, small shifts of
sea level made great differences to the shape and land connections of
the British Isles, France and the Low Countries.

To get a true picture of the magnitude of earth surface features, sea
level must be disregarded. The true relief of the group of volcanoes
composing the island of Hawaii, rising abruptly from the deep central
Pacific floor, is actually about twice as much as shows above sea level,
making it a giant mountain with a relief of about 9 kilometres. It
greatly exceeds Everest, which rises only 6 kilometres above the general
level of adjacent valley floors. In south-western New Zealand, where
there is little or no marginal shelf, the real relief of the mountains of
Fiordland is around 4 kilometres, a figure not exceeded by Mounts
Cook and Tasman because the offshore slopes of Westland are
comparatively gentle.

The relief of Hawaii sounds enormous if one is thinking of climbing
a steep slope 9 kilometres high, but again a sense of proportion is
needed. Neglecting the slight differences between the polar and
equatorial measurements, the earth's radius is approximately 6400
kilometres. Even the greatest ups and downs of its surface depart only

insignificantly from the generalised level of the surface of the crust. If the earth is represented in profile by drawing a circle 15 cm in diameter, the altitude of Everest above sea level is contained within the thickness of a fine pencil line and could not be shown on that scale. The extreme range of levels, from the height of Everest to the depth of the deepest known ocean trench near the Mariana Islands in the north-west Pacific is nearly 20 km but that is only one part in 320 of the earth's radius. It would be difficult to 'feel' the Southern Alps on an earth scaled down to billiard-ball size.

Outlining what are quite trifling bulges of the earth's crust, the present boundaries between land and sea which seem so important in the affairs of men are transient things by no means fixed in position through time. The present ratio of land to ocean area is 29:71; ten thousand years ago, when water was still locked up in Ice Age glaciers and ice-caps not yet melted, the proportion of land was substantially greater. Going further back in time, earth materials have been moved about and the relative levels of crustal surface and ocean surface have changed greatly, but in relation to the bulk of the earth neither the quantities of material moved nor the forces involved should seem incredible.

One further point. When the percentages of crustal surface areas standing at varying distances above and below sea level are plotted on a frequency curve, it is found that extreme ranges of mountain altitude and ocean deep are confined to a very small fraction of the total . surface. The mountain ranges of New Zealand seem to take up a good deal of our land, but we must bear in mind that the really high land is of limited extent compared with the whole of the broad, largely submerged crustal bulge of which New Zealand is just the biggest emergent portion.

Defining New Zealand Geologically

In earlier days sovereignty over lands ended at the coast. There were advantages if a nation could dominate adjacent seas for purposes of defence or trade, but no one would have thought about owning the sea bottom. Now there are international laws and conventions, 'territorial waters', the Three Mile Limit, and most recently, because of new defence factors and competitive interest in the food and other marine resources, the Two Hundred Mile Limit—if you can enforce it! Such limits are completely arbitrary, so we may inquire whether there is any logical basis upon which a 'geological New Zealand' could be defined. It could become a practical question in the not-too-distant future.

Like the coastlines of most other lands, ours has fluctuated widely in the geologically recent past. Its present position is of no particular geological significance. More important, perhaps, is the broad 'crustal excrescence' mentioned above. Bounded by a distinct drop to deep ocean floors, this bulge includes the continental shoal areas where they exist, which is around most of the mainland coasts, together with ridges reaching across the north Tasman Sea towards Lord Howe and Norfolk Islands, eastwards to the Chathams and south to the Bountys and Campbell Island. Though only a faint crustal bulge, it is a separate

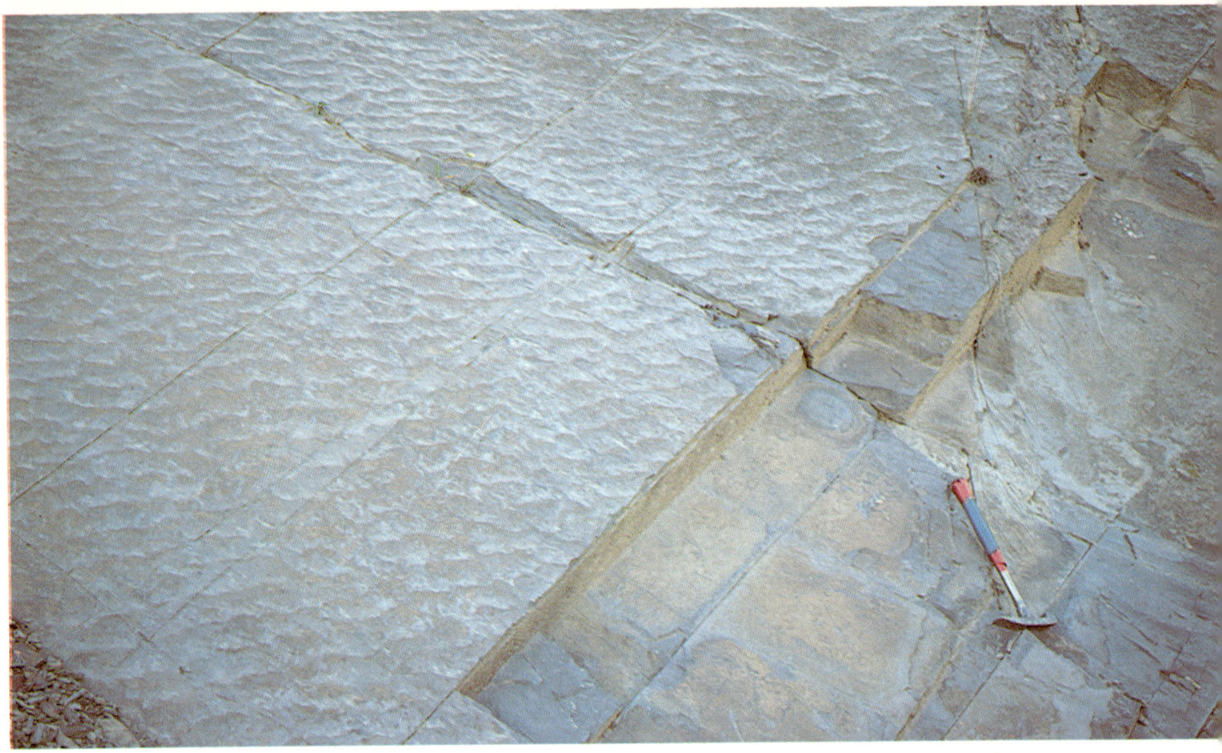

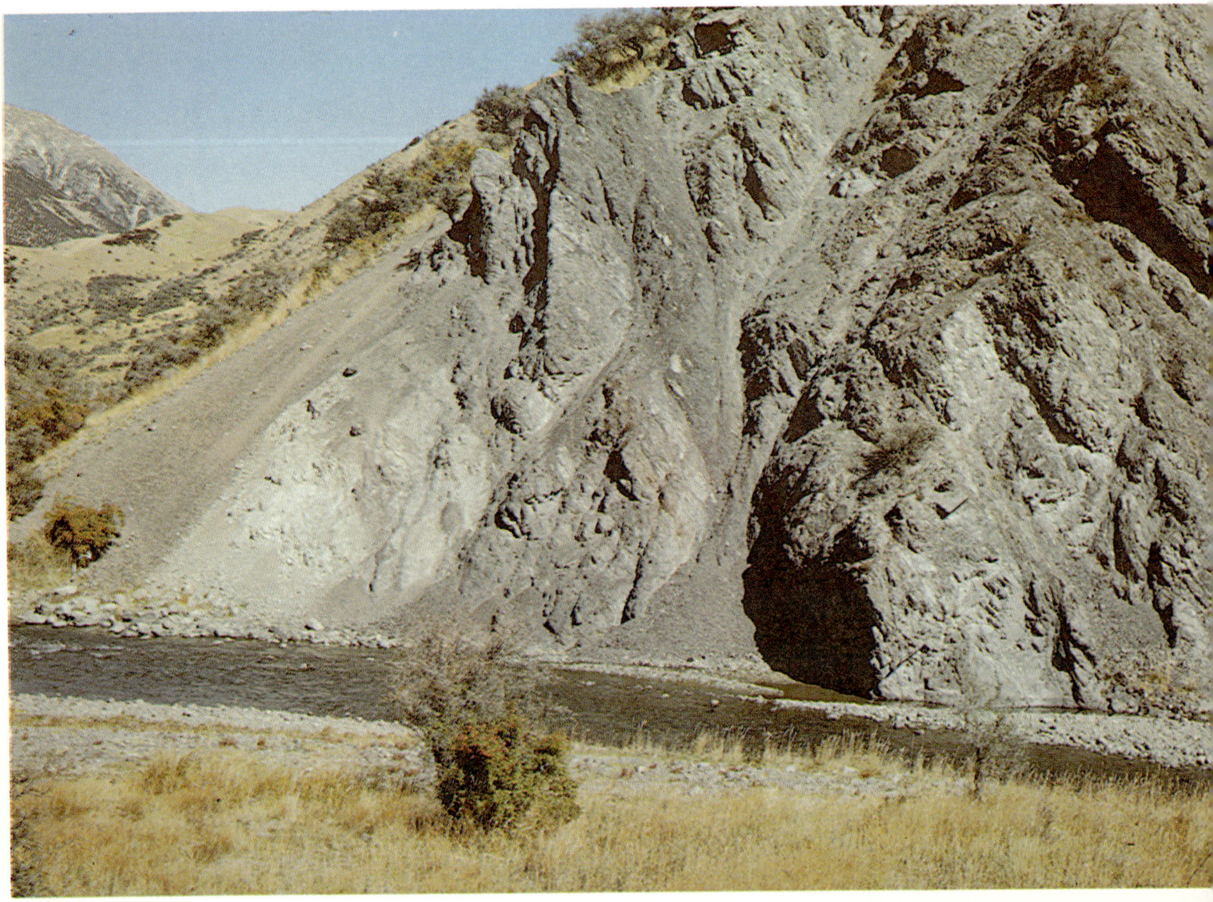

5. Exposed along a bedding surface, this purple-coloured laminated siltstone of the Waiua Formation (Permian) in the Wairoa valley, Nelson, shows fossilised ripples that were caused by gentle seabed currents while the sediment was accumulating. Breakage of the rock has been guided by a simple, almost rectangular system of joints, perpendicular to the bedding (p. 104).

6. Torlesse Supergroup rocks, including reddish basaltic tuff beds, have been crushed and sheared by movements on a major fault in the valley of Edwards River, a tributary of the Clarence River in inland Marlborough.

7. The Torlesse Supergroup, as exposed here in a deep canyon of the Waimakariri River, Canterbury, comprises thick beds of hard, greywacke sandstone, thinner sandstone beds and dark argillite or siltstone, all involved in complex folding and faulting (p. 112).

entity distinct from the continents of Australia and Antarctica and the eastern margin of Asia. The feature has been referred to by geographers and geologists as the 'New Zealand Platform', a piece of continental crust which swung away from the Austral-Asian platform when the Tasman Sea wedge opened up about seventy million years ago. Many attempts have been made to picture how all the continental bulges once fitted together. New Zealand has sometimes been omitted altogether from these reconstructions or represented by its present shape, the substantial New Zealand continental platform being neglected. Geologically the submerged portions are just as much part of New Zealand as the south Atlantic continental shelves are parts of Africa and America.

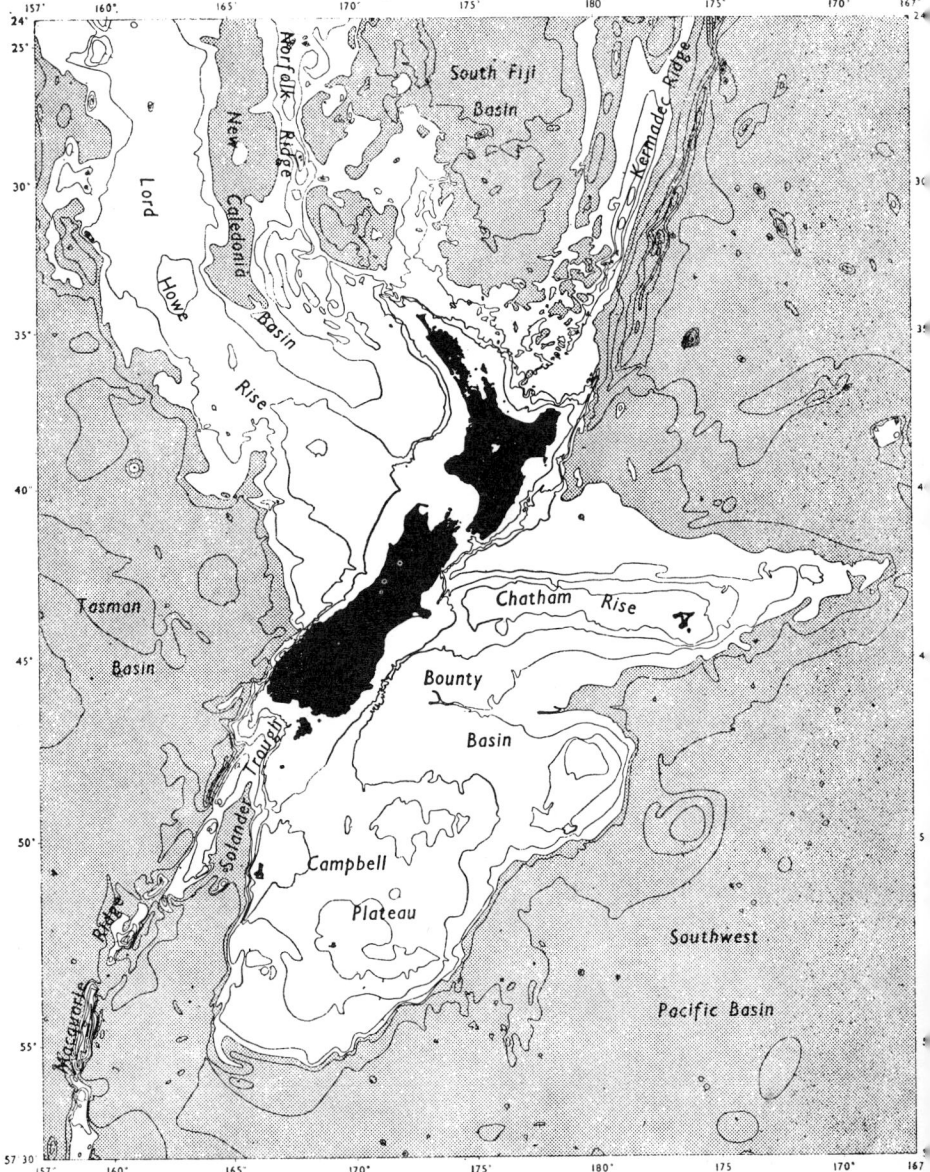

Fig. 2.1
In this bathymetric map of New Zealand and the surrounding sea bed, the contour line representing a level 4000 m below the present ocean surface has been chosen to outline the New Zealand Platform and its north-westward continental connections. The north-north-eastward and south-south-westward ridges have a different significance (Chapter Ten). (From C. A. Fleming, 1970, after P. Lawrence, 1967)

How Long has New Zealand had its Present Shape?

It is difficult to think of our country looking any other way than it does on maps today—three main islands of distinctive shape and a number of smaller ones at distances of up to 50 km offshore from the main ones. Even if its present shape does not delineate the 'geological' New Zealand, at least it provides a figure of reference when we come to consider the outline of former lands that have existed on and around this site, though only for comparatively recent geological times. Its usefulness for that purpose diminishes as we go further back in time because we have to take account of (and mentally undo) large horizontal displacements of the crust. In the undoing, the reference figure has become unrecognisably distorted by the time we have gone back 100 million years.

It would be simpler if the vertical movements of the crust had been uniform over the whole region, but New Zealand, close to the Pacific Ocean border, is in an area where tilting and differential changes of crust level have accompanied the oscillations of ocean level through our later geological past.

For simplicity, let us begin by considering the effects of rising or falling sea level alone. It is easy to get an impression of what the effects of a modest rise of sea level would be by tracing out the contour lines on a topographic map. Suppose the sea were to rise by 100 m (or 300 ft), the new coastal outline would be roughly that of the equivalent contour line. Unfortunately, the more readily available bathymetric maps and charts of the surrounding sea floor do not usually have suitable contour lines to enable the same thing to be done for a sudden fall of sea level. Although in reality coastal processes would quickly modify the picture, it is obvious that a 100-metre shift of sea level, either way, would have a drastic effect on New Zealand's outline, out of all proportion to the significance of 100 m relative to the earth's 6400-km radius. A fall of 100 m would increase the land area by about a quarter, Cook and Foveaux straits would vanish, and the consequential changes in circulation of coastal waters would affect the climates of inland areas. (Fig. 2.2.)

This is no pointless exercise. Sea level was, indeed, about 100 m lower than at present as recently as 10,000 years ago, and approximately 100,000 years ago it was relatively higher by 50 to 100 metres. There have of course been erosional modifications to the landscape detail over that longer period. The point being emphasised, of course, is that the outline of New Zealand was substantially different only yesterday, geologically speaking.

What was its Shape in the Past?

We are now getting into 'paleogeography', a term that is self-explanatory if it is recalled that 'paleo' comes from the Greek, meaning ancient.

One of the big difficulties in answering the question is caused by the gross horizontal distortion of the crust suffered by this region, contributed to largely by lateral shifts along the Alpine Fault and several other great, transverse displacements which cross New Zealand

8. Lake Wakatipu is in a setting of Haast Schist mountains, the lower slopes having been model- led by glacial scour and the upper slopes and crests fretted by physical weath- ering agencies, especially frost (p. 317).

Fig. 2.2
Approximate outline of the continuous land of New Zealand when sea level was still low towards the end of the last Ice Age, during the Otira Glaciation (Chapter Nine). (After Fleming)

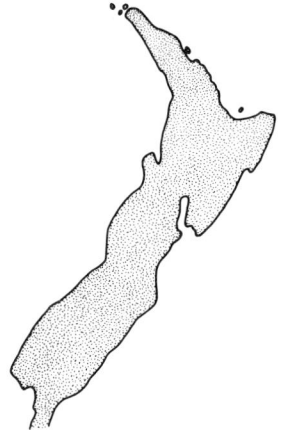

Fig. 2.3
Approximate outline of New Zealand early in Pleistocene time, neglecting the effect of offsetting along major faults, which in any case probably would not have greatly altered the shape from that shown in this sketch. (After Fleming)

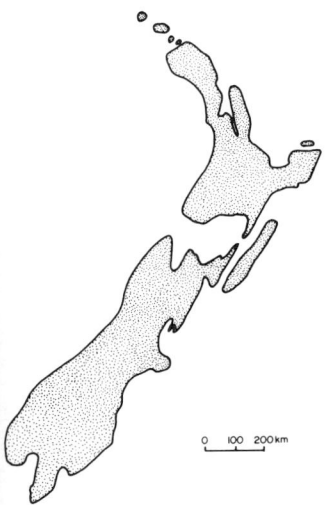

obliquely. So great, indeed, that in travelling not very far back into the geological past we must forget about the present shape altogether and use the outline of the New Zealand Platform as the reference figure.

A convenient stop on this journey backwards through time is at about 2 million years ago. (A widely accepted date for the beginning of the Pleistocene Period is 1.8 million years ago.) At this point we are already too far back to be able to think of changes merely in terms of simple, uniform submergence or emergence of a land similar in form to the present. There is ample evidence of major differential movements, upwards in some places and downwards in others, since that time. In Taranaki and Hawke's Bay, for example, wide areas are underlain by sandstones and gravels, rich in places with fossilised marine life, which were deposited under the sea during that period and later uplifted to make new land. While these areas were being depressed, other areas were rising.

Fig. 2.3 is a paleogeographic outline of New Zealand in early Pleistocene time as envisaged by C. A. Fleming. It neglects the effect of lateral fault displacements since then, but such displacements have not greatly affected the picture over the last 2 million years. The outstanding difference from the present is that land is inferred to have joined the Marlborough Sounds area with south-western Wellington, while a broad seaway connected Manawatu with eastern seas then lapping the embryo Ruahine and Tararua ranges. The alpine axis of the South Island was rising and probably already had a relief of the order of 1000 m but many subsidiary ranges had not yet acquired the stature of mountains.

Having made full allowance for the errors and inadequacies of these pictures, it is safe to say that the shape of the land at the beginning of the Pleistocene Period, a little under 2 million years ago, bore little resemblance to the present one. And that, after all, is not very far back; our oldest known rocks are 300 times older.

Reconstructions for periods before the Pleistocene Period are completely independent of modern surface outlines. Mostly they have been worked out by putting together all available clues as to where sedimentary strata were being deposited during the period in question and where erosion of land must have been going on to supply some or all of that sediment. Here and there, however, disconnected scraps of ancient land surface, sometimes still carrying soil, in other cases exhumed from under younger deposits, are recognisable. An intriguing example can be seen at the western end of Motutapu Island, Hauraki Gulf, where the present-day sea-cliff cuts obliquely across an earlier cliff buried under marine sandstone of Miocene age. It is uncommon to know so exactly where the coastline was located as far back as 25 million years ago.*

Fig. 4.10 (Chapter Four) will take us still farther back in time with the aid of a series of paleogeographic maps of the New Zealand region from about 250 million years ago,

* *Acknowledgement.* I wish to thank Mr Ian Keyes of the N.Z. Geological Survey, Lower Hutt, for arranging to have Fig. 2.4 (after Fleming, 1962) re-drawn for me by Mrs P. Williams, also of the Geological Survey.

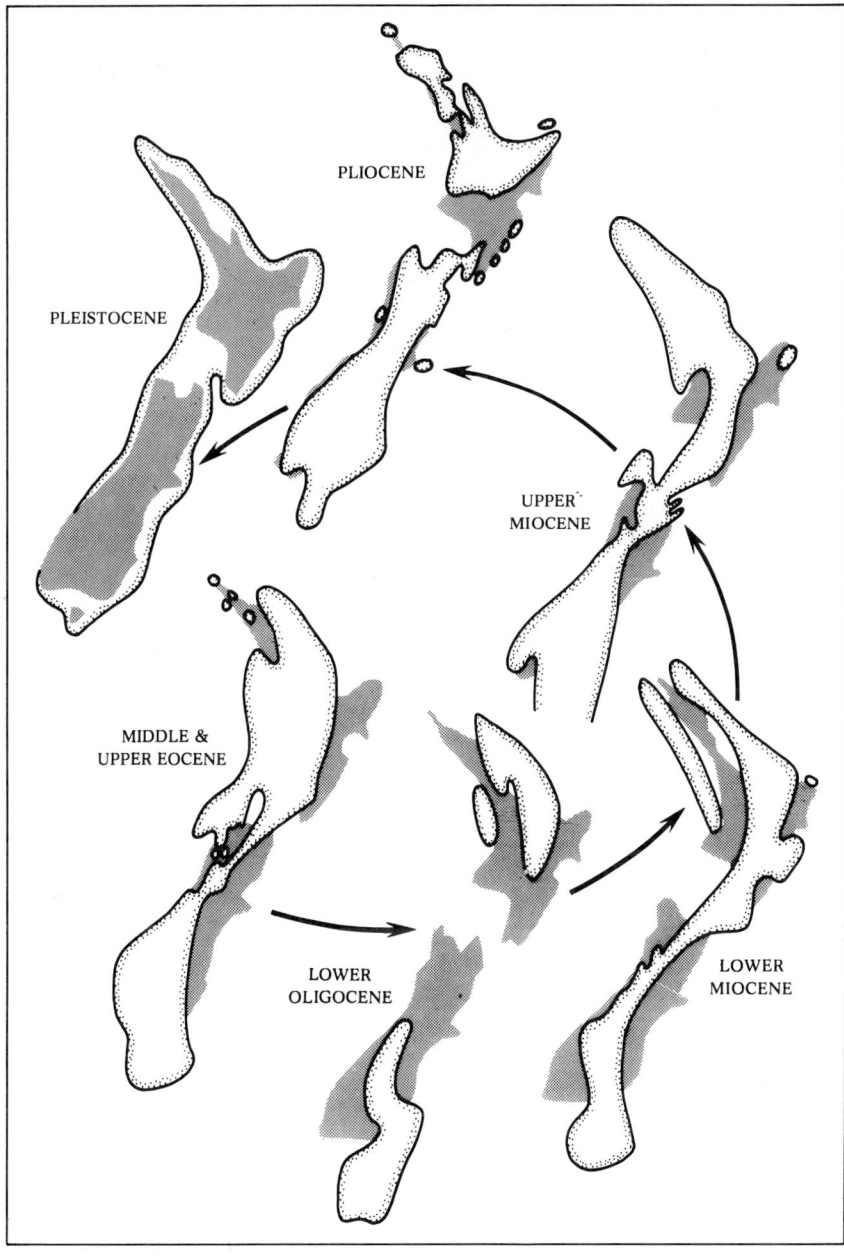

PLIOCENE

PLEISTOCENE

UPPER
MIOCENE

MIDDLE &
UPPER EOCENE

LOWER
OLIGOCENE

LOWER
MIOCENE

Fig. 2.4
The outline of New Zealand at different stages of our geological history shown by means of highly generalised and conjectural paleogeographic maps. A major difficulty is to provide a suitable basis of comparison. In this attempt the present outline of New Zealand is the basis, but it must be remembered that there have been offsets along the great north-easterly-trending faults which traverse the country obliquely, and that the effect of these offsets is disregarded in this series of maps, which were compiled for the New Zealand Encyclopedia *in 1966. (After Fleming. Crown copyright; reproduced with permission)*

built up from interpretations of the sedimentary rock record as indicated above. Rocks to represent all intervening ages are not present (or accessible) everywhere, neither do all those that are available for study yield unambiguous information about how they were formed, so these maps can only be generalised and highly conjectural. But they are worth making, with all their shortcomings, because there is no better way of representing an overall view of what is known about the geographical conditions of any region at particular times in the geological past.

For still earlier periods our knowledge is too vague and incomplete to

9. These exposures of Haast Schist at the car park in the Waiho valley downstream from Franz Josef Glacier were deeply covered by moving ice less than twenty years ago. Large blocks and smaller fragments detached along joint planes (e.g. the evenly sloping surface) and along the surfaces of foliation of the schist (facing the camera) and were plucked away by the ice. The schist belongs in the Garnet Zone of progressive regional metamorphism (p. 121). The photograph was taken in 1977.

10. Separation Point Granite outcrops at the southern end of Tatas Beach, Golden Bay. The sparkling, golden colour of the popular beaches of Golden Bay, Totaranui and Kaiteriteri is mainly due to an abundance of mica flakes released by the decay of this granite. Most New Zealand granites are much affected by joints and fractures as in this picture; they limit the size of solid blocks that can be quarried, and this is the main reason for their restricted use for building or ornamental stonework (p. 140).

11. *Spheroidal concretions in muddy siltstone of late Cretaceous age in the Waipara River, North Canterbury. Note that there are outer, incompletely cemented shells, obviously much softer than the core. Concretions from the same area have been found to contain, as central nuclei, pieces of fossil wood and reptilian bones (p. 179).*

12. *Blue-grey mudstone and siltstone and brown-weathering sandstone make up a large proportion of the late Tertiary marine strata in many areas, popularly going by such names as 'papa' in the North Island and 'Blue Bottom' (old miners' term) in Westland. This Miocene mudstone-sandstone above the limestone quarry at Tarakohe, Golden Bay, is typical (p. 186).*

justify preparing this type of paleogeographic map. Also the absence of an easily recognisable frame of reference to take the place of the present outlines, the limitations of which have been stressed already, becomes an increasing problem in the more remote past. Discussions about paleogeography are also confused by the fact that the shifts known to have occurred in the positions of crustal segments and in the direction of the axis of rotation relative to the body of the earth affect the applications of the terms 'north' and 'south' to New Zealand's geological past.

How Old is New Zealand?

The question now being asked is: 'At what stage in the geological evolution of this piece of earth crust can we justifiably begin calling it "New Zealand"?' The present outline has little relevance because land certainly existed here for a long time before it acquired its modern shape. The present form has in fact developed progressively from an archipelago of low islands back in the Oligocene Period. Few would object to these being regarded as an embryonic New Zealand.

It is less clear whether the name should be applied still further back in time to a low-lying land of sub-continental extent, which existed in this region of the globe from late in the Cretaceous Period some 70 million years ago until it became almost submerged during the Oligocene Period (Chapter Seven), or to the mountainous terrain from which it was developed by long-continued erosion. For some 150 million years prior to the emergence of that ancient land there had been a great ocean basin on this site which geologists call the 'New Zealand Geosyncline'. Then adjoining ancestral Australia, it accumulated enormous quantities of sediment destined to be the raw materials for the basement of much of New Zealand (Chapter Four). Though there are indications of still earlier land near at hand when our oldest known rocks were being formed, this could in no way be regarded as ancestral New Zealand.

It seems therefore that New Zealand can be said to date from the Oligocene Period about 30 million years ago, and less justifiably from the Late Cretaceous Period, more than twice as far back.

It emerges from the foregoing discussion that the original question was ambiguous, although geologists are often asked it in just those terms. There is a world of difference between 'How long has the present land been here, looking as it does today?', 'When did land first appear on this site?' and 'How old are the oldest rocks beneath the present land?' Clearly, it is easier to answer the last question than the others.

Our Place in the Pacific

In later chapters we will discuss the information that geophysicists have discovered about the nature of the earth's crust in the New Zealand region, its thickness and its history. To round off this chapter it will be useful to note that the region includes a segment of what has been termed poetically the 'Pacific Ring of Fire', a rather sensational designation for the zone of geologically young mountains, young or recently active volcanoes and prevalent strong earthquakes which almost encircles the ocean. More prosaically, it is known as the Circum-Pacific Seismic Belt (or Mobile Belt) within which occur not only a large proportion of all major shocks, but also all those of exceptionally deep origin. The reason for the prevalence of earthquakes and volcanoes especially around the margins of the ocean has become clearer in recent years, in the light of the well-supported hypothesis that the floors of the Pacific and certain other oceans are spreading from the centre and under-riding the edges of adjoining continents. The theory has become very popular because of the variety of observations and relationships which it can account for, leaving fewer anomalies unexplained than its competitors. (Further references, Chapters Ten and Twelve.)

As applied to this sector of the Pacific, the hypothesis suggests that the ocean floor is under-riding the eastern edge of the ancient continental slab to which the New Zealand Platform belongs. It was observed earlier in this century that we are near an important and very distinct geological boundary. This boundary is sometimes identified as the Marshall Line after the late Professor P. Marshall who had a lot to do with its recognition. The line runs northwards from New Zealand alongside the Kermadec Trench and separates the almost entirely basalt-producing volcanoes of the Central Pacific from the tract of continental character on the western side, including granite, ancient sedimentary and metamorphic rocks, and many andesite and rhyolite volcanoes.

It is now appreciated that hundreds of kilometres of sideways displacement of the crust has occurred along this line since the Cretaceous Period. Within New Zealand the displacement was mainly along the Alpine Fault and related transcurrent faults. Although this book is not the place for a full treatment of the sea-floor-spreading hypothesis, its application to New Zealand may be noted briefly here. According to recent versions (and it must be remembered that the theory is relatively new and still being developed) the basaltic floor of this south-western segment of the Pacific Ocean tends to thrust beneath our continental margin, driven by a slow westwards flow or creep of sub-crustal material, while at the same time a similar sub-crustal flow moves north-westwards on the western side of the southern extension of the Marshall Line. The hypothesis has found favour among many New Zealand geologists. It seems to fit with most of the facts of our later geological history as well as explaining many features of the sub-oceanic topography around us.

New Zealand is now in a temperate zone of latitude. According to some good authorities, the region has most likely been roughly within the same latitude range as it now occupies, despite other continental wanderings, since at least as far back as when this segment is supposed to have swung away from Australia in the Cretaceous Period. In earlier times, when our oldest strata were being formed, this region may have occupied other positions relative to the Poles.

13. Tertiary strata are very thick in the Murchison district. These almost-vertical late Miocene beds of hard sandstone, conglomerate and shale near Longford are part of the Longford Formation which exceeds 3000 m in thickness (p. 188).

14. Basaltic scoria making up part of one of the summit cones of Rangitoto Volcano, Auckland. The layers of this airfall ash deposit lie at the angle of repose of loose, dry debris. (Compare with Plate 15.)

15. This fine basaltic debris from submarine eruptions in the Oligocene Period would probably have been in horizontal or gently dipping layers when deposited on the sea-floor. The present steepness is the result of later crustal movements. (Compare with Plate 14.) The locality is near Avoca station on the Midland Railway in Canterbury.

Chapter Three

OUR MOST ANCIENT ROCKS

'No nehe noa atu'*

This chapter sets out upon the main task of this book, which is to present a survey of what has been learnt in substantially over a century of study of the rocks of New Zealand. It always seems logical in such surveys to proceed as far as possible chronologically, dealing in turn with successive episodes of regional geological history. However, strict chronological treatment is sometimes neither practicable nor kindest to the reader, and in this case it was found better to bring together as separate topics some types of geological events which went on continually or repeatedly.

The known geological history of New Zealand spans 600 million years and there are scraps of information from twice as far back as that. Our well-founded geological story divides naturally enough into three main geological cycles, each beginning with the formation of sedimentary rocks and ending with the distortion of those rocks and the raising of mountains, and separated by episodes wherein fluid granitic magmas were generated in the upper crust along with metamorphic rocks and profound erosion occurred. This sequence provides a central theme for the remainder of the book. It is broken, as hinted above, to bring together the story of our granitic and metamorphic rocks, volcanic activity of all ages, and the record of climatic events over the last two million years. The structural framework, though developed and modified during and between the successive cycles, is best dealt with in one place, before describing the comparatively recent modelling of the face of New Zealand and the pattern of earthquake distribution. The structure of New Zealand and its origin have an important bearing on these last topics. Finally, we look at the usefulness of all this information to New Zealanders.

Though there is now a substantial consensus of opinion on the broader issues and differences are mainly at the level of a detailed correlation of strata between areas, it would be a mistake to think that there are no disagreements at all. Geologists in New Zealand, no less than anywhere else, are quite capable of differing irreconcilably about how to interpret the one set of observations. Introducing an outline of the geology of the Canterbury region I once wrote: '. . . the author must deal with unresolved disputes and irreconcilable viewpoints by

---

* 'In most ancient times . . .'

making his own choice [on the reader's behalf] and if the selection
appears arbitrary it should be remembered that the available space
makes it impracticable to attempt a full discussion of conflicting views.'*
These remarks are no less relevant here.

When mentioning localities where examples of geological
phenomena or strata representing various periods may easily be seen, I
take it for granted that readers who want to see these examples for
themselves will try to obtain the published geological maps of the areas.
The whole of New Zealand is now covered by geological maps on the
1:250,000 scale (roughly 2.5 km per centimetre, or 4 miles to an inch)
and on larger scales, in every case with summaries of the geology of the
area printed with the map or in a separate pamphlet.† Many on the
scale of 1:63,360 (roughly 630 m per centimetre, or one mile to an inch)
accompany Geological Survey *Bulletins* which have been appearing at
frequent intervals since 1906. These usually contain a wealth of
detailed information, but unfortunately, many of the earlier issues are
out of print and not easy to come by.

The Age of the Oldest New Zealand Rocks

In the introductory chapter it was noted that rock in the geological
context consists of aggregates of mineral grains or particles of
pre-existing rock, or both. We now add the further thought that
whereas many rock components may have originated or at least
acquired their present form at the same time or only a little before the
rock itself was formed, other components had already been in existence
for varying lengths of time. All the same, when speaking of the *age* of
rock what is usually meant is not the ages of the components but the
length of time since the components were assembled. Ages, moreover,
have usually been given in terms of the conventional geological time
divisions, though it is becoming increasingly common to find them
expressed in whichever multiple is appropriate (ten-thousands,
millions, etc.) of the basic, secular time unit, the year. When dealing
with mineral components whose ages have been estimated using one of
the methods involving measurement of radioactive decay-products, the
radiometric age is the logical one to use.

The greatest age yet determined for a component of a New Zealand
rock sample is approximately 1300 million years for mica flakes in the
hard greywacke or quartzite strata of western South Island districts
known as Greenland Group. This date gives us merely a maximum
possible age for some constituents of the sediment which went to make
up the Greenland rocks, whereas, as recently indicated by a new fossil
discovery near Reefton, at least some of the strata mapped as
Greenland Group were deposited in the Ordovician Period, and are

---

* *Natural History of Canterbury*, G. A. Knox, ed., 1969, p. 25.

† As part of the national adoption of metric measures, future regional geological
mapping will be on scales related to the basic 1:50,000 scale with a metric grid instead of
a thousand yard grid, to be used in future by the Lands and Survey Department. The
Director of the N.Z. Geological Survey advises that stocks of the 1:250,000 geological
map series will be available for some years to come but that no further revisions or new
editions (e.g. Sheet 21, Christchurch, 1973) are contemplated. Ultimately, a new
1:250,000 series, based on the metric grid, will be prepared.

16. Ngauruhoe, photographed after the 1954 eruption which produced the first flow of lava beyond the crater seen since the coming of the pakeha. The flow is distinguishable as a dark band on the outer slopes of this composite cone (p. 226). (Photo: R. H. Clark)

Figs. 3.1 and 2
Roadside outcrops of the Greenland Group (p. 81) greywacke and slate near the Seventeen Mile Bluff between Greymouth and Westport. The upper picture shows a type of rock cleavage (not uncommon in these rocks) in which the cleavage planes cutting diagonally across the strata are distorted into an open 'S' form (hence sometimes called 'sigmoidal cleavage'). 'Gash veins' of quartz, supposedly filling tension cracks, are also visible.

therefore only about one-third as old as the oldest components. This also makes the Greenland beds younger than rocks in the Cobb Valley, north-west Nelson, which belong to the Haupiri Group containing fossils of Cambrian age—the oldest in New Zealand for which we have a reliable geological date of formation.

When systematic geological exploration of New Zealand was beginning more than one hundred years ago the men involved had acquired their previous experience and formal scientific education (if any) in the British Isles, Western Europe and North America. In those countries sedimentary strata of comparatively simple structure and dating from the Paleozoic Era contain abundant, well-preserved fossils and are much easier to study than rocks of corresponding age in New Zealand. The widespread forest cover and difficulties of travel here were severe hindrances in the earliest attempts at stratigraphic studies. Our early geologists were frustrated by the comparative scarcity of determinable fossils in what they believed to be the oldest sedimentary rocks, and surprised by their hardness ('induration') and structural complexity. Pioneer workers were thus inclined to regard the older, apparently barren sedimentary rocks along with the metamorphic schists and gneisses as being older than Cambrian, that being the Period in which abundant and varied animal life is first recorded by fossils. There may also have been a lingering influence of the eighteenth-century doctrine which tended to identify the crystalline granitic and metamorphic rocks as relics of the earth's primitive state.

The older sedimentary strata became important in the early colonial days because of the gold and other valuable minerals they contained, and the achievements of the handful of early New Zealand geologists in view of the obstacles mentioned above were quite remarkable. Mistakes were made, however, in trying to identify New Zealand fossils with similar-looking ones in the northern hemisphere, and there is little point in tracing out here the involved and confused story of initial attempts to classify and date our oldest rocks. For those who would like to follow the development of ideas through to modern times, a rather cynical but well-documented review by J. B. Waterhouse, written for the occasion of the New Zealand Geological Survey centennial celebrations, provides a good beginning.*

Before ending this section it would be as well to point out that lower parts of the otherwise wholly unfossiliferous Greenland Group strata of south-west Nelson and Westland, though indistinguishable from the rock which yielded Ordovician fossils recently at one locality near Reefton, might yet prove to be older and possibly even of Precambrian age, as shown on the New Zealand Geological Survey 1 : 1,000,000 scale geological map of the South Island published in 1973.

Vanished Ancient Terrains

Every sedimentary rock is haunted by ghosts—properly attuned observers can detect spectral glimpses of its ancestry, of the sources of

---

* J. B. Waterhouse, 'A Historical Survey of the Pre-Cretaceous Geology of New Zealand', *N.Z. Journal of Geology and Geophysics*, vol. 8, no. 6, pp. 931–98; 1965.

its components, and of the environmental conditions existing while these were derived, transported and finally put together. Sediments that accumulated on mid-ocean floors are made up very largely of organic remains, inorganic chemical products and minute particles of volcanic or perhaps cosmic origin, but the greater thicknesses of strata of all ages include large proportions of grains, pebbles, particles of any size derived from erosion of some pre-existing land. These are the clastic or detrital class of sedimentary components—giving the useful word 'clast' for an individual detrital fragment, regardless of its size. Where volcanoes were erupting at the time, volcanogenic (or pyroclastic) components of all sizes may constitute part or all of the sediment supply. Naturally, the larger and more varied the component fragments, the greater the range of information yielded by them, though much can be learnt from the shapes and sizes of grains of a single mineral, such as quartz, which may make up practically all of a sandstone.

Rounded, clastic pebbles making up a conglomerate bed give direct information about the kinds of rock which underlay the surface of some area being eroded at the time of its formation, and may point to streams or currents as the likely agents of transportation. The kind of stratification within the conglomerate bed may even give a clue as to the direction from which the sediments came, and with great good fortune fossils inside the pebbles or radiometric age determinations may make it possible to 'date' the source terrain. Some distance and duration of previous stream-transport or working by waves and currents on a beach must have occurred to round the edges of rock fragments which were angular when first broken from the parent rock. Therefore, the angular shape of breccia components may suggest that the source area was not far enough away from the site of accumulation of the breccia to give an opportunity for rock fragments to be worn into rounded shapes. That is unless there are also some special textural or structural features in the breccia, and associated beds indicate the need to look into the possibility that the clasts were transported to the site of deposition in a mudflow, or by a glacier—agents which are generally less effective than running water in the rounding of rock fragments.

A conglomerate bed composed mostly of cobbles and pebbles of granite must have been deposited not too far from vigorously eroding mountains of granite, metamorphic gneiss, or something of the sort. Sandstone beds made up of feldspar crystal debris and mica flakes and angular quartz grains also require an eroding granite terrain; without the feldspar crystals but with some clay instead, the implication would be that a gentler, deeply-weathered granitic landscape was the source. Seldom will there be conclusive evidence as to the age either of the eroding surface, or of the bedrock underneath, except that the granite must have been sufficiently older than the sedimentary rock product to give time for erosion to have exposed the crystalline rock types that could not have been formed at less than a few kilometres below the surface.

I may have succeeded in making the study of paleogeography seem delightfully simple, whereas in fact there are always uncertainties and possible alternative intepretations. Nevertheless a good deal has been discovered about the sources of materials making up our oldest sedimentary rocks in early Paleozoic times. Evidence from the Haupiri Group suggests that the source was an ancient, vanished land of appreciable topographic relief underlain by quartzite, granite, schist and gneiss, and that for parts of the time at least, volcanoes were erupting; some of these may have been on dry land, but perhaps the majority broke out under the sea in which Haupiri sediments were accumulating.

Apart from the possible volcanoes, what would the land have looked like? It must have been partly mountainous, lofty and steep enough to promote physical breakdown of rock by weather, and export of pebbles and cobbles of debris by streams. There was little of land vegetation in early Paleozoic times, except algae and perhaps other primitive plant orders with poor survival-value as fossils which would have exerted little moderation over the erosional processes. This makes it all the more difficult to guess at what the climate was like. The physical weathering agencies were active and running water in the form of substantial streams and rivers was available to transport the products to the sea, so extreme climatic conditions seem to be ruled out. Yet perhaps the closest modern parallel would be with the barren mountains of sub-equatorial desert belts in Africa, Australia and south-western United States, where physical weathering is vigorous and rain falls rarely, though in torrents when it does so.

Next question: Where were these early Paleozoic land sources located relative to where rocks of these ages now occur in New Zealand? To begin with, our wedge-shaped piece of continental crust lay alongside the Austral-Asian mass before the break-up, in the Cretaceous Period, of an ancient assemblage of continents usually referred to as 'Gondwanaland', but although the open ocean on the other side of us is unpromising as a source of sediment in early Paleozoic times, it would still be hasty to assume that the material for our oldest sedimentary rocks simply came from 'Australia'. Australia had not then achieved a separate geological identity by drifting away from the mother continent, and our area was still very much involved in the events leading to the formation of what was *to become* Eastern Australia. On the other hand, there are no indications that the clastic sediments (as distinct from what was contributed by contemporary volcanoes) came from any other direction. Most probably the ancient crystalline and sedimentary rocks of Precambrian age which are now the core of the Australian land mass formed the borderland and supply-source ('foreland') for the trough in which the early Paleozoic sediments of both Eastern Australia and New Zealand were deposited. These sediments in their turn, eventually became the rocks of the foreland for the trough of deposition for our Permian, Triassic and Jurassic strata in the next succeeding cycle.

We are, in fact, discussing a well-recognised process termed 'continental accretion'. The New Zealand rocks of early Paleozoic age represent a strip that was added on to the eastern fringe of our segment of the primordial continent; another strip was added later during the New Zealand Geosyncline episode* before the Gondwanaland assemblage of continents finally broke up and dispersed.

The term 'Tasman Geosyncline' has been applied by Australian geologists to events in Eastern Australia which lasted about 150 million years and affected New Zealand geology as well. A long-enduring, complex system of subsiding troughs made room for the thick early Paleozoic strata found today in the east and south east of Australia and Tasmania and the even thicker sequences which occur in the west and the south of the South Island. In both regions the sediments deposited at this time now make up distinctive groups of sedimentary and metamorphic rocks. In 1967 the name 'Buller Geosyncline' was given by C. A. Landis and D. S. Coombs to the part of the system in which the New Zealand strata were formed; recently the whole chapter of events has been described as the 'Early Geosynclinal Cycle' (*Geology of New Zealand* vol. 1; R. P. Suggate, ed., 1978).

In general the geological contrasts between Australia and New Zealand are more noticeable than the resemblances, but strong affinities do exist between parts of our older rock succession and equivalent strata in Eastern Australia. Similarity is closest of all as regards strata of Ordovician age in New Zealand and in Central Victoria, to the extent that our rocks can be classified and dated geologically according to a scheme developed for Victoria. Working with the Australians, R. A. Cooper has made detailed comparisons across the Tasman with some surprising results. For example, he claims that as regards both types of rocks and fossil successions, similarity between Central Victoria and New Zealand is closer than that between Central Victoria and the rest of Victoria, New South Wales and Tasmania, which resemble one another. Cooper is making a bold attempt to explain this peculiar situation in terms of a series of horizontal offset displacements during the Silurian and Devonian Periods. The present Tasman Sea is not in any sense a direct descendant of the early Paleozoic troughs of sedimentation.

R. A. Cooper, G. W. Grindley and M. G. Laird notably among New Zealand geologists have been working over recent years to produce a clearer picture of the situation in which our early Paleozoic rocks were formed. They see them as having been laid down in two quite different kinds of marine environment, one of which was dominated by the activities of a string of volcanoes, perhaps rather like the 'island-arc' festoons of modern oceanic volcanoes in the western Pacific, Indonesia, the Caribbean region and elsewhere, whereas the other one was almost free of volcanic material. The volcanoes erupted repeatedly through the Cambrian and Ordovician Periods, contributing a substantial portion of what went into an 'Eastern Belt' of sediments which include also lenses of limestone (now altered to marble) and conglomerate. The 'Western Belt' of non-volcanic sediments consists of quartz sandstones (now quartzite), greywacke and slate derived from a continental source which as we have seen must have lain to the west. That is, of course, in

---

* The geosyncline concept is explained more fully in Chapter Four.

terms of present-day compass bearings. When trying to comprehend what such directions would have meant in the past it has to be remembered that subsequent displacements and rotations of segments of the crust will have affected these.

Deposition of the Western Belt sediments began and ended earlier than in the Eastern Belt, wherein the youngest strata are of Devonian age. The granite rock of the present mountains of west Nelson and Westland is believed to lie along what was then the axial zone of the Western Belt rocks, from which the granites were formed later in late Devonian or Carboniferous times (some perhaps earlier as well). The eastern border of the Eastern Belt coincides with an important boundary dividing New Zealand almost from end to end, separating areas that have experienced different tectonic and metamorphic histories. This 'Median Tectonic Line' was identified and named by Landis and Coombs, and will be dealt with more fully in later chapters. Some early Paleozoic sedimentary rocks occur in Fiordland, but by far the most complete sedimentary record is in central and western Nelson.

Users of *Geology of New Zealand* will find that it introduces a change in the meaning of 'Buller Geosyncline'. The term attaches to the Western Belt only, and it is referred to as a 'miogeosyncline' to indicate that the sediments are believed to have accumulated in a less vigorously subsiding trough relatively free from volcanic activity, while the Eastern Belt is separately named the 'Anatoki Eugeosyncline' to imply that it had a more vigorous history in which volcanoes were involved. Time and experience will show whether the change has been worth the confusion and ambiguity which inevitably result from such modifications. Meanwhile, it will be convenient in this book to use 'Buller Geosyncline' in the way it was introduced by Landis and Coombs in 1967, and so avoid having to go on repeating 'the early Paleozoic geosynclinal sedimentary phase' or some such cumbersome expression.

Paleozoic Complex of North-west Nelson

The word 'complex' is applied in this context where a number of rock units occur together but not in a simple or consistent way. Never was it more highly justified than when A. R. Lillie used it in 1959 in a lexicon of New Zealand stratigraphic names to embrace the various groups of sedimentary, igneous and metamorphic rocks of early Paleozoic age in north-west Nelson. Since the whole assemblage seems now satisfactorily classified into groups of formations on the basis of rock-type coupled with position in the supposed succession of strata, we may soon forget that from the earliest days of geological exploration until recently these rocks were the subject of continual argument about the order of sequence of beds, their ages and their correlations (i.e. identifications with other strata elsewhere).

Seldom is it clear from a quick inspection on the ground how the identifiable formations and groups are related to one another. Recognisable fossils, though locally abundant, are not easily seen by the unpractised eye and are limited to a few formations only. All of the strata are involved in complicated folding, so that as often as not they are now upside-down, while in places older strata have been thrust

Fig. 3.3
Rugged crests formed of
Paleozoic quartzite and
schist in the central
mountains of Nelson. The
mountain in the fore-
ground is Mt Cobb. Tarns
nestle in hollows excavated
by small glaciers during
the Pleistocene Period.
(Photo: N.Z. Geological
Survey)

above younger. As if these difficulties were not enough, most if not all of the sedimentary and igneous rocks are to some extent sheared and distorted, and parts of them have been changed into schist by metamorphism.

A great deal has been added to our knowledge of these rocks since 1959, and a more refined classification has emerged, but the structural picture remains far from simple. Not surprisingly, perhaps, since we are talking about the part of New Zealand that has the longest geological history of all, and an eventful history at that; yet it would be a mistake to assume that geological age and geological complexity are always directly related. In the present chapter we will be concerned mainly with the less-altered sedimentary rocks, but these will include conglomerates in which the component pebbles have been drawn out into elongated forms, fossils distorted in shape, and limestone converted into crystalline marble. The major stratigraphic groupings used on the published geological maps of the region (Haupiri, Aorere, Mount Arthur and Baton River Groups) provide logical sub-headings for the descriptions which follow.

The Nelson rocks of Paleozoic age received attention in the very early days. They were examined in 1859 by Ferdinand von Hochstetter and again by Julius von Haast in 1860. Largely because of the many mineral occurrences in Nelson Province, every one of the original Geological Survey geologists and a few mining engineers as well studied and reported on the older strata, the granites and the metamorphic rocks during the next few decades.

In their first attempts to sort out the Nelson Paleozoic succession, the old Geological Survey (i.e. the organisation established under James Hector in 1865) classified the beds now embodied in the Haupiri Group of Cambrian age as the 'Te Anau System', chiefly because they contained volcanic debris and in other ways resembled the late Paleozoic strata of Southland. Strata of the Baton River Group which had been found to contain fossil corals and brachiopods near Wangapeka were called 'Upper Silurian' (they are now considered Devonian, so the original guess was not too bad), while the underlying Mount Arthur rocks and the 'Aorere Slates' containing graptolite fossils which eventually gave their true age as Ordovician were classed 'Lower Silurian' (again, creditably close to the mark).

Many changes, however, were to follow during the ensuing eighty years, and one of the central causes of prolonged confusion and dispute was the Haupiri Group. Although instinctively suspected by sound observers to be relatively old, Haupiri beds had the habit of turning up in positions apparently high in the succession, and were therefore given various guessed ages ranging between Devonian and Permian. The problem finally became acute in 1948, when W. N. Benson found trilobite fossils of Cambrian age in the Cobb Valley. Successive expeditions then undertook the challenging task of solving the structural obscurities of north-west Nelson from a fresh standpoint, making use of much structural information that had been ignored previously but which indicated the places where the strata are now upside-down. A bold solution was conceived by G. W. Grindley about 1960. According to this, the originally lower Haupiri volcanic sandstones, conglomerates, slate and limstone now seem to be *above* rocks of Ordovician age because they form part of great overriding thrust sheets, squeezed out when the tightly compressed isoclinal-fold structures collapsed and sheared under long-continued stress. This was an heroic attempt (not the first, however) to apply to a New Zealand problem the kind of approach which resolved complicated geology in the western European Alps earlier in the century. In its turn, Grindley's theory may fall victim to further fossil discoveries and detailed mapping.

The following is the order of succession and the ranges of geological age now generally accepted for the early Paleozoic rock groups in Nelson, age increasing downwards.*

---

* A long established and universally observed convention is to set out the names and descriptions of a succession of strata in tables or diagrammatic 'columns' with the oldest rocks at the bottom and the youngest at the top. Likewise, it is usual in writing about them to deal in turn with successively younger rock units. These practices follow logically from the basic stratigraphic 'Law of Superposition' which declares that in any sequence of strata that have not been inverted (and there are independent ways of detecting overturning) each bed is a little older than the one above it; and vice versa.

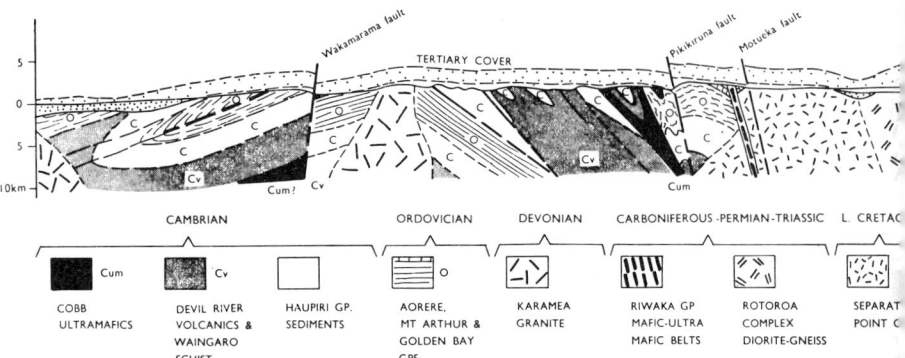

Fig. 3.4
Hypothetical section across west Nelson from the upper Takaka depression north-eastwards to Kaiteriteri, to show G. W. Grindley's view as to how the various rock groups are related to one another. (After Grindley, 1974)

Baton River Group   —Late Silurian to Early Devonian.
Mount Arthur Group—Late Ordovician to Silurian (?)
Golden Bay Group ⎫
Aorere Group   ⎬ —Early and Middle Ordovician
                    ⎭
Haupiri Group   —Middle to Late Cambrian

The following paragraphs will give a brief account of the rocks making up each of these groups and their major divisions.

Haupiri Group

When the New Zealand Geological Survey undertook to map the whole country at a scale of 1 : 250,000, some rather ugly chickens came home to roost. Problems of the kind discussed in the preceding section could no longer be put aside. Assigned the Golden Bay Sheet (13), Grindley grappled with the many-sided puzzles of the early Paleozoic rocks, and came up with a new classification of the Haupiri sedimentary beds into three formations, treating the associated metamorphic and intruded igneous formations separately. The Cobb Valley fossils which first indicated a Cambrian age are in limestone lenses within the middle sedimentary unit. Sheet 13 (Golden Bay) was issued in 1961. Further work by Grindley and others over the next ten years is incorporated with modifications and in finer detail in the 1 : 63,360 geological maps ('Mile to an inch') for the Takaka, Kahurangi and Cape Farewell-Collingwood areas. Table 3.1 (opposite) summarises how the Haupiri Group is subdivided for the mapping of the Golden Bay area.

The volcanic material making up the bulk of Devil River Formation, though considerably altered, can still be seen to have been originally lava flows, intruded sheets parallel with the bedding layers (sills), fragments of lava more or less welded together (agglomerate) and large amounts of finely fragmental and initially glassy volcanic debris (vitric tuff). The composition was originally andesite for the most part, but in a recent revision sills of basalt previously included in the Cobb Intrusives are counted in with the Devil River Volcanics. The alteration is both metamorphic and metasomatic (Chapter Five); volcanic mineral components are now reconstituted into chlorite, epidote, actinolite and other minerals indicative of low-grade metamorphism. In their present altered state the rocks might not be recognised as of volcanic origin

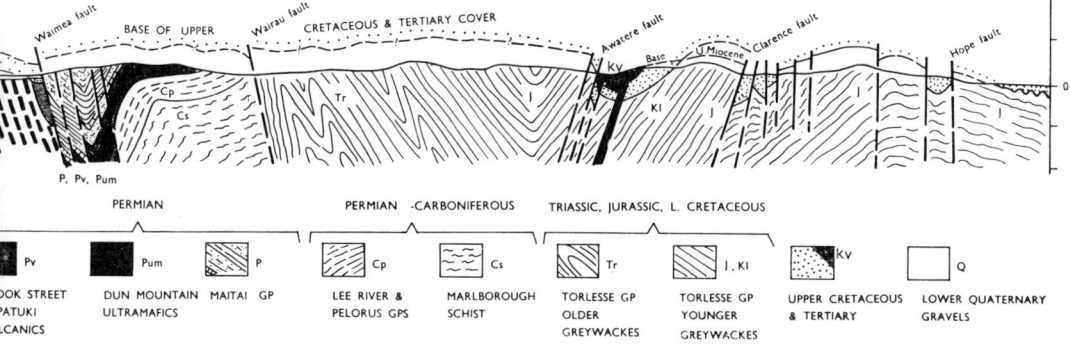

because of recrystallization and the imprint of fractures and cleavage systems and replacement of typically volcanic minerals. With a little experience, however, they are picked out readily enough by their characteristic greenish tinge on freshly broken surfaces of unweathered rock.

The volcanic sources were eruptions either on the floor of the eastern trough or along the eastern margin, or both. The Balloon Formation is made up of grey sandstone, dark argillite (hardened mudstone) with a micaceous sheen, marble, chert and green schist, the latter attributed to metamorphism of layers that were rich in fine, volcanic material. The sediments may have been laid down at the same time as the Devil River Volcanics but farther from the eruptive centres. They amount to about 330 m in thickness.*

The Buller Geosyncline event suffered an interruption at the end of the main period of volcanic activity in mid-Cambrian times; volcanic material is still recognisable in the Tasman Formation, but the

Table 3–1	SEDIMENTARY FORMATIONS	EQUIVALENT METAMORPHIC FORMATIONS	INTRUDED IGNEOUS ROCKS
HAUPIRI GROUP (Cambrian)	Sandhills Creek		
	Anatoki	Wakamarama Schist	
	Lockett Conglomerate		
	Tasman		Cobb Intrusives
	Balloon		
	Devil River Volcanics	Waingaro Schist	

(Adapted from notes accompanying Geological Map of New Zealand, *Sheet 8–Takaka; G. W. Grindley, 1971.)*

---

* Recent paleontological work bringing in some new groups of fossil micro-organisms suggests that the Balloon Formation may be of Precambrian age, and that the stratigraphy of the Mount Arthur rocks will be revised.

conditions had changed to the extent that a rich fauna of brachiopods, trilobites and other typically marine life could survive and be preserved in the eastern belt. Uplift of the floor of the geosynclinal trough raised fault-bounded blocks of country above sea level for a time and thick, coarse gravel fans were built out from them. These constitute the Lockett Conglomerate, made up of fragments of all sizes up to about 1 m across, some rounded but mostly angular in shape. They are mostly volcanic in some places, while in others there are recognisable pieces of the underlying Tasman Formation.

The uppermost division of the Haupiri Group as defined by Grindley in 1961, though somewhat metamorphosed, retains enough of the original sedimentary bedding features to show that sedimentation was going on again in geosynclinal fashion. The Anatoki Formation before it was altered would have looked very like the greywacke, argillite and greenish tuff beds with limestone lenses here and there that are typical of the preliminary geosynclinal phase of the mountain-building cycle. The Anatoki Formation has been identified as far south as Maruia Springs Junction in the southern corner of Nelson Province, that is, over a total length of 180 km from north to south. In the mapping of the Kahurangi Sheet, far west Nelson, a distinctive, bright green, spotted sandstone with slate or phyllite has been separated from the Anatoki Formation and called Sandhills Creek Formation. How it relates to the rest of the group is not fully clear.

More highly metamorphosed equivalents of Haupiri sedimentary rocks, mapped separately as Waingaro Schist and Wakamarama Schist, are considered along with the other metamorphic rocks of Nelson in Chapter Five. Similarly complex, injected masses of dark-coloured igneous rocks (Cobb Intrusives) are now counted as part of the group.

Haupiri Group rocks are mainly to be found in the mountainous, rather inaccessible terrain between the upper Takaka Valley and the headwaters of the Aorere and Karamea rivers, and in the Wakamarama Range north west of Bainham. Undoubtedly, the easiest Haupiri localities to visit are in the neighbourhood of the Cobb hydro-electricity station. The road from the powerhouse to the dam traverses Devil River, Waingaro Schist and Cobb Instrusives outcrops, while Tasman, Anatoki and Balloon formations appear in the Cobb Valley upstream from the reservoir. Anatoki River and Track provide a cross-section 11 km long showing most of the formations. Smaller Haupiri areas are crossed by roads and creeks in the Baton and Wangapeka catchments, and by the Wangapeka Track. Sheet 13 (Golden Bay) of the 1:250,000 map and the Takaka Sheet (S.8) of the 1:63,360 series will be found useful guides for field study of this distinctive group of our oldest rocks.

'Aorere'—a Stratigraphic Shambles

The codes and conventions accepted by geologists in various countries differ in detail but are all in agreement on one rule. Every geologist who wants to give a formal name to any assemblage of rock of any type is required to provide a clear, diagnostic description and definition, giving both the lithology (i.e. all the descriptive characters such as mineral composition, size and shape of components, colour, stratification, if any, etc.) and a type locality. The latter is vitally important. It enables other geologists, if they wish, to check the diagnostic description and assess the validity of the unit to their own satisfaction or to make their own direct comparisons with samples collected elsewhere. These excellent practices have grown up only over recent decades, for the need was not seen in earlier days. With few notable exceptions the earliest published descriptions and localities are given in vague terms. Much of the controversy and occasional disharmony among geologists through the years sprang from ambiguity in original descriptions and localities, which led to different interpretations and applications by subsequent workers. It can be difficult and confusing, and often unrewarding, to try to trace the origins and conflicting past uses of old terms.

A notorious example is provided by 'Aorere', a name attached to older greywacke and slate rocks in western South Island districts which attracted a good deal of attention by being the country-rock for gold-bearing quartz veins. It was first used in print in 1878 by James Hector, then Director of the original Geological Survey of New Zealand, in a bare list of fossil localities and with no more information than 'Aorere slates: graptolite slates, Collingwood'. Through many subsequent variations, the meanings of 'Aorere' became a confused tangle. Besides showing what can happen as a result of a poor initial definition, its story points also to the dangers both in continually redefining names based purely on descriptive rock characters, and in extending the application of rock-units thus defined too far away from the original locality.

When first introduced, 'Aorere' presumably (though we cannot now be certain) was meant to include *only* the graptolite-bearing slates found at several places near Collingwood. A few years later (1883) S. H. Cox in the first long report on the geology of the Collingwood area gave the earliest description of a sequence of greywacke sandstone, argillite, chert and blue slate beds underlying what Hector had called 'Mount Arthur Series'. By 1907 the now-reconstituted Geological Survey when describing the Parapara district in its *Bulletin No. 3* extended 'Aorere' to take in schist, marble and what was termed 'complex carbonate rock' that had been part of the 'Mount Arthur Series' of Hector and some later authors, as well as the blue graptolite-bearing slates of the original 'Aorere'. But this was not all. The authors of *Bulletin 3* went on to include barren (i.e. unfossiliferous) greywacke rocks in the western part of the region.

'Aorere' thus distorted from its original meaning was bad enough, but it would not have been so serious if it had remained in the Nelson area. Unfortunately, as systematic geological surveys reached southwards during the next three decades through the Buller catchments into North Westland, it was the barren greywacke-slate sequences which carried the name 'Aorere Series' eventually as far south as the Big Grey River. At this point its complete synonymy, when used this way, with the 'Greenland Series' of Westland became obvious and embarrassing. After 1948 this grossly distorted application of

'Aorere' was officially dropped. Back in its home territory it was therefore restricted to something nearer the original use. At the same time 'Series' was changed to 'Formation' or 'Group' to conform with modern stratigraphic language, since 'Series' now has a special, formal meaning involving geological time divisions as well as position in the rock succession.

With such a sorry background, it might have been better if the term 'Aorere' had been discarded completely. However, G. W. Grindley and D. G. Bishop within the last ten years have continued to use 'Aorere Group' with further redefinitions during re-mapping of western parts of north-west Nelson. This now embodies several rather similar sets of strata comprising well-bedded grey sandstones, phyllites, and spotted schists lacking the volcanic material that gives a distinctly greenish cast to the underlying Haupiri rocks, in stratigraphic order as below:

AORERE GROUP (early-mid-Ordovician)	Patriarch Formation
	Roaring Lion Formation
	Aorangi Mine Formation
	Webb Formation

The rocks of the Patriarch Formation were part of the Mount Arthur Group on Sheet 13 (Golden Bay) of the 1:250,000 map series. The graptolite-bearing slates to which Hector gave the name 'Aorere' are now part of Aorangi Mine Formation which also, as the name suggests, is the host rock for gold quartz veins which were mined for several decades prior to the First World War.

If they were all present in one area (which they are not), Aorere Group rocks would amount to a total thickness of at least 5 km. Together with the Karamea Granite they form the greater part of the basement rocks in the far west of Nelson. Trampers on the Heaphy Track traverse Aorere rocks for more than 20 km at the eastern end of the walk, and if returning by way of Wangapeka Track they see them again on both approaches to Wangapeka Saddle.

Golden Bay Group

This comparative newcomer arose from a regrouping for which no reasons have been published of formations previously classified with the Mount Arthur Group. Of the three sedimentary formations now in the Golden Bay Group, the oldest and the youngest (Leslie and Peel) date from 1929, when W. N. Benson and R. A. Keble first published detailed studies on the graptolite beds of Ordovician age in Nelson. An intervening succession of unfossiliferous strata was named Douglas Formation by Grindley in 1961, for the Golden Bay Sheet, and the supposed metamorphosed equivalents of these sedimentary rocks are recognised as Bay Schist. The sedimentary formations are described in similar terms in the booklet with the Takāka Sheet as successions of light-coloured quartzite beds with phyllites and green and dark graptolite-rich slaty rocks. The last have yielded some of our best-known Ordovician graptolite faunas. Despite complex folding and thrusting, the position of the Golden Bay strata below Mount Arthur rocks seems established. They form a well-defined belt southwards through the interior mountains of west Nelson to the Wangapeka

Valley, and are probably most accessible from the Flora Track-Leslie River route into the upper Karamea Valley and in the Cobb River headwaters. The total thickness is about 2 km, and the whole group is of late Ordovician age.

Mount Arthur Group

A dominating feature of the western skyline as seen by Nelson City residents is the broad-shouldered summit of Mount Arthur, popularly (and correctly) identified with marble crags and deep caverns. The geological associations of the name go back to the earliest days of the province, for it was mentioned by Haast in a report of his 1860 explorations on behalf of the provincial government and also by Ferdinand von Hochstetter in *Geologie von Neu-Seeland* (1864) in connection with fossil localities. As early as 1870 'Mount Arthur Series' appeared as a formal name (though without description) in a table of formations drawn up for the old Geological Survey by Hector, but it was not until 1879 that the rocks referred to were described by Alexander McKay.

There seems to have been no doubt in the minds of Hector and McKay that the slates in which Haast found fossils in 1860 were beneath (stratigraphically speaking, that is) the crystalline limestones of Mount Arthur itself, yet there was to be much confusion later as to how the marble was related to the groups of strata later called 'Aorere' and 'Haupiri'. S. H. Cox in 1883 described these as though lying above the marble, but in 1907, in the early days of the present Geological Survey, J. M. Bell and his co-authors of the Parapara *Bulletin* included the 'complex carbonate rock' and marble of that district *within* the 'Aorere Series', with the 'Haupiri Series' on top. Various other versions of the sequence were suggested by later authors and the true succession was not understood until after the discovery of Cambrian fossils in the Cobb Valley in 1948. These forced recognition of the fact, already suspected by some geologists, that the Haupiri was really older. There was also difficulty in finding some continuous sections in the field, not complicated by folding or faults, to link the unfossiliferous strata above the marble south of Mount Arthur with unfossiliferous strata below the Baton River fossil beds, then believed to be of Silurian age (now Devonian). In later mapping the difficulty was resolved by introducing a Wangapeka Formation as a 'buffer' between them (see p. 79).

Anyone who has made more than a casual study of the older rocks of north-west Nelson can appreciate that it was going to need advances in paleontology, new techniques in structural mapping and bold interpretations of the complicated structures before the habit of Mount Arthur Marble of turning up in the wrong place in the succession would be explained. This state of affairs did not come about until midway through the twentieth century. At the outset of the new 1:250,000 mapping of New Zealand in 1957, almost a century after Haast's fossil discovery on Mount Arthur, it seemed that interpretations of the early Paleozoic succession had at last settled down. Thus, according to Grindley in 1961 (the legend for Sheet 13—Golden Bay) the content and ages of the Mount Arthur Group were:

Fig. 3.5
Quartzite and metamor-
phosed mafic igneous
rocks of the Onekaka
Schist formation at the
roadside on Takaka Hill
(eastern side).

Fig. 3.6 (right)
Roadside outcrop of
Pikikiruna Schist on the
western slope of the
Pikikiruna Range below
outcrops of Mount Arthur
Marble. It is thought to be
the metamorphosed equi-
valent of the Flora For-
mation (late Ordovician).

Wangapeka Formation }	late Ordovician-Silurian
Mount Arthur Marble }	
Flora Formation }	
Peel Formation }	middle Ordovician
Douglas Formation }	
Leslie Formation }	
Patriarch Formation	early Ordovician

The subsequent 1:63,360 mapping in the Golden Bay region by
Grindley and Bishop introduced a further regrouping and brought in
the supposed metamorphic equivalents along with the
less-metamorphosed sedimentary units, thus:

	Sedimentary:	*Metamorphic:*
MOUNT ARTHUR GROUP (Upper Ordovician-Silurian)	{ Wangapeka Formation { Arthur Marble { Flora Formation	(Onekaka Schist { Arthur Marble (Pikikiruna Schist

As noted earlier, Patriarch Formation was transferred to the Aorere
Group and the Leslie, Douglas and Peel formations to the
newly-established Golden Bay Group, but reasons for the changes are
not given. Let us hope they are the last for a long time.*

Flora Formation is described as a dark phyllite and argillite unit
similar to the Leslie and Douglas Formations. Arthur Marble (prefix
'Mount' eliminated so as not to clash with the group name) varies from
a recrystallised limestone still retaining traces of original sedimentary
structures to complex, distorted masses of marble showing evidence
that it has 'flowed' in the solid state under pressure. It is believed to
have been formed originally as many separate lens-like sheets of limey
debris, which might indeed have been reefs built on top of shoal areas.
The colour varies from almost white to a medium grey, the latter
attributed to finely-divided graphite dispersed between the calcite
crystals. At Mount Burnett, near Collingwood, the normal calcite

---

* These stratigraphic revisions are given some prominence here because some users of
the maps are likely to be puzzled by the unexplained differences between the legends of
the 1:250,000 and 1:63,360 versions. Future editions are likely to show further
amendments arising from current work by R. A. Cooper.

marble is partially or wholly replaced by the calcium-magnesium carbonate rock dolomite—valued for refractory furnace linings and as a source of magnesium. Near Kaiteriteri, where it is in contact with granite, the even more prized refractory mineral wollastonite is found. The 'complex carbonate rocks' referred to in connection with Aorere Group include marble lenses, but these belong to the Haupiri Group. As 'Takaka Marble', it has been a well-known ornamental stone.

Wangapeka Formation comprises quartzite and argillite with no particular distinction. This lack of distinction might be applied also to the Pikikiruna and Onekaka schist formations, identified on the basis of field relations with Flora and Wangapeka sedimentary formations respectively. Mount Arthur Group as a whole outcrops extensively along a belt between Pakawau and Springs Junction. Marble Hill, alongside Highway 7 where it crosses the Alpine Fault, is near the southern extremity. For obvious reasons, the thickness of the group is hard to estimate; Wangapeka Formation is estimated at about 3 km, and Arthur Marble at up to 1000 m, allowing plenty of bulk for the cave labyrinths and deep potholes in Abel Tasman National Park.

Mount Arthur graptolites compare well with the standard sequences in Victoria. There are also trilobites and conodonts (the latter are rather puzzling, minute tooth-like structures of an unknown animal). The marble is confidently given an Ordovician age, and the Wangapeka Formation is late Ordovician to Silurian.

Early Paleozoic Rocks Farther South

The Nelson belt of early Paleozoic strata tapers southwards as far as the Owen Valley, beyond which they appear less frequently from beneath the younger cover of Tertiary strata. Underneath, they are very probably continuous as far as Maruia Springs Junction, just south of which the Alpine Fault, cutting obliquely across the grain of the Paleozoic rocks, brings them in contact with greywacke and schist rocks of the New Zealand Geosyncline that have a more north-easterly trend.

But surely that cannot be the end of them; what happens farther south still? Before the notion of a great lateral displacement along the

Fig. 3.7 (left)
The relatively pure calcite marbles of the Mount Arthur Group weather mainly by solution, yielding little insoluble waste to form the basis for soil. Marble terrain, as in this view from alongside the road over the Pikikiruna Range to Takaka, is notoriously rough, unsuited to agriculture and often dangerous for grazing animals because of hidden cavities.

Fig. 3.8 (right)
A cliff of the dolomite rock (calcium-magnesium carbonate) which makes up part of the Mount Arthur Marble formation of Ordovician age on the slopes of Mount Burnett near Collingwood. Dolomite is used in the manufacture of some fertilizers, refractory furnace-linings and in glass-making.

Alpine Fault was put forward by H. W. Wellman in 1949 there had
been no consistent views as to the real nature of this major geological
boundary, though a few more perceptive members of the older
generation, including A. McKay and P. G. Morgan had referred to it in
terms of faulting on a large scale. It was Wellman also who pointed out
that the surface trace of the Alpine Fault could not be so remarkably
straight (as anyone can see when flying across the central Alps, given
clear weather in the west) if, as some had suggested, it was a great
eastwards-dipping thrust plane upon which the alpine chain had been
pushed up. Certainly, if such an east-dipping plane does slope
downwards into the crust under the western flank of the Alps it could
perhaps be hiding a strip of older Paleozoic rocks between Maruia and
the western corner of Fiordland where a small patch of Ordovician
strata has been known to exist for a long time. The late J. T. Kingma
was one of the proponents of this view, but the present consensus of
opinion is that a large amount of lateral off-setting along a steep fault
fits better with all the data. (See Chapter Ten.)

Precipitous fiord walls and craggy ridge-tops in far-western
Fiordland provide splendid opportunities for very agile people assisted
by helicopters to study clean, freshly ice-worn rock surfaces. It is ideal
for the energetic petrologist but frustrating for the stratigrapher
because the great bulk of older rocks in Fiordland consist of granites
and highly metamorphosed sediments from which he can make only
indirect guesses about the original character and ages of the rocks.
Only in the south-west corner are they sufficiently unaltered for fossils
still to be recognisable. The probable continuation of the Nelson early
Paleozoic sediments appears between Cape Providence at the entrance
to Chalky Inlet and Long Point. Over about 20 km the coastline
provides good exposures of only slightly metamorphosed greywacke,
quartzite and slate of the Preservation Formation which has yielded
graptolite fossils of early Ordovician age. This age corresponds with
that of the Aorangi Mine Formation of the Aorere Group in north-west
Nelson, and indeed the lithological character is similar too.

The Seaview Formation is partly the metamorphosed equivalent of
Preservation Formation, still showing traces of conglomerate pebbles,
stratification and other sedimentary features. It has been mapped
extensively in the central Fiordland mountains. B. L. Wood, who
compiled the Fiord Sheet (27) of the 1:250,000 series points out that
the structure and succession of Preservation Formation strata are not
clear except for the graptolite-bearing members, and that the
relationship of this formation with the other Paleozoic rocks is obscured
by fault boundaries. Nevertheless, the rocks of central and northern
Fiordland are now divided up into about a dozen formations, allotted
ages ranging from Cambrian to Devonian, and the whole mass is seen
as warped up in a broad, complicated arch structure ('anticlinorium')
and seamed with many intrusive sheets of granitic rocks dating from
both before and after structures and textures of metamorphism were
imprinted upon the Paleozoic sediments. The total thickness, though
impossible to determine closely, must be of the order of many
kilometres.

It has been known since at least as far back as 1875 that old sedimentary rocks existed in the south-western corner of the South Island. F. W. Hutton in his early *Geology of Otago* guessed the age of these to be late Paleozoic, perhaps because they were not very much altered. Fossils do not seem to have been collected from them until 1895, when a prospector named William Docherty collected samples with Alexander McKay, who had been sent to this inaccessible and stormy region to report on gold deposits in Preservation Inlet and Wilson River for the Mines Department. The graptolite fossils, then in a private collection, were examined later and compared with those in the Ordovician strata of Victoria by T. S. Hall in 1915. Further localities were discovered at Cape Providence by James Park in 1921 and reported in 1922. The first intensive study was made by a party led by W. N. Benson in January, 1933. Benson was able later to make comparisons with the Lower Ordovician strata of north-west Nelson.

Very likely, other north-west Nelson early Paleozoic formations have their equivalents among the metamorphic rocks of Fiordland; indeed, the 1:1,000,000 geological map of the South Island gives a general picture of the whole of the interior of Fiordland as being composed of metamorphosed early Paleozoic rocks.

Greenland Group—Paleozoic or Older?

At no time have geologists in New Zealand been unduly oppressed by worship of great authorities. They have been free to follow the perpetual urge of scepticism and curiosity, to question and to research, always in the hope of getting a little nearer to truth, yet never certain of having found a final answer. From the outside their endeavours may look like an endless succession of changes of mind, whereas in fact one is seeing the outcome of continual natural curiosity, that essential ingredient in scientific progress in all fields.

The principle has been well illustrated over the past century by the many changes of opinion as to the stratigraphic position and age of what we now call the Greenland Group. Particular importance was given to this thick sequence of strongly-folded, hard quartzitic sandstone and silky, light-coloured slates in Westland and south-west Nelson, because they are the host rocks of gold-bearing quartz veins in the Reefton and other formerly productive mining fields, and the ultimate source of much alluvial gold as well. It is an involved tale, linked with the 'Aorere' story told earlier, of complex structure, rarity of fossils, premature and incorrect correlations, conflicting interpretations of the evidence, all set against a background of continually evolving geological philosophy.

Since the rocks were first examined geologically by Haast in 1860 the whole succession, or parts of it, has been ascribed ages, based, in the absence of fossils, on indirect evidence, ranging from Precambrian to Triassic. The original urge to discover the age and relations of the Greenland rocks was economic—they yielded gold—but this has not been an important incentive since the beginning of World War II and the decline in interest in gold-quartz mining as an investment. Pure scientific curiosity was the direct cause of the latest upset which seemed

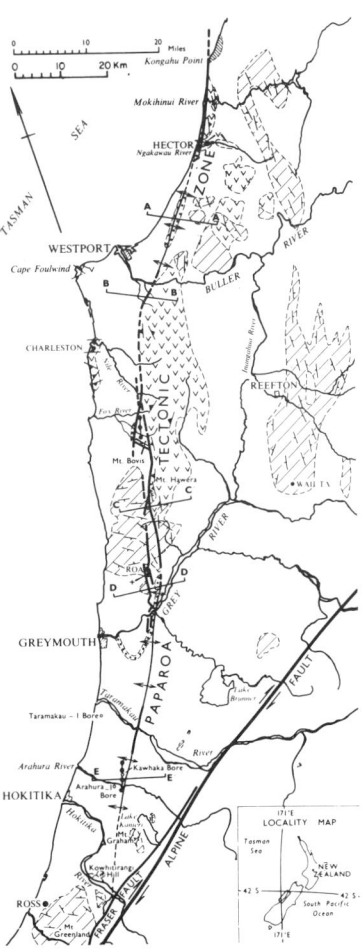

Fig. 3.9
Part of a map by M. G.
Laird which brings out the
contrast of structural trend
between the Greenland
and Waiuta Groups (both
areas shown by diagonal
ruling) on either side of
the Grey-Inangahua
depression. Structure is
indicated by conventional
geological map symbols for
strike and dip (see Appendix I). The v-symbols
indicate areas of granite
and gneiss.

to make it necessary to abandon a conclusion reached more than twenty years ago that the Greenland Group was very probably of Precambrian age.

Neglecting the earlier history, it will suffice here to say that from about 1955 until recently there had been general agreement on the following points:

1. The Greenland Group as first properly described (though under another name) is at least as old as early Paleozoic but probably Precambrian; at any rate pre-Devonian.
2. The marked difference in structural trend between the western belt of these rocks, to which the name was first applied, and the inland belt passing north through the Reefton goldfield justified giving the latter a separate name, Waiuta Group, after the locality of the Blackwater Mine which successfully mined gold from it for half a century.
3. It was not justified to use any of the formation names given to early Paleozoic strata in north-west Nelson, though certain similarities are noted.

The question of the Greenland Group and its associations seemed to be at rest. Then came the upset referred to above. The discovery of early Ordovician graptolites by R. A. Cooper in 1974 in the Waitahu Valley near Reefton was a surprise, and afforded great credit to the discoverer because for more than a century geologists and interested goldminers had been searching diligently and hopefully over rock outcrops, river bed gravel and mine dumps.

The discovery also seems to have stimulated renewed interest in the group, and with the aid of new techniques a lot more has been found out in the past few years. The story is instructive and well worth following further. Geologists with particular West Coast interests welcomed the Waitahu discovery as showing that at least part of the Waiuta Group could at last be given a definite age, i.e. Ordovician, while others who on the grounds of petrological similarity had rejected the need to separate Waiuta Group from Greenland Group were ready to attach an Ordovician age to the latter group as well. Yet it has to be remembered that apart from being very similar kinds of greywacke and slate rock, the two groups lack internal evidence of being the *same* age, and have distinctly different structural trends. The Greenland Group could still be of different age but for that principle of philosophy which declares that the simplest hypothesis that fits the known facts is better than a more complicated one. So, if the two groups are identical in all respects but the age of only one is known for sure, it is better to accept that they are identical unless and until some conflicting piece of evidence turns up. On the other hand, neither the top nor the bottom of the Waiuta Group is exposed anywhere, there is so far only one fossil locality, and its position within the group is uncertain. Lingering doubts must remain, therefore, that parts of either group could still prove to be younger than (or older than) Ordovician—perhaps even Precambrian.

The Ordovician graptolite discovery was welcomed especially by M. G. Laird and D. Shelley, whose careful stratigraphic and structural studies had already convinced them that the distinction of two groups was unnecessary and that the Greenland rocks are really deeper-water sediments laid down in the same seas and over the same period as some of the early Paleozoic rocks of north-west Nelson. This correlation was similar to that implied by J. Henderson in 1917 when he called the gold-bearing Reefton rocks 'Aorere', but it was more precisely stated and now had some support from the fossil evidence.

Radiometric dating both of the rock as a whole and of selected minerals in it had been attempted but the results were difficult to interpret until recently, when fresh evidence from an independent source offered a solution to the anomalies. It had been known for nearly ten years that particles of the mineral zircon in a sample of Greenland Group sandstone gave radiometric ages of almost 1500 million years, which is well back in Precambrian time. Zircon is a normal though minor component of granite and related crystalline rocks of plutonic origin, so the result pointed to the ultimate source of some Greenland material having been in some very ancient terrain of granitic rocks that had cooled down prior to that date. Other assays of the radio-isotope ratios (potassium/argon, strontium/rubidium, lead/uranium) in whole-rock samples, however, were showing that Greenland rocks had suffered a 'heating-up' and minor metamorphism during two or three separate episodes between roughly 500 and 300 million years ago. The various mineral components therefore had been assembled into a sedimentary rock long enough *before* the 500 million year date to allow for deep burial, heating-up, recrystallization, cooling and the winding up of the radiometric 'clock'. This is not in conflict with an early Ordovician date of deposition for part of the Greenland Group, but it does not rule out other parts being older.

Recent studies by S. Nathan on the chemical and mineral nature of Greenland sediments show that the bulk of the material did not come *directly* from a granitic terrain—for one thing, there are no feldspar grains—but rather from an intermediate source which must have been a land underlain by sedimentary sandstones and slates similar to the Greenland rocks derived from them. But how old was the earlier suite of sandstones and slates? They could well have been Precambrian, and this would account for the otherwise anomalous whole-rock radiometric age of 870 million years obtained from rocks in South Westland that had been mapped as Greenland Group. Important in this connection is the recent discovery by B. M. Hume of rocks indistinguishable from Greenland resting unconformably upon granitic gneiss in the central Paparoa Range. If his interpretation of the field evidence is correct, it would be very unlikely that sedimentary rocks in such a situation would *not* contain grains of feldspar.

It seems, therefore, that while we might not be able to distinguish, as rocks, between Precambrian Greenland and Ordovician Greenland, there could however be structural differences between them. Still farther back into the remote Precambrian past we can 'glimpse' the ancient granitic source of it all which, about 1500 million years ago, crystallised with zircon among its more durable components, and at the same time wound up its 'clock'.

The Waitahu discovery came too late to be incorporated in the 1:1,000,000 *Geological Map of South Island* (1972), which shows these with the oldest sedimentary rocks as 'Upper Precambrian?' from the Little Wanganui Valley near Karamea all the way south to Milford Sound.

West of the South Island main divide there are innumerable opportunities to examine Greenland Group rocks together with their metamorphosed products, hornfels and spotted slates, close by granite intrusions. They crop out on roadsides, in river gorges and banks, but nowhere better than on the coast north of Greymouth between Thirteen Mile Creek and Barrytown.

Do We Have Silurian Rocks in New Zealand?

Suggestions of Silurian age were made as early as 1864, when Hochstetter, following Haast, referred slates and greywackes at Mount Arthur to that Period on the basis of coral, trilobite and brachiopod fossils. Silurian also appears in a revised Geological Survey classification by Hector in 1877 to cover the 'Mount Arthur beds'. 'Lower Silurian' was the age given by McKay in 1879 to the Mount Arthur rocks, but it must be remembered that this was very shortly after the Ordovician had been recognised as a separate period in Britain. Also in 1879 McKay classified the fossiliferous beds found in the Baton River by gold prospectors a few years before as 'Upper Silurian'. Hochstetter had been considerably farther off the mark in suggesting a Silurian age for some rocks in the Auckland region.

The Baton River beds were always regarded as late Silurian in age until a large collection was sent to the British paleontologist J. Shirley, who caused a stir by reporting in 1938 that the Baton fossils were not Silurian, but Devonian and similar in age to the 'Reefton Series' (now Group; q.v.). Looking at both the character of the fauna and the kind of sediment containing it, Shirley concluded that the Baton River beds had formed in the comparatively deep water of an outer continental shelf, whereas the Reefton fossil beds were deposited closer to a shoreline. Shirley's findings were accepted, and 'Silurian' disappeared from New Zealand maps and classifications for some thirty years.

It has now been restored by R. A. Cooper, who in 1965 found grey quartzite with brachiopod fossils resting conformably (i.e. in unbroken sequence) on top of Mount Arthur marble on the Pikikiruna Range at the head of Riwaka River. Comparative studies with Silurian fossils overseas had, by 1970, confirmed an early Silurian age for the Hailes Knob Formation, as the quartzite is now called. This has raised the chances that lower, unfossiliferous parts of the underlying Ellis Formation are Silurian as well. Compilers of the 1:1,000,000 geological map took the plunge, and showed large areas of Silurian rocks west of the Motueka River.

The answer, then, to the question with which this section opened, is 'Yes'.

The Devonian—Further Puzzles

Outcropping Devonian rocks in New Zealand are confined to western and southern Nelson. The earliest fossil finds were in quartzite, limestone and shale, still to be seen alongside Highway 7 about 9 km east of Reefton. Haast saw the rocks during his penetrations of the southern Nelson hinterland in 1860, but the discovery of fossils is attributed to a prospector named Theodor Ranft working in branch valleys south of the Inangahua River in 1872. Hector looked at the fossils in 1873 and in the following year dispatched S. H. Cox and A. McKay to make collections and study the succession of strata. Ten years were to pass before the first detailed report appeared. Since then the fossil beds have been known variously as Devonian Series, Devonian Formation, Reefton Series, Reefton Formation, and now as Reefton Group at Reefton and Baton River Group in west Nelson.

From the outset controversy surrounded the proper interpretation of the Devonian beds and their relationship with other rocks at Reefton. This was inevitable, because the many contacts between the Devonian rocks and the adjoining Waiuta Group are either obscured or faulted. McKay and Hector correctly gave a Devonian age, but others later considered the Reefton beds to be Silurian because the fossils were similar to those at Baton River—a correlation that was to be confirmed much later, though with a Devonian age attached instead. The next piece of logic was not so good. Because the gold-bearing, but unfossiliferous, greywackes and slates had been correlated with the 'Maitai System', then considered Carboniferous, or Permian, or even Triassic, the Devonian fossiliferous beds being supposedly older, just *had* to be stratigraphically below them. And that was the 'official' Geological Survey view at the time. It then seemed strange that the 'older' Devonian strata had failed to become mineralised along with the gold-bearing rocks. This, however, would no longer be a problem today since it has become understood that permeating solutions may or may not deposit ore, depending upon the nature of the country-rock. None of the Devonian rock is very favourable country-rock.

J. Henderson in 1917, after a comprehensive study of Reefton geology, saw the weight of evidence as being in favour of the 'Aorere Series', as he called the auriferous rocks, occurring below the Devonian. Even more detailed mapping in 1935–7, in connection with geophysical exploration for hidden gold veins, followed by yet another remapping and revision in 1957, could come up with no better solution. Not until the Ordovician fossils were found in the Waitahu Valley in 1974 did it become virtually certain. Further detailed mapping and fossil-collecting in 1962 by I. Willis, followed by comparative study of the fossils in museums abroad where type specimens are available, has clarified the situation also of the Devonian strata at Baton River.

In modern classifications Reefton Group contains the best-known Devonian successions, long familiar to collectors at the roadside outcrops in the Inangahua Valley, in Lankeys and Stony Creeks where specimens of brachiopods, corals, and more rarely of molluscs and trilobites can still be found notwithstanding a century of collecting. Other localities can be found with the help of *N.Z. Geological Survey Bulletin 56* (R. P. Suggate, 1957) and accompanying maps, but one must be prepared to endure dense bush and undergrowth, sandflies, slippery rocks and waterfalls in gorgey smaller streams.

The lithological character of the Reefton Group is adequately summarised by the four formations defined by Suggate; Lower Reefton Quartzite, Reefton Mudstone, Reefton Limestone, and Upper Reefton Quartzite. The sequence and thickness, however, seem to vary from place to place. The rocks are less strongly fractured and sheared than the adjacent Waiuta Group, but the structure is nonetheless puzzling, and the succession, especially that in the roadside section, is incomplete and misleading as a result of faulting. Since there is neither a clear top nor a clear base to the group, only a minimum thickness can be estimated, and this according to Suggate is 1.2 km.

The Baton River Group also has been a Paleozoic fossil-hunting ground for a very long time. It, too, was discovered by gold prospectors and, as at Reefton, fossils occur almost invariably as casts or impressions. Outcrops are easily accessible in the Wangapeka River and its branches. The fossil-bearing Baton Formation is mainly of blue and green hardened mudstone (argillite) at the base, grading upwards into more sandy rocks and laterally into limestone and quartzite. The rocks are thus of a different sedimentary type or 'facies' from those at Reefton. The fossils show affinities with Devonian faunas in Australia, whereas a higher proportion of the Reefton ones are described as 'endemic', meaning that they have not been found away from New Zealand. There is also a small difference of age. Both groups are early Devonian, but the Baton River rocks are a little older in terms of the European standard succession.

South of Mount Arthur the Baton Formation is separated from Ordovician rocks by an unconformity, whereas in other places the unfossiliferous quartzite and quartz conglomerate of the Ellis Formation continues below with no apparent intervening break in the sequence. Accordingly, the Ellis Formation may span the Silurian-Devonian time boundary, and part of it could be equivalent to the Hailes Knob Quartzite.

Deposited apparently in a relatively near-shore environment, the Devonian rocks of Reefton reflect a slower tempo of sediment accumulation than do the older Paleozoic greywackes. Soon after the Reefton rocks had been deposited subsidence must have ceased, for they have suffered neither very deep burial under later sediments of the same cycle nor invasion by granitic intrusions.

Despite the unsolved structural problems, the Reefton Devonian area is, as it were, close to a beaten track, and it will continue to attract the interest of paleontologists and collectors. The list of Devonian fossils from the Reefton Group will be increased substantially as a result of work by Margaret Bradshaw still in progress at Reefton. She informs me that molluscs are turning out to be a more important element in the Reefton faunas, relative to brachiopods and corals, than the older literature would suggest.

End of the First Episode—Tuhua Orogeny

Long episodes of geosyncline sedimentation are sometimes interrupted by mild uplift, reducing the capacity of the trough to accommodate sediment; or by vigorous uplift and erosion of adjoining land areas, or by both. In any case, conglomerates or non-marine sediments (these more easily recognisable after the appearance of land plants and animals) take the place of typical geosynclinal greywacke and unconformable breaks interrupt the succession of strata. Such events may occur midway through the geosynclinal episode and do not necessarily lead into the orogenic phase with invasion of granite plutons into higher levels.

Development of the early Paleozoic Buller Geosyncline was thus interrupted, first in late Cambrian time to bring about the formation of the Lockett and Anatoki conglomerates as well as local breaks in the Haupiri succession. This event has been given the name 'Haupiri Disturbance'. A further interruption in late Ordovician or early Silurian times caused the Baton River and Reefton Group strata to be separated from older Paleozoic rocks by unconformity. This later event, and not the Haupiri Disturbance, is regarded as an initial phase of the Tuhua Orogeny which brings to a close the early Paleozoic cycle.

The main phase of the Tuhua movements affected the whole mass of Buller Geosyncline rocks, with folding, metamorphism and, in part, intrusion of the Tuhua Granites (Chapter Six). It is marked by a substantial time-gap in our sedimentary record. Roughly one hundred million years, from middle Devonian to late Carboniferous, are missing, making a clear separation between the events described so far and those connected with the New Zealand Geosyncline (Chapter Four).

As well as bringing to a close the early Paleozoic cycle of sedimentation, the Tuhua Orogeny laid foundations for events that were to follow. The tectonic squeezing of geosynclinal deposits, their metamorphism and invasion by granite 'froze' the former mobility of the sediments and added them to the already rigid eastern margin of the Austral-Asian slab. In tectonic language the Buller Geosyncline sediments were 'cratonised'. The accompanying upwards bulging of the crust, which has been termed the 'New Zealand Geanticline', provided the foreland which was to be an important source area for sediment when renewed subsidence offset to the east initiated the New Zealand Geosyncline.

Chapter Four

NEW ZEALAND IN MIDDLE AGE

(Geologically Speaking)

The 'Greywacke'

We cannot take up the next part of the story without some discussion of the kind of hard rock which is most commonly seen by New Zealanders in nearly all parts of the country except the volcanic regions and the metamorphic rock terrains of the south. This is the bluish-grey or greenish-grey sandstone known even by many non-geologists as 'greywacke'. We see it as gravel on the river beds and beaches, as metal on the macadam roads, as shingle on the mountain scree slopes and as aggregate going into concrete mixers. It is a pity that as a formal petrographic term this familiar rock name has its weaknesses which result from its long use in different countries and in different contexts before its meaning was properly defined. The word had its origin in the language of eighteenth-century German mineralogists. Hochstetter first used it for New Zealand rocks in 1859. Many geological papers and articles have been produced in the course of attempts to arrive at a consistent, generally acceptable meaning, or to replace the word with several new words specially coined as part of the complex systematic terminology now favoured by some petrographers. Meanwhile, some avoid using the term but many geologists in New Zealand, Australia and western Europe go on applying 'greywacke' according to local custom and tradition despite its ambiguity and in face of objections by purists—sometimes acknowledging their slackness in so doing by writing it at least once in inverted commas, as I have done above!

Refinements and sub-classifications of greywacke in terms of mineral compostion have never caught on, except for the recognition of two main varieties (the names of which we need not include here) and intermediates between them. In one, quartz grains occur in a fine matrix of flaky minerals derived from the alteration of original clay in the sediment; in the other feldspar partly substitutes for quartz along with a substantial amount of debris from pre-existing igneous rocks, either granite or contemporary volcanic material. The greywackes that were mentioned when describing the early Paleozoic rocks of New Zealand are of both types, whereas those referred to in the sections that follow contain some quartz but include much material of igneous origin, both as mineral grains and as rock particles.

Some of our greywacke sequences of both types consist mainly of thick beds of sandstone, but usually the sandstone strata alternate with mudstone or siltstone beds of varying thicknesses. In our older literature particularly, these finer units were often called slate or argillite to draw attention to the beginnings of metamorphic alteration, or merely hardness ('induration') due to compaction and minor mineral changes. Hard, siliceous chert often occurs with the greywacke sequences, and where it is present volcanic rocks are generally found as well. Sandstone units show regular systems of planar fractures ('joints'), while the finer ones may also show a slaty sheen on fractured surfaces, sometimes cutting across the stratification. And, last but not least, individual beds may change in texture regularly from coarse at the base to fine at the top with an abrupt change to coarse again at the bottom of the next layer. This is what is meant by 'graded bedding' and it provides one way of telling where strata have been overturned by folding.

There has been much diversity of opinion also about the exact nature of the sedimentary processes which produce greywacke, especially about how the sediment was brought to the place where it accumulated. On the other hand, although there is a possibility that some greywackes could have originated in shallow seas or even in lakes, there is general agreement about the intimate relationship between thick greywacke sequences and geosynclines. They are universally accepted as typical deposits of the early phases of subsidence before the orogenic phase begins and while the foreland source-area is rising vigorously and volcanoes are active on or close to the margin of a geosynclinal trough.

From the very earliest days of the study of New Zealand geology notice was taken of the widespread occurrence of similar-looking greywacke rocks, mostly devoid of useful fossils, in the main mountain chains of both islands. The interest aroused by this phenomenon is seen, for example, in the simple classification of the supposedly older strata used by James Hector for the pioneering 1869 and 1879 geological maps of New Zealand. Not only because fossils were scarce, but also because of the failure of individual beds to allow themselves to be traced and mapped over any distance, the formal stratigraphic classification of the 'greywackes' caused a great deal of trouble to successive generations of investigators. At times the fashion has been to try to sub-divide and distinguish the greywacke formations of the main ranges more finely, making formal units on the basis of such lithological differences as could be detected, and on fossils in the restricted areas where these were abundant. At other times the tendency has been to try to meet the needs of regional mapping by 'lumping' into larger units, generalising and correlating sequences between one area and another, with or without the guidance of fossils.

The high peak of the generalising trend came early in the present century, and is sometimes identified, if not quite justly because others were involved, with Patrick Marshall and his widely read little *Geology of New Zealand* of 1912. Under the name 'Maitai System' Marshall lumped together, as being of Triassic and Jurassic age or 'Trias-Jura' for short, all the sedimentary rocks of the mountain chains of both islands; all the sedimentary basement rocks showing from beneath the younger Cretaceous and Tertiary covering strata; and even the foliated (i.e., layered) schists; in other words, virtually all the basement rocks of New Zealand apart from the fossiliferous older Paleozoics, the granites and the crystalline gneisses. The rocks thus embraced in the 'Trias-Jura'

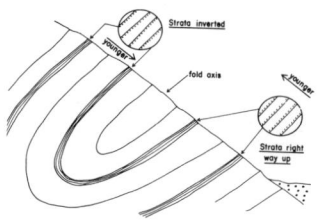

Fig. 4.1
Where graded bedding is present, the gradation normally would have been from coarser to finer upwards when the sediment was deposited. This can give a valuable clue to cases where the strata are no longer lying in the order in which they were laid down, but have subsequently been inverted. In this imaginary example, if it were not for the evidence from graded bedding, showing that the beds in the upper part of the outcrop are inverted, the presence of a tightly folded syncline could be missed.

Maitai System according to Marshall actually range in age from late
Carboniferous to early Cretaceous.

Innumerable errors, mis-correlations, mis-identifications of fossils
and a mountain of dogma were embodied in what J. B. Waterhouse has
called 'Marshall's vast Maitai empire'. It was defended with a certain
amount of authoritarian bluff for some years after an independent and
cynical British paleontologist named C. T. Trechmann in 1917 showed
that some of the supposed Triassic-Jurassic beds in Nelson were of
Permian age, and lingered until J. Marwick and M. Ongley were able to
produce field evidence which convinced everyone that late Paleozoic
strata intervened between Triassic strata and metamorphic rocks in
South Otago. W. N. Benson gave wide publicity to the revolution and
the overthrow of authoritarianism in his presidential address to the
Australasian Association for the Advancement of Science at Sydney in
1921.

Yet, despite all the error and confusion, for which Marshall was not
alone to blame, the obsolete 'Maitai System', stripped of its age
limitations, did express the undoubted unity of the sedimentary,
contemporary volcanic and metamorphic rocks generated within one
distinct episode of New Zealand's geological history. The truth of this is
borne out by the fact that the sedimentary rocks of the main ranges
were still being spoken of loosely as 'the greywacke', meaning that they
were a basement beneath younger rocks, until after World War II. In
short, the extended 'Maitai' concept defined a unity that has come to be
recognised anew in the last twenty years in the context of the New
Zealand Geosyncline and its deposits.*

The New Zealand Geosyncline

The 'geosyncline' concept is more than one hundred years old. It was
introduced to emphasise the importance of the discovery that
sedimentary strata representing a given span of geological time may
increase in thickness enormously within well-defined, usually elongated
belts on the regional scale. One of the early examples noted was in
North America; early Paleozoic strata which are no more than a few
hundred metres thick over the mid-continent region increase abruptly
to many kilometres as the eastern Appalachian mountain belt is
approached. It was recognised at the same time that along with this
thickening, the rocks became typically of the greywacke-argillite kind
with distinctive bedding features together with conglomerates and
volcanic rocks as well, and that towards the axis of the mountain belt
the sedimentary rocks were replaced by schists and crystalline rocks,
and were severely deformed by folding and faulting. It was noted also
that the boundary of the deformed and metamorphosed belt was
usually marked by one or more great thrust-planes.

Also recognised at an early stage was the implication that crustal
furrows or troughs must have been deepening over a long span of time
so as to keep on making room for such enormous thicknesses to

---

* Some readers may be interested to read J. B. Waterhouse's comments on the 'Maitai'
and 'Trias-Jura' story in *N.Z. Journal of Geology and Geophysics*, vol. 8, pp. 963–5, 1965.

accumulate below sea level. When first introduced the term 'geosyncline' focussed attention upon this inference that the earth's crust had been 'folded down' locally; later, its use conveyed two additional notions: firstly that of a distinct phase in the cycle of mountain building and a notable event in regional geological histories, as outlined in Chapter One; and secondly the term suggested the bulky 'wedge' of sedimentary and volcanic rocks originally as much as tens of kilometres thick but now greatly reduced as a result of subsequent uplift and erosion—in other words, the 'contents' of the downfold and the evidence for the geosynclinal event.

Geological exploration and the study of fossils in New Zealand had by the late 1920s reached a stage where estimates of the thicknesses of Mesozoic and Tertiary strata could be made with some confidence, even if most of them had to be minimum estimates because either the tops or the bottoms of the sequences measured, or both, were obscured or inaccessible. Moreover, it was seldom possible to match together sequences measured in different areas with exactness, if at all. Nevertheless, estimates were attempted.

Successions of Mesozoic strata studied in the North Island, though clearly incomplete, aggregated thicknesses of more than 8.5 km for the early Mesozoic and 3 km for the late Mesozoic (without taking account of great expansions of the Cretaceous portions subsequently determined during oil explorations). The full significance of these thick sequences seems not to have been appreciated when the estimates first became available. Credit is usually given to the bold and fertile mind of E. O. Macpherson for first perceiving that the stratigraphic and structural information which the Geological Survey and the oil exploration companies had been putting together during the years between the world wars could be synthesised into a slightly unusual history of geosynclinal trough-sinking, contemporary volcanic activity, subsequent folding—a story which links incidentally with the Mesozoic history of New Caledonia.

The full presentation of Macpherson's ideas in 1946, though complicated and rather vague, did for the first time emphasise the geosynclinal character of the basement rocks over most of New Zealand. He envisaged two parallel, linear troughs with associated chains of active volcanoes which he likened to modern, oceanic island-arc volcanoes; an older 'western' one dating from late Paleozoic to early Mesozoic, and a younger 'eastern' one from mid-Cretaceous to early Tertiary in age. These were thought to be separated by a 'median ridge' which persistently rose. From time to time the ridge interrupted the deposition of sediment along a central zone, introducing local minor breaks and unconformities. The whole system was subsequently to be distorted into a reversed 'S' form, but surprisingly, Macpherson failed to link any of this with the great dislocation which we now call the Alpine Fault. Despite its vagueness, its incompleteness, its complexity, Macpherson's scheme was imaginative and bold. It certainly provoked much thought and discussion over the next few years.

Ten years later the geosynclinal interpretation was fully accepted and had been placed on a stronger footing by H. W. Wellman. The coining of the expression 'New Zealand Geosyncline' is attributed to A. R. Mutch, but it first appeared in print in 1956 in Wellman's *Structural Outline of New Zealand*. This broad synthesis built upon the sounder aspects of Macpherson's ideas and took into account the many post-war advances, to which Wellman's own research had contributed very significantly. Since then, there have been successive revisions and refinements of the original concepts, bringing in new philosophies of crustal mobility and ocean floor substructure. A valuable summing-up of the situation was presented by C. A. Fleming in 1968 when he delivered the William Smith Memorial Lecture before the Geological Society of London.

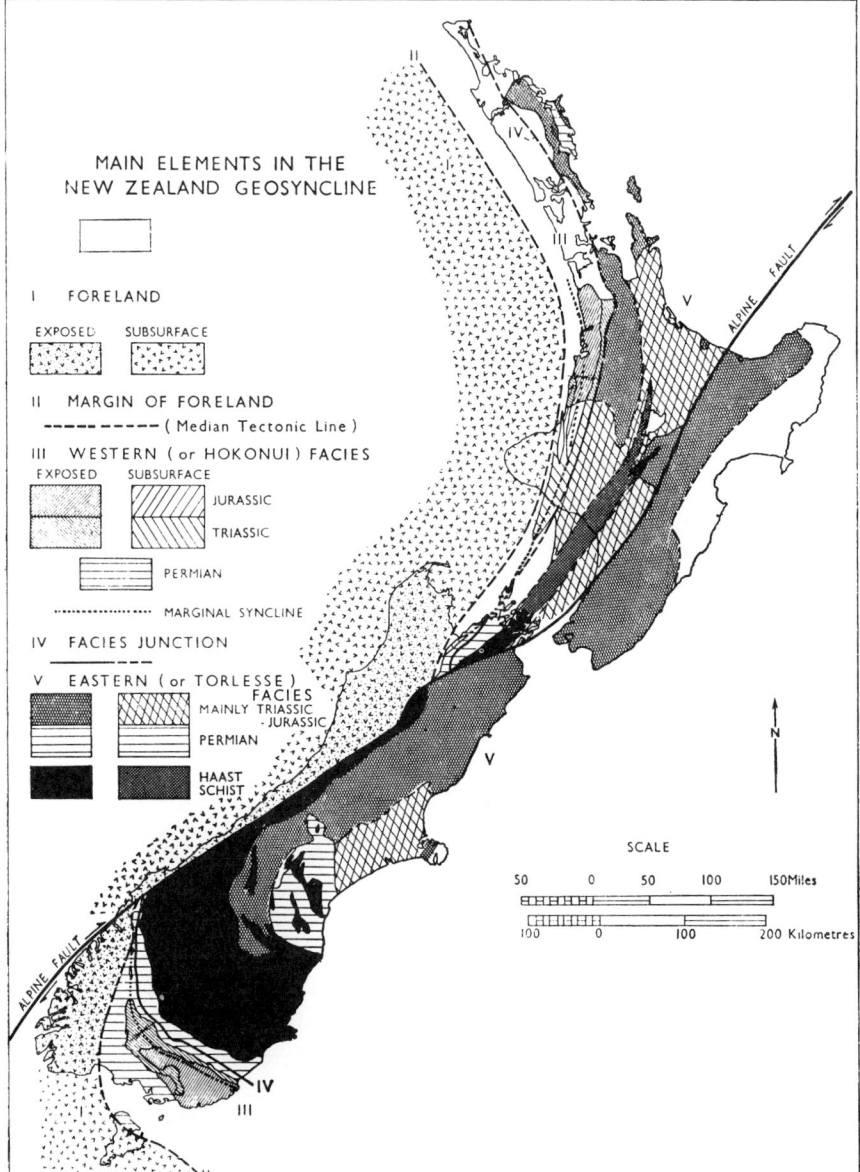

Fig. 4.2
The main elements of the New Zealand Geosyncline are outlined in this map by C. A. Fleming (1970) which distinguishes between the areas where the geosynclinal rocks are directly observable at the surface and those where they are inferred to exist beneath younger rock cover.

Modern studies of the New Zealand Geosyncline continue to distinguish between a western and an eastern belt and to emphasise the contrasting types of sediment deposited in them, but there are differences of interpretation of the structural evidence and some confusing differences in the use of terms by different authors as a result. The terms introduced by Wellman in 1952 to denote the contrasting rock types, 'Hokonui Facies' for those of the western trough and 'Alpine Facies' for those of the eastern trough, are still understood and frequently heard. The words 'marginal' and 'axial' also indicate the general belief that the western trough was close to land at the time, and the eastern trough nearer the axis of the geosyncline as a whole, though probably not identical with it. In addition, there are separate names for the strata that accumulated in the troughs, e.g., 'Murihiku Group' (or Supergroup) for part of the marginal trough contents in Southland and Otago; 'Torlesse Group' (or Supergroup) for Alpine Facies rocks everywhere. However, at the time of writing the naming of these units is again under discussion, and it remains to be seen whether any of the new names that have recently been proposed will come into general use.

To return to 'Hokonui Facies' and 'Alpine Facies', the distinctions emphasised by these terms as originally used by Wellman were supposed to be applicable to the whole of what we are now calling the New Zealand Geosyncline rocks of all ages spanned by that event. The main points of contrast were:

Hokonui. Abundant volcanic material as lavas, tuffs, and volcanic breccias; sandstone and conglomerate predominant over mudstone or argillite; limestone beds in the parts of Permian age; fossils common; relatively simple fold structures.

Alpine. Very thick; sediments of volcanic origin less common than quartz-rich greywacke and argillite; limestone rare; fossils in widely scattered localities, though abundant where found.

It has proved to be an oversimplification. In the quarter-century since these names were introduced a tremendous amount of additional stratigraphic and structural information has been gathered from the New Zealand Geosyncline rocks throughout the country and it has become increasingly difficult to make generalisations like those

Fig. 4.3 (left) Greywacke sandstone and siltstone of the Torlesse Supergroup, in places including conglomerate and basaltic volcanic material erupted beneath the sea, make up much of the mountainous country of both the North and South Islands. A fair range of Torlesse rock types is well displayed in the sea cliffs and rocky tidal platforms of the southern Wellington coasts. In many places (as for example here, west of Ohiro Bay) they are severely folded and fractured. Some of the deformation probably occurred while the material was accumulating in the geosynclinal trough, but the fracturing and crushing occurred mainly during the Rangitata Orogeny.

Fig. 4.4 (right) A Brunton-type compass is being used to observe the strike (the direction of the line of intersection between rock layers and the horizontal) in Torlesse greywacke rock near Island Bay, Wellington. The same instrument can be used in another way to measure the dip, that is, the deviation of the strata from the vertical.

embodied in the two-facies concept which apply consistently to both
Permian and post-Permian portions. More fossil localities continue to
be found in the Alpine Facies, while also the boundary between
dominantly volcanic-derived sandstones and dominantly
granite-derived greywackes was not consistent throughout the history
of the geosyncline. Admittedly (as J. D. Campbell has urged in a recent
letter to me) it may be unwise to go on using Wellman's facies names
because, in the light of later knowledge, they tend to exaggerate the
differences and obscure uncertainties about the true position of the
geographical boundary between them at different times. Yet to
abandon the two-facies idea altogether would amount to ignoring a
distinction which is real enough in a general way, and which must have
a significance that cannot be overlooked when working out the history
of the geosyncline.

There are difficulties also with regard to the alternative pairs of
terms, namely 'marginal' and 'axial' (because it is now clearly an
oversimplification to suppose that the trough consistently subsided
more rapidly along one fixed, central line); and 'western' and 'eastern'
(which runs into complications in the south because the general facies
boundary swings almost to an east-west direction). For want of a better
general term at the moment I have decided to use 'Hokonui' and
'Alpine' in this book, hoping that it has now been made clear to readers
that the contrast is not as sharp, nor is its application as simple as it
seemed to be when the terms were introduced.

The boundary between the two facies belts, spoken of now as the
'facies junction' has come in for a good deal of scrutiny recently. In the
Auckland district it is associated not only with a change of sediment
type but also with a change eastwards from softer to harder rocks.
Farther south, geophysicists have shown that a hidden belt of rocks
with strong magnetic properties runs beneath younger rock cover from
Kawhia to the Taranaki Bight, thence across Cook Strait to join up with
igneous rocks in Nelson that terminate against the Alpine Fault. The
western boundary of the Alpine Facies should lie to the east of this
igneous belt, and might have been expected to reappear on the other
side of the Alpine Fault in western Otago, but in fact, although
theoretically the facies junction should be marked by a belt of igneous
rocks of Permian age sweeping eastwards across Otago, there are no
typical Alpine Facies greywacke rocks between this belt and the schist
rocks of Central Otago.

Sedimentary strata of the Alpine Facies grade laterally into
metamorphic rocks of the Haast Schist belt (Chapter Five). Contrary to
the beliefs of Macpherson and Wellman, the main belt of metamorphic
rocks appears to lie not between the facies belts, but rather to the west
of the facies junction line.

In most areas the deeper parts of the geosynclinal contents have been
too much altered by metamorphism for any basal contact with older
rocks to be recognisable. W. R. Lauder, however, has suggested that a
complex of granitic rocks exposed on the eastern shore of Tasman Bay
between Wakapuaka and Pepin Island might conceivably be a surviving
scrap of older basement beneath the western flank of the geosyncline.

Still farther west, in the Golden Bay area, strata of Permian age overlap on to the early Paleozoic rocks that were part of the 'foreland', but they cannot be considered to have anything to do with the geosyncline.

Double-handling of the Sediment

Geosynclinal deposits commonly show a diverse range of types of stratification and texture which are not easy to reconcile with a simple picture of progressive accumulation of sediment on the floor of a subsiding trough. These features are usually explained by supposing the sediment reached the final site of deposition only after many stages and by different modes of transportation across the ocean floor. In the first stage mud, sand and gravel were deposited directly from currents at or near river mouths or other major bulky sources of sediment such as erupting volcanoes, but only temporarily. Most of it soon moved farther out as soon as the oozy, water-saturated sediment accumulation

Fig. 4.5
Greywacke and slate beds of the Torlesse Supergroup on the Pioneer Ridge at the head of Fox Glacier now dip almost vertically as a result of the Rangitata and Kaikoura orogenic movements combined. The small-scale folding, puckering and squeezing were produced before the sediments had consolidated, probably while they were slumping and sliding down-slope towards the trough of the New Zealand Geosyncline. (Photo: F. Barta)

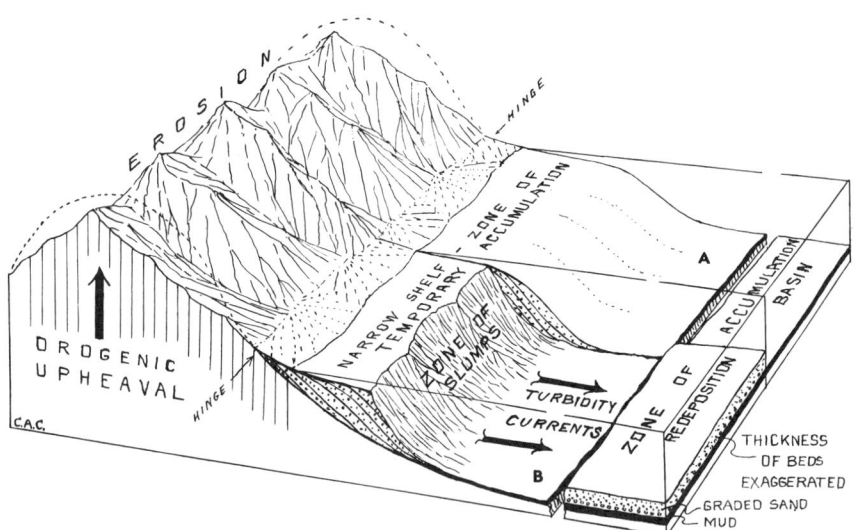

Fig. 4.6
The vertical scale is exaggerated in this diagram by C. A. Cotton, to illustrate the double-handling of sediments; slopes, especially submarine, are in fact gentler and horizontal distances greater than they appear here. (Note particularly that there is a break, representing perhaps many kilometres of sea-floor distance, towards the right of the diagram.) The point being emphasised is that floods of sediment derived from erosion of a rising landmass near the coast are usually deposited first in near-shore zones or on the continental shelf and slope. Later, these unstable deposits are likely to slump, slide and flow seawards in turbidity currents, to be deposited finally in well-bedded sequences. Most of the typical geosynclinal sequences of sediments are now thought to have experienced such double-handling.

became too unstable to remain even on gentle slopes of one or two degrees; then, when the resulting submarine slide or mudflow came to a rest, much of the sand and mud was carried on farther in a succession of pulses of highly turbid water-and-sediment flows spreading out across the bottom until finally the deposits from both mudflows and turbid-water flows came to rest permanently in more stable situations.

Individual beds caused by the settling of each distinct mudflow have the characteristic greywacke mixture of different particle sizes, whereas the layers formed by the settling of successive tongues of turbid water beyond the limits of the mudflows which generated them developed the gradations from coarser to finer sediment within each layer which is described as 'graded bedding'. C. A. Cotton applied the term 're-deposited beds' to our greywacke-argillite sequences showing combinations of the above features along with evidence of scouring and scraping of the soft sea bottom by the passage of both kinds of flows over it.

The double-handling (or perhaps 'multi-handling') mechanism helps to explain many characteristics of geosynclinal strata, including the beds of fine breccia containing chips and flakes of mudstones indistinguishable from argillite layers within the same sequence and pebbly beds which for various reasons seem not to have been deposited near to a land source. Another expression commonly used in connection with redeposited sediments is 'turbidite facies', a term which draws attention to the important role played in their formation by sediment suspended in dense, mud-laden water flowing across the ocean floor. Many sedimentologists, however, think the importance of so-called 'turbidity-currents' has been exaggerated and prefer to use a purely descriptive term, 'flysch', a German word originating among investigators of the European Alps in the nineteenth century.

Time-span of the New Zealand Geosyncline
Fossils are certainly abundant in the western, marginal belt of the New Zealand Geosyncline. Although there were many minor, local breaks in

the deposition of sediment, due to minor pulses of uplift and erosion, they did not occur everywhere at the same time, and the Hokonui Facies sediments as a whole contain a fairly complete record. Painstaking work on the fossil faunas for over a century, aided by continually improving techniques and many re-examinations of the containing rock sequences, has evolved what promises to be a stable classification linked by reliable correlations with overseas geological time standards.

The situation is not quite so good with regard to the Alpine Facies because of the greater thickness of barren strata and the structurally more complicated tracts which lie between widely spaced fossil localities. There are virtually no extensive, highly distinctive beds by which one succession can be connected with another elsewhere. Again, however, dedicated stratigraphers refuse to be discouraged and continue to search diligently for fossils in the Torlesse Supergroup, which embodies the Alpine Facies sediments making up the greater bulk of unmetamorphosed New Zealand Geosyncline material. Substantial progress has been made with working out a sequence and structure for the greywacke strata of the central Southern Alps, where the mountaineer-geologist can now reach critical areas, sometimes with helicopter support.

The difficulty about making a simple statement as to the age range of the New Zealand Geosyncline is that the rocks in it have neither an obvious stratigraphic base nor a clearly defined top. For the beginning we can cite the oldest strata not too much metamorphosed for their fossils to be recognisable; for the end, it becomes a question of trying to agree which are the youngest beds belonging to the geosyncline.

It is true that everywhere except on the East Coast of the North Island and in the north-east of the South Island New Zealand Geosyncline rocks are planed off at the top by an erosion surface ('Late Cretaceous Peneplain'; Chapter Seven), making an important unconformity and time-gap separating them clearly from the younger covering strata. Yet in these eastern areas the interruption was of short duration. Some geologists indeed insist that in those special areas marine sediment was continuing to accumulate with only minor interruption and in very much the New Zealand Geosyncline style until well into Cretaceous time, thus overlapping in age the youngest beds of the final cycle of rock-formation (Chapter Seven).* There is no doubt, however, that the unconformity marks a major interruption over most of the country.

Since Trechmann revised the Maitai rocks and their fossils in Nelson it has become certain that the great sedimentation episode spoken of in Marshall's day as the 'Maitai System' of 'Trias-Jura' age was already well under way early in the Permian Period. Suspicions of an earlier beginning in the Carboniferous Period, held by Wellman and others

---

* According to this view the New Zealand Geosyncline sedimentation continued on into the Tertiary Era, as Macpherson claimed for his 'eastern trough', and the Rangitata Orogeny (to be discussed at the end of this chapter) was the first phase of crustal movements culminating in the Kaikoura Orogeny (Chapter Seven). Though not widely supported, this view is consistent with the fact that these last two orogenies are not separated by an important geosynclinal event.

but generally disbelieved, were confirmed correct in 1971, when D. G. and T. B. Jenkins extracted some conodonts from supposedly Permian rocks at Kakahu, South Canterbury, which showed them to be of Carboniferous age. Coming after R. A. Cooper's determination of a Silurian age for rocks in west Nelson, the Kakahu discovery means that every one of the major geological time-divisions, from Cambrian onwards, is represented in New Zealand.

Considering now the upper limit, a thorough analysis of the evidence based on his own acquaintance with critical sections in the field had convinced I. G. Speden by 1973 that our oldest Cretaceous beds are, after all, separated from the Torlesse Supergroup by a significant unconformity, some late Jurassic and early Cretaceous time not being represented by any surviving strata in New Zealand. The time-span of the New Zealand Geosyncline can therefore be given as from late Carboniferous to late Jurassic, an interval of some 150 million years.

Whence Came the Sediment?
As noted earlier, the geosyncline concept is built around the idea of a subsiding crustal furrow, the case for which rested originally upon the need to account for exceptionally thick sedimentary strata occurring in elongated belts. Great mid-ocean deeps such as the Kermadec Trench and Mindanao Deep have been called 'empty geosynclines', being too far from major land masses to receive much sediment other than that which is of organic or chemical origin or produced by nearby volcanoes. There is good evidence from the nature of its sediments that the New Zealand Geosyncline was kept well supplied for most of the time. Where was the sediment coming from?

Again, of course, it is the origin of the non-volcanic sediment that mainly concerns us. In the days before the two contrasting facies of sediment had been recognised and claimed to be near-shore and outer trough deposits, there were already vague ideas that the 'foreland' source of most of the sediment lay to the west. This seemed obvious enough because more ancient rocks existed in the west of the South Island and in Australia, while there is no trace of any land east of the Chatham Rise, nor any evidence from the way earthquake vibrations arrive here to suggest that ancient submerged land lies beneath the Pacific floor beyond. Macpherson's original concept of western and eastern troughs with contrasting sedimentary facies as well as his belief that the western trough developed and terminated sooner than the eastern trough are both consistent also with a westerly source. Indeed, it was fully in line also with the philosophy then already popular that the Australian continent has had another strip of sedimentary and metamorphic rock added to its south-eastern perimeter by each orogenic cycle, and so has been encroaching upon the Pacific basin since at least as far back as early Paleozoic time. In modern language, the Buller Geosyncline sediments are said to have been 'cratonised', i.e. made rigid, during the Tuhua Orogeny to become ready to provide the 'foreland' for the New Zealand Geosyncline, and this would have been before the Tasman Sea was opened up by the New Zealand segment swinging away from Australia in the Cretaceous Period.

There have always been a few anomalies, however. For instance, it is hard to see how sediments of sand grade or coarser could continue to reach the eastern, axial part of the geosyncline from the west after a 'median ridge' began to rise, and there are indications that one had begun to do so in Permian times, early in the history of the geosyncline, and that it rose again in the middle of the Jurassic Period. Among many ingenious suggestions to get around this difficulty, one from J. C. Schofield is to the effect that sediment was fed in at either end of the subsidiary trenches. In the North Island particularly it seems likely that subsidiary trenches and ridges changed position repeatedly, with a good deal of sediment being transferred to and fro across the geosyncline in the process.

None of these ideas entirely obviates the need to account for some non-volcanic sediment supply from the east. The work of J. T. Kingma, J. D. Bradshaw, P. B. Andrews and others in Canterbury shows that conglomerate members of the eastern trough sediments are more common on the eastern side but become thinner, finer-textured and rarer towards the west, that is, towards the supposed position of the trough axis. Moreover, it is difficult to explain the composition of the conglomerate pebbles if they have to be perceived as deriving from erosion of the median ridge alone. Kingma was probably close to the mark when he interpreted the conglomerates as material which had slumped from the fronts of former river deltas farther out from the shore—*but* from a shore that must have faced in the opposite direction from the present coast! Connected with this problem is the apparently abrupt ending of the Haast Schist belt (and thus also of the 'median ridge') at the Chatham Islands.

During the past decade acceptance and better understanding of the concept of major lateral displacements of the continents has opened the question of whether New Zealand could have been alongside Antarctica during the periods concerned. Attempts have indeed been made to justify some such reconstruction, though not very convincingly. Yet another suggestion is that the vanished land source has been engulfed by the denser rocks of the sub-crustal mantle while the expanding Pacific Ocean floor has been underriding the eastern margin of our piece of the Austral-Asian continent. An attractive speculation, perhaps, but the problem really remains unsolved.

Volcanoes Near and Beneath the Sea

No problem arises as to the source of sediment of the Hokonui Facies which went into the marginal trough. This lay close to the old 'foreland' composed of early Paleozoic sedimentary rocks and of the granites and gneisses derived from them, and in any case a large proportion of the sediment was of volcanic origin. Lesser amounts of volcanic material are present in places in the Alpine Facies.

Molten rock ('magma') was injected into the sediments while they were accumulating, and more was erupted from volcanoes near the shore and very likely under the sea as well. Volcanic activity was certainly going on almost from the outset, if not from earlier, in the south. The oldest known New Zealand Geosyncline sediments from

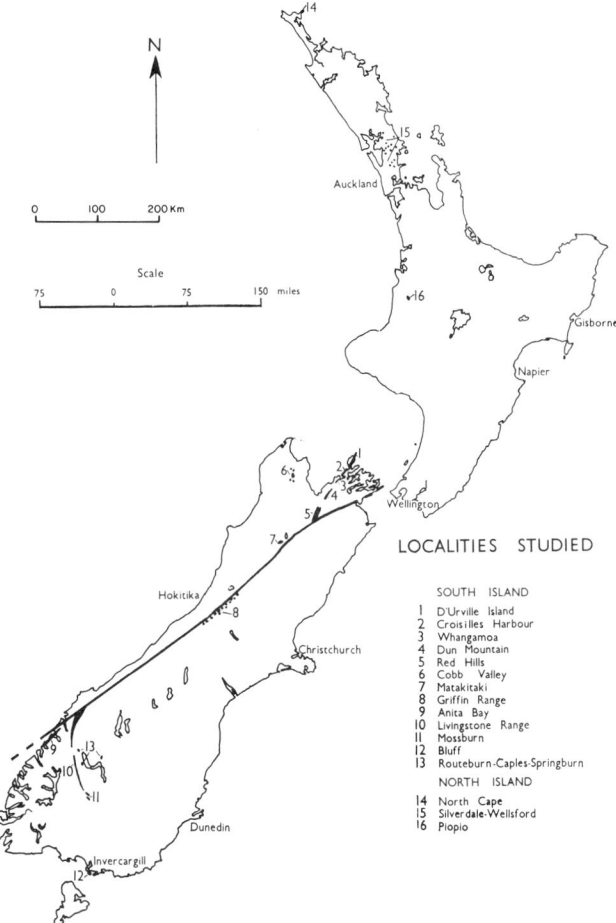

LOCALITIES STUDIED

SOUTH ISLAND
1   D'Urville Island
2   Croisilles Harbour
3   Whangamoa
4   Dun Mountain
5   Red Hills
6   Cobb   Valley
7   Matakitaki
8   Griffin Range
9   Anita Bay
10 Livingstone Range
11 Mossburn
12 Bluff
13 Routeburn-Caples-Springburn

NORTH ISLAND
14 North Cape
15 Silverdale-Wellsford
16 Piopio

Fig. 4.7
Some basalt erupted under the sea during the development of the New Zealand Geosyncline disintegrated into shards on contact with water, forming distinctive beds of dark-grey basaltic tuff; in other places the lava squeezed into bulbous masses enveloped by glassy skins of quickly-chilled lava, forming 'lava pillows'. Good examples can be seen on the coastal rock platform at Red Rocks Point, west of Ohiro Bay, Wellington.

Kakahu, South Canterbury, include the silica rock called chert, which is usually formed in association with undersea volcanic action. The Longwood Group and Caples Group of Southland and Otago, both Permian in age, contain volcanic debris, as does the Pelorus Group of eastern Nelson.

With regard to Nelson, however, according to W. R. Lauder the 'median ridge' dividing the geosyncline into two distinct troughs came into existence when volcanic eruptions were beginning in earnest. It must be remembered that we are speaking of Permian times when the Nelson and Otago areas were closer together than they are now, for the great Alpine Fault offsetting had still to come about.

Volcanic activity in the New Zealand Geosyncline will be discussed further in Chapter Eight.

The 'Ultramafics'—Dunite and Serpentine*

Far less well understood than the rest of the igneous rock family, though usually classed with them, are the 'ultramafic' types. The second half of the word is made up to emphasise their richness in magnesium (ma) and iron (f) relative to sodium, potassium and silica, and it synonymises with 'basic' in the context of igneous rocks. The ultramafic

* See also Chapters Six and Eight.

rocks are thus 'more basic than basalt'. Olivine is a characteristic mineral component of both mafic and ultramafic types; peridot is the gemstone variety and an old alternative name for olivine; hence 'peridotite', one of the more common, broad groups of ultramafic rocks.

One of the ways in which Hochstetter's early explorations put Nelson very much in the geological picture was his discovery of a peculiar, essentially olivine rock which he named dunite, after Dun Mountain. Among the usual products from alteration of dunite and other ultramafic igneous rocks by heat and pressure in a chemical environment including water, serpentine is the most familiar; serpentinites thus are altered ultramafic igneous rocks. A convenient group name for the wide variety of mafic and ultramafic igneous rocks and their alteration products that occur together in geosynclinal deposits is ophiolite. Though not a new term, it has become very fashionable in recent years to use it in the context of modern tectonic views as to the evolution of geosynclines. As layers, lens-shaped masses and thicker 'pods', peridotites and serpentinites occur within well-defined zones in the rocks of the New Zealand Geosyncline. Strips of ultramafic rocks emplaced in the Haast Schist of central Westland and aligned generally parallel with the trend of the schist have long been known as the Pounamu Formation, the name recalling the Maori word for nephrite (i.e. New Zealand jade or 'greenstone') which some of them contain. (See below, p. 102.)

Formerly regarded simply as intruded igneous bodies of magma containing a high concentration of mafic components, peridotites seemed at first to present no problem as regards their emplacement as igneous rocks and their subsequent alteration to serpentinite. With more precise mapping in New Zealand and elsewhere, and a truer appreciation of their relationships in the field with other rocks, it became increasingly difficult to find a general explanation that would fit the circumstances of all occurrences of serpentinites. The literature on the subject is now very extensive, and I could not undertake to outline it here, except to note that according to some of the suggested modes of origin, peridotites and serpentinites scarcely belong in the 'igneous' category at all. Modern tectonic theories have, incidentally, made it easier to see how ultramafic rock materials derived from the deeper crustal or upper mantle levels in the earth can have been injected into the lower zones of geosynclinal troughs, and why they (and their alteration products) now occur discontinuously, with much disruption and displacement from their original positions.

Rocks of the 'ophiolite' association occur at the surface in Nelson for 150 km north of the Alpine Fault, and again south of the fault for another 185 km in Otago and Southland. They have been studied extensively in recent years, and in a general account published in 1976, D. S. Coombs and several collaborators have established the name 'Dun Mountain Ophiolite Belt' for the whole assemblage. Geophysical evidence suggests that it actually extends hundreds of kilometres further, in both directions. Indeed, the otherwise anomalous presence of serpentinite in Tertiary covering strata at Piopio, near Te Kuiti, and

Fig. 4.8 (opposite right) The locations of the important areas of ultramafic rocks. Except probably for area 6, all are connected with the evolution of the New Zealand Geosyncline. (R. G. Coleman, 1966)

in Northland, may be explained by a slice of this rather slippery rock having been injected mechanically, in the cold state, from the ophiolite belt in the basement.

Most rockhounds are aware that greenstone country (i.e. in the New Zealand sense of 'greenstone', meaning jade or nephrite; geologically, it has usually had other meanings, but is almost obsolete) is also serpentine country. Our nephrite is obtained only from strongly metamorphosed parts of the ophiolite belt east of the Alpine Fault. The peridotites attracted mining interest in Nelson because they contain chromite and copper, and in west Otago where they also contain the native nickel-iron alloy mineral, awaruite.

The Gross Thickness—What Would it Mean?

I write 'would' because we cannot obtain even an approximate figure for the total thickness of deposits trapped in the New Zealand Geosyncline during the 150 million years of its existence. But supposing we could add up the maximum thicknesses deposited during each successive interval of geological time spanned by the geosyncline, how would we interpret the result? A ridiculous question, perhaps, but the reasons why it is ridiculous are instructive.

In the first place, although the total thickness could not be less than some tens of kilometres, this would *not* mean that such a colossal thickness of sediment could ever have accumulated in any one area. The maximum thicknesses estimated for each period were measured in different places. Next, the many visible minor unconformities interrupting the sequences of strata, the breccia and conglomerate pebbles derived from what appear to be older deposits of the same geosyncline, the apparently conflicting evidence as to the direction of transport of sediment all give ample support for the view that accumulation was not continuous and progressive in any one place. On the contrary, indications are that at different times and in different places, strips of the main trough ceased to subside and in fact probably rose and were eroded to supply sediment to adjacent strips then sinking. In other words, there was 'cannibalism' aplenty in this geosyncline, as in all others of the same type. Furthermore, the gross aggregate would still be merely a minimum indication of the quantity because the trough was evidently kept well filled and the amount retained was therefore limited by the rate of subsidence of the trough floor. The surplus would have been exported, but where to we cannot know. Again on the question of making room for the accumulation, progressive burial ever more deeply under later deposits squeezed out water, compacted the particles and eventually changed the sedimentary minerals into higher-density equivalents that take up less volume. Thus, to provide space for the last hundred metres of thickness at the top of the column would not have required the crust locally to subside as much as a hundred metres.

A further important reason why the estimation of total thicknesses is an impossible exercise arises from the belief now widely held that the orogenic phase of the cycle, in which the geosyncline contents are squeezed upwards and outwards ('everted'), need not await until the sedimentation phase is completely finished, but can begin quite early along the axial zone with the elevation of a median ridge and the beginning of metamorphism of the crumpled basal layers.

These conclusions were formerly hard to explain. How could the sinking trough be making room for sediment to accumulate in one place while squeezing up the lower parts in another place at the same time? The anomaly is removed in the light of plate-tectonics theory, which sees geosynclinal troughs being generated where the crust beneath the continually spreading ocean floor pushes under the adjoining continental plate, squeezing the contents of the marginal furrow and carrying the lower parts down into deeper zones of metamorphism and plutonism at the same time, while still making room for more sediment. How fast? Deepening at the rate of centimetres per year would be in order; there were plenty of years.

Carboniferous and Permian Sedimentary Rocks

Not much is known about conditions here during the Carboniferous Period. Part of it was taken up by the Tuhua Orogeny. Only at Kakahu have sediments of the New Zealand Geosyncline so far been shown to be as old as Carboniferous. Generally the deeper parts are either barren of fossils or too much altered by metamorphism to tell us much about the environments in which they were formed.

In 1878 some of the earliest fossil collections made in Nelson enabled James Hector to date the Permian rocks fairly closely as 'Carboniferous'. Hector had previously noted peculiar fossil shell fragments with an internal fibrous structure at right angles to the shell surface, and not unlike an unrelated bivalve fossil genus, *Inoceramus*, numerous species of which are common in marine Jurassic and Cretaceous beds throughout the world. He was not put off the scent by this resemblance. Others were being confounded also by the complexity of structure near Nelson, particularly by the fact that a steeply-dipping part of the sequence at the best known localities has been rotated past the vertical (i.e. inverted). Limestone containing the prismatic shell fragments thus appears to be on top of instead of underneath strata with good fossils of Triassic age.

This inversion was the origin of confusion, wrangling and mistakes in correlation not only at Nelson but all over the country for the next forty years; it was the basic error which led first to the construction and eventually to the collapse of Marshall's 'Trias-Jura'. The fossil in question has since gone by various names besides *Inoceramus*, including *Maitaia* and *Aphanaia*; it is now assigned to the genus *Atomodesma*. This fossil is seldom absent from pieces of the lower Maitai limestone at Nelson and is often recognisable in other associated beds in Nelson and in Southland and Otago as well. The most common 'sign' is small rectangular windows on weathered rock surfaces, empty holes from which the shell material has been dissolved out.

J. B. Waterhouse has claimed that New Zealand can produce the thickest and most complete record of the Permian Period in the world. Our Permian sequences vary considerably between regions and accordingly there are numerous local formation names. Only the broader divisions are named and described below.* On the 1:250,000 geological map series, Permian beds appear mainly on sheets 14, 15, 22, 23, 24 and 25.

A Permian age for parts of the greywacke basement rocks of the North Island was first established nearly thirty years ago when blocks of limestone entangled in volcanic lavas near Whangaroa, North Auckland, were found to contain corals and microfossils. It is now acknowledged that parts of the unfossiliferous axial-range greywackes of the North Island are probably Permian too. Red and green argillites in South Canterbury, some showing traces of *Atomodesma*, suggest the presence of Permian rocks there, and this has been confirmed by microfossil samples collected near Benmore Dam. The chief Permian areas, however, are Nelson, Otago and Southland.

In eastern Nelson the Hokonui Facies beds are folded into the complex Nelson syncline. Greenish volcanic sandstones and argillites at the top of the Pelorus Group, though unfossiliferous, are most probably of Permian age, but the known Permian succession really begins above with the Lee River Group of volcanic-derived sedimentary rocks, mafic and ultramafic igneous rocks totalling perhaps as much as 4 km in thickness. This group includes the serpentinite and dunite of Dun Mountain and also the Brook Street Igneous Formation consisting mainly of volcanic sediment along with sodium-enriched varieties of basalt and trachyte, known respectively as spilite and keratophyre. The Permian age of the Lee River Group is generally accepted as correct, although it depends mainly upon the evidence of poorly preserved *Atomodesma* fragments in the Brook Street beds.

The succeeding Maitai Group (not to be confused with the obsolete 'Maitai system') embraces five formations as listed below, type sections for the lower two being located in the classic Nelson area, and the upper three on D'Urville Island. Wooded Peak Limestone has also been called 'Maitai Limestone', and the latter name has also been used for limestone beds of Permian age away from Nelson, but fortunately this practice has almost ceased. Divisions within the Maitai Group are:

	(top)	
	Stephens Formation	— igneous conglomerate; 1200 m.
	Waiua Formation	— banded maroon/purple mudstone and grey siltstone; 600 m.
MAITAI GROUP	Greville Formation	— banded grey sandstone, siltstone; 1200 m.
	Tramway Sandstone	— well-bedded red/green sandstones; 152–415 m.
	Wooded Peak Limestone	— fine-grained, dark limestone, emits foetid odour when struck; 152–335 m.
	(base)	

---

* More detailed accounts can be found in Waterhouse's 'Permian Stratigraphy and Faunas of New Zealand', *N.Z. Geological Survey Bulletin 72*, 1964; and 'Proposal of Series and Stages for the Permian of New Zealand', *Transactions, Royal Soc. N.Z. (Geology)*, vol.5, pp. 161-80, 1967; for Nelson in particular see 'Geology of Nelson Urban Area', *N.Z. Geological Survey Urban Series Map 1*, M. R. Johnston, 1979.

Places where these units may be looked for alongside Highway 6 and on roads and tracks near Nelson city can be discovered on Sheet 14 of the 1 : 250,000 Geological Map. (See footnote, p. 104.)

Southland can justly claim to be the greatest showplace for the Permian of New Zealand. On the whole relationships with rocks of other ages are less complicated here than in Nelson and there are many accessible and well-exposed sections of both sedimentary and igneous. Good exposures may be seen at many places where hard rocks crop out on the coast between Orepuki and Bluff. Permian strata also come to the surface inland along the southern flank of a broad southeast-northwest belt in which the Hokonui Facies strata are downfolded in the Southland Syncline; on the northern flank of it most of the Permian outcrops are in southern and western Otago. On the whole they all resemble the Nelson Permian rocks, although the succession differs there to the extent that the same group names cannot be used in both regions. Different formations are, in fact, recognised on the northern and southern flanks of the Southland Syncline.

Because of the above-mentioned differences, the very old name 'Te Anau Group' (or Supergroup; formerly 'System'), long used to unify supposedly Permian strata throughout the South Island, has been dropped from classifications used by the Geological Survey.

The major divisions in Southland are:

(top)

PRODUCTUS CREEK GROUP:	limestones; volcanic-derived sandstones, siltstones and argillites; good fossils.
TAKITIMU GROUP:	red/green volcanic breccias; red/green/grey sandstones and mudstones; basalt and andesite lava rocks.

(transition downwards into Longwood Complex of mafic igneous rocks with interpolated sedimentary beds and intrusions of granite, etc.)

Units recognised in South Otago on the northern flank of the syncline, not closely equivalent to the above, are:

(top)

ARTHURTON GROUP:	conglomerate at base; then siltstone and argillite with thick sandstone beds; limestone with *Atomodesma*, corals and other fossils; sediment derived mainly from mafic igneous rocks with some granite and schist fragments.
WAIPAHI GROUP:	reddish-brown coarse greywacke sandstone; tuff; argillite and basalt lavas towards the top.
CAPLES GROUP:	greenish greywacke, argillite, conglomerate; grades northwards into Haast Schist.

(base)

The early and middle Permian Period is represented by the Takitimu and Waipahi Groups, and the late Permian by the Productus Creek and Arthurton Groups. While the Caples Group contains no internal evidence of Permian age, it is usually included together with Permian rocks in geological maps. Late Permian is also represented by the Mataura Island Group of fossiliferous limestone, sandstone and mudstone formations east of the Mataura River between Mataura Island and Te Peka, a distance of about 10 km. Also called the Kuriwao Group, it is made up of three named formations. In Nelson the Lee River Group probably spans early and middle Permian time, whereas the Maitai Group is late Permian.

A separate, concluding note is warranted for the Permian fossils found in north-west Nelson in 1963, far to the west of previously known rocks of that age. The fossils occur in a pebbly sandstone below the ridge leading south towards Parapara Peak, west of Takaka. They were discovered by chance in loose boulders on the bed of Pariwhakaoho Stream. The source rocks had previously been mapped with the Haupiri Group, but they are actually above Ordovician slates and their age is equivalent to the upper part of the Productus Creek Group of Southland. The fossils have affinities with an eastern Australian fauna that is supposed to have inhabited cold seas offshore from 'Gondwanaland', which in the Permian was still largely under the glaciers of the Carboniferous-Permian Ice Age. (Later work on our Permian faunas has suggested that the water in the New Zealand Geosyncline was alternately cool and warm throughout the period.) Three formations have been defined:

PARAPARA GROUP
- Walker Quartzite
- Flowers Formation (sandstone etc., fossils)
- Pupu Conglomerate.

These beds rest on what is perhaps an old peneplain carved across the early Paleozoic rocks. They contrast strongly with all the rest of the New Zealand Permian, which elsewhere are typically geosynclinal types. Parapara Peak is the only place where strata of Permian age are known to overlap on to the old land generated during the Tuhua Orogeny (the 'New Zealand Geanticline' of some authors), and are underlain by rocks of the Buller Geosyncline.

Stages and Series; Formations, Groups and Systems

Users of the 1 : 250,000 Geological Map of New Zealand are sometimes puzzled by a change to a different kind of classification of the sedimentary strata which comes in half way up the succession we are now considering under the heading 'New Zealand Geosyncline'. The two different ways of handling the task of subdividing sequences of strata and the time intervals they represent were introduced in Chapter One. We will now look more closely into the scheme by which the different stratigraphic categories are shown on New Zealand maps.

The 1 : 250,000 scale maps show the older sedimentary strata in terms only of formations and groups, e.g. Anatoki Formation, Wooded Peak Limestone, Maitai Group. Strata of Triassic or younger age, on the other hand, are mapped as stages and series, though formations and groups are mentioned too under the corresponding headings in the accompanying texts. On all twenty-seven sheets of this map Permian and older sedimentary strata and the igneous and metamorphic rocks of all ages are mapped and described as formations and groups, these being units distinguished by the character of the rock itself and its position relative to other rock units in the local sequence;

hence 'rock-stratigraphic' or 'lithostratigraphic' classification and naming. In this case the map colours and boundary lines show the extent on the ground of each rock-type unit. For Triassic and younger sedimentary strata the type of stratigraphic unit used for these maps is one defined by the position of the strata in the geological time sequence and not by the kind of rock; hence, 'time-stratigraphic' or 'chronostratigraphic' classification. The map colours and boundaries therefore mean something different for the Triassic and younger sedimentary strata. Further, because in this case it is the succession of fossil life that provides the measure of time, the classification is 'biostratigraphic'. The map legends show how the stages of this classification are grouped together successively into series. But beware! In maps and publications issued prior to about 1950 'series' was still being used in New Zealand in an obsolete way for what we now call formations.

All very confusing, but the double system adopted for this particular map series was an inevitable compromise for which there are a number of practical reasons. For one thing, when the 1 : 250,000 mapping programme was launched, paleontological knowledge of the older strata was not up to the standard for setting up a biostratigraphic classification for strata of all ages; for another, the numerous local formations and groups into which the Cretaceous and Tertiary strata have been divided up in each area could not have been shown clearly on the 1 : 250,000 scale. Larger scale maps mostly show formations, which are the kind of unit most users of these maps can recognise in the field from lithological descriptions and without the need to find and identify fossils.

By 1967, when the 1 : 250,000 mapping programme was well on the way to completion, biostratigraphic classification of our Permian strata had been achieved, and since then stages and series have been proposed to cover still more of the Paleozoic Era in New Zealand. This does not mean that these would necessarily be used to replace formations and groups for the older sediments in any future remapping on the same scale, for the lithostratigraphic divisions remain the more practical kind for the majority of users of these maps.

Letter symbols to identify the stages and series on our geological maps (as well as colour) have been standardised for more than thirty years. An inital capital letter (now two for the Permian units) denotes series, followed by one or two lower-case letters denoting stage (e.g. Lwh : Landon Series, Whaingaroan Stage, one of the divisions of the Oligocene strata in New Zealand.

'System' is not used on our 1 : 250,000-scale maps but it does appear in many older papers and Geological Survey *Bulletins* as a major time-rock unit, such as 'Te Anau System' which once embraced all of our Permian rocks. Its proper use is international, to wrap up all the rocks everywhere that belong in one of the major geological time-rock divisions, such as the 'Devonian System'.

Time-stratigraphic Classification of the New Zealand Permian Strata

J. B. Waterhouse in 1967 set up six stages based on characteristic fossil assemblages including brachiopods, molluscs and corals, and grouped them into series named Aparima and D'Urville (Table 4-1). He made it clear that although these divisions correspond roughly with international standard divisions of Permian time, based on sections in the Ural Mountains, the agreement is not exact.

NEW ZEALAND PERMIAN UNITS Table 4–1

D'URVILLE SERIES	Makarewan Stage	LATE PERMIAN
	Waiitian Stage	
	Puruhauan Stage	
APARIMA SERIES	Braxtonian Stage	EARLY PERMIAN
	Mangapirian Stage	
	Telfordian Stage	

Triassic and Jurassic Rocks

The earliest discovery of fossils of Triassic age in New Zealand was made by Hochstetter in 1859 in the hills east of Richmond, Nelson, but the first detailed work on our Triassic and Jurassic rocks was undertaken in Southland and South Otago. Hochstetter also saw Jurassic beds at Waikato Heads but thought at the time they were early Cretaceous. The southern areas were attractive because fossils are abundant and structures simple, and a further incentive was Hector's mis-identification of some fossils which led him to hope the equivalents of coal-bearing rocks in New South Wales might be found there. There were other initial mistakes, but the Southland succession still provides stage names and reference localities for our Triassic classification, some terms echoing Hector's original scheme of a hundred years ago. The present understanding of our Jurassic succession grew mainly from work in the south-west Auckland area, from where Hochstetter proudly announced the first New Zealand discovery of ammonite fossils (ancient, extinct relatives of nautilus), but is also based partly on the Southland region.

As with the Permian, the Hokonui Facies sediments of the marginal geosyncline trough lend themselves more readily to fine subdivision and precise, detailed mapping. The Alpine Facies is more difficult to map, but there are many, scattered fossil localities. J. D. Campbell and G. Warren in 1965 compiled a list of 350 localities in the South Island and I. G. Speden has recently presented a further list of 110 in the North Island. More are continually turning up.

In the times of the great 'Trias-Jura' controversy it became important to establish whether the deposition of what we are calling the New Zealand Geosyncline rocks was interrupted. Marshall's categorical declaration that there were no breaks within the 'Maitai System' was unacceptable to some of his contemporaries, and an incentive to examine all known sections very carefully. Much of the North Island basement is obscured by younger rocks, and fossils are harder to find, while in Nelson the relation of Triassic with Permian beds is uncertain in the classic area because a major fault separates them. So it was in

Southland again that an answer was found. While remapping there and in South Otago in 1953 B. L. Wood confirmed that there was a time-gap between Permian and Triassic strata marked by a slight angular difference in the attitude of beds above and below the boundary ('angular unconformity'). This shows that mild crustal warping and uplift were affecting the sediments in the western, marginal trough even at this early stage in its history, but this was a minor disturbance compared with pre-Permian upheavals during the Tuhua Orogeny, or with the Rangitata Orogeny which terminated the New Zealand Geosyncline. Minor conglomerate layers and unconformable contacts between beds become less common away from the southern trough margin, but on the whole the differences between the opposite flanks of the main east-west syncline are less marked than they are in the Permian beds. Coarse sandstone containing volcanic tuff, pebbly beds and blue/grey banded siltstones are the dominant sedimentary types.

The middle Jurassic upper part of the Southland succession shows a progressive change upwards into coarser sandstones, breccias and conglomerates containing fossil leaves and thin coal beds but no marine fossils. This was the time when the trees grew and left their roots fossilised to form the well-known 'fossil forest' at Curio Bay, Waikawa. Volcanic sediment was still being supplied, but some of it no doubt was being eroded from earlier coastal volcanoes along with material from the old land beyond.

In the Auckland region the Hokonui Facies of the Jurassic strata is notable for shaly mudstone beds hundreds of metres thick. Basaltic lava and breccias occur in the far north, but again much of the volcanic content of Auckland may have come from erosion of older volcanic terrains. As in Southland the uppermost beds include conglomerate, coarse sandstones, fossil wood and coaly layers, announcing the end of the geosynclinal phase and the formation of an increasing extent of land nearby.

Chiefly from Hokonui Facies sediments in south-west Auckland and Southland, fossils have now been found representing all of the main international divisions of Triassic and Jurassic time except for one at the very end of the Jurassic Period. Early Triassic was believed missing until 1959, when it was determined that ammonites from near Wairaki, Southland, are of that age. The two periods are now represented by thirteen biostratigraphic stage units assembled into five series. These are listed in Table 4-2, with their approximate ages according to the international scale. Quite independently, the strata have also been divided up on the basis of rock type and sequence into formations and groups, the names of some of which are mentioned in the legends and texts which go with the 1 : 250,000-scale maps. Many more are described in regional accounts of the geology in published papers and Geological Survey *Bulletins*.

'Portmanteau' Names—Hokonui, Torlesse, Murihiku and Others
Applied to a piece of travellers' luggage the word 'portmanteau' is obsolete but it is still useful metaphorically to denote a large, imaginary

NEW ZEALAND TRIASSIC AND JURASSIC UNITS Table 4–2		
OTEKE SERIES	(possible time-gap in New Zealand) Puaroan Stage	LATE JURASSIC
KAWHIA SERIES	Ohauan Stage Heterian Stage	
	(time-gap in N.Z.) Temaikan Stage	MIDDLE JURASSIC
HERANGI SERIES	Ururoan Stage Aratauran Stage	EARLY JURASSIC
BALFOUR SERIES	Otapirian Stage Warepan Stage Otamitan Stage	LATE TRIASSIC
GORE SERIES	Oretian Stage Kaihikuan Stage Etalian Stage Malakovian Stage	EARLY TRIASSIC

bag in which to put many oddments not yet sorted out. When F. W. Hutton introduced 'Hokonui System' in 1885 it was by no means a loose, disordered package, but actually a clearly-conceived group name for all of the fossiliferous formations (then called 'series') of Triassic and Jurassic age in New Zealand. Some had been described from Hokonui Hills and Mataura, Southland; some from Wairoa River, Nelson; some from Waikato Heads, Auckland; some by himself but mostly by other early investigators. It has to be remembered when mentioning old names like 'Hokonui' that the difference between dividing strata into units of rock type and marking off sequences of strata to represent spans of time was not generally appreciated in the past; the distinction was not made until much later. Nevertheless, it is clear enough what Hutton intended to be the composition of the Hokonui System.

At first this name was applied almost entirely to the better-dated parts of what is now recognised as the western marginal belt of the New Zealand Geosyncline. There was then no consistent naming of the sparsely-fossiliferous, eastern or axial trough greywackes, etc. forming most of the main ranges and basement rocks of both islands. Gradually, and especially after the 'Maitai' confusion had been cleared up, it became increasingly likely on indirect evidence that the bulk of these rocks were Triassic and Jurassic, so the habit grew up of calling these Hokonui System too. It was a bad habit. For one thing, Hokonui System was introduced originally to embody a sequence of strata distinguished by their geological age as indicated by fossils, whereas no such units could be recognised in the largely unfossiliferous bulk of the axial trough greywackes. As the rocks themselves were also of a distinctly different type, as well as being of uncertain age, it was never really justifiable to extend the name 'Hokonui' to them.

The problem was noticed (and backed away from) in 1947 when, for the first time since Marshall's *Geology of New Zealand* came out in 1912, a coloured geological map of New Zealand was being prepared on the occasion of the Seventh Pacific Science Congress being held in this country. For this map the axial trough strata of the New Zealand Geosyncline were grouped under the legend name 'Undifferentiated Permian, Triassic and Jurassic'. By 1952 it was no longer possible to use the old term in this extended sense because H. W. Wellman had begun what has since become the common practice of recognising the special sedimentary character of the marginal trough rocks as 'Hokonui Facies'. When the new 1 : 250,000-scale mapping programme was being designed in 1957 it was quickly appreciated that a new unit to embody the axial trough facies of the New Zealand Geosyncline rocks was an imperative need.

Taking up a casual suggestion by B. H. Mason, the stratigraphic co-ordinator for the new map series, R. P. Suggate, announced in 1961 the intention to revive another old term introduced almost a century earlier by Haast for the greywackes and argillites of the Southern Alps and Canterbury foothills ranges—'Mount Torlesse Formation'—

Fig. 4.9
Greywacke strata of the Torlesse Supergroup folded tightly into a syncline in the Malte Brun Range, Southern Alps. (Photo: N.Z. Geological Survey)

shortening it to 'Torlesse' and giving it group status. It was not used, however, as a mapping division on the 1 : 250,000 series, these showing instead the compilers' 'best guesses' of the ages of the axial trough 'Alpine Facies' rocks in terms of stages and series.

Torlesse Group was not used outside the South Island until 1963. By general consensus it has since 1971 been raised to Supergroup rank because of the need to incorporate whole groups of formations within it. Southerners were roused into action to remedy the lack of a similar lithostratigraphic 'portmanteau' to replace Hokonui in Southland and Otago when talking about the marginal-trough Triassic and Jurassic beds as rocks rather than as time-units. And what better word for the prefix than the old Maori name for those southern districts? Hence, the 'Murihiku Supergroup' introduced by J. D. Campbell and D. S. Coombs in 1966. It applies, however, only to the Mesozoic parts of the New Zealand Geosyncline in that region. There is no corresponding term, yet, for the Nelson equivalents, unless as Campbell suggested, Hochstetter's old name 'Richmond Sandstone' can be revived as a group for the late Triassic rocks of the Nelson ranges.

The Torlesse Group (or Supergroup) is made up of great thicknesses of Alpine Facies greywacke and argillite and their associated cherts, jaspillites, etc. as described at the beginning of this chapter. The use of 'argillite' has been criticised here because the finer beds tend to be silty rather than of clay grade (Fr. *argile*: clay). Towards the east the sequences are diversified with the appearance of thin limestone and conglomerate beds and masses of basaltic lava usually associated with chert and jaspillite. One of the largest occurrences of this association in one place is in the head of Camp Stream below 'The Notch' on the crest of the Torlesse Range.

Fossils are rare and few animal groups are represented, but the list continually grows and attention is also paid to the trails and castings of unidentified sea-bottom dwellers. Most frequently found is the distinctive, radially-ribbed bivalve mollusc of the genus *Monotis* (or rather, casts and internal moulds of shells; shell material rarely has survived). This creature seems to have thrived in the sediment-charged ocean waters and in its life habit was not unduly discouraged by the unstable, flowing and slumping bottom conditions; it may have avoided unstable places, or perhaps was not a bottom dweller at all. Also frequently seen are the silica tubes of a polychaete worm known as *Torlessia*, flattened by compaction where lying in the bedding planes, but circular where perpendicular to the bedding. Other groups represented locally are ammonites, belemnites, brachiopods, rare marine reptilian bones and land plants. The fossil plant beds of Canterbury, e.g. at Tank Gully near Mount Potts, at Clent Hills and near Whitecliffs, are well known to local amateur geologists. It is questionable whether the Jurassic plant beds at Whitecliffs belong to the Torlesse Group or not. R. Speight mapped them there as 'Wakaepa Series' after the Maori name for Selwyn River; separated from the main bulk of Torlesse rocks by unconformity, and less indurated, they are nevertheless as much a part of the New Zealand Geosyncline in its waning stages as are the Jurassic plant beds south of Port Waikato.

There are other problems and some unsatisfactory aspects of the above use of Torlesse Group and its relation to the Haast Schist Group. A number of papers have appeared recently on this subject, and a number of new names for newly-conceived categories of stratigraphic data have been suggested. Unfortunately it cannot be said that these have done much to clarify the issues so far, and it would be premature to include them in this book. I feel safe in assuming that 'Torlesse Supergroup' will continue to be found an adequate and useful definition for some time to come.

The lists of Torlesse fossil localities mentioned above may be found in papers by J. D. Campbell and G. Warren for the South Island in *Transactions of the Royal Society of N.Z. (Geology)* vol.3, pp.99-117, 1965, and by I. G. Speden for the North Island in *Journal of the Royal Society of N.Z.* vol.6, pp.73-91, 1976.

Haast Schist—Metamorphic Spine of the South Island

The metamorphic rocks of New Zealand will be discussed as a whole in the following chapter.

While proposing to revive 'Torlesse' for the Alpine Facies Triassic and Jurassic rocks, R. P. Suggate in 1961 took the opportunity to suggest a unifying name also for the middle belt of schists which lies to the west of the facies junction (see earlier in this chapter). The suggestion was accepted and almost universally implemented so that we now use the term 'Haast Schist Group' for what was formerly called 'Marlborough Schist' in the north, 'Alpine Schist' in the west and 'Otago Schist' in the south of the South Island, and by other names as well in the older literature. Some of the rocks included are on the borderline between sedimentary and metamorphic, as for example Caples Group and Tuapeka Group of South Otago; others can only be shown to have started on the road to metamorphism by virtue of minor mineral changes detectable under the microscope.

Sustaining the metaphor a little longer, the 'spine' is actually dislocated, with a gap about 70 km long due to displacement where the Wairau Fault cuts obliquely across the schist belt. It has slipped a disc or two also at the Awatere, Ahaura and Taramakau-Hope faults.

Essentially a rock-type unit, the Haast Schist Group does not need to have age limitations. In fact, it is limited by its definition to the metamorphosed equivalents of the Torlesse Supergroup strata, which means that its parent material ranged in age from Carboniferous to Jurassic. The date of metamorphism is another matter. It may have begun before the onset of the Rangitata Orogeny, but additional tectonic features were imposed upon the schists during the late Tertiary to early Pleistocene Kaikoura Orogeny.

Rangitata Orogeny—Final Phase of the New Zealand Geosyncline

A number of incidental references have already been made to the events which brought the New Zealand Geosyncline episode to a close, raised up mountains which, though destined to be almost levelled by erosion and nearly submerged by the sea once more, may be considered a 'proto-New Zealand'. (See also p. 169.)

Exactly when the Rangitata Orogeny began is not easy to tell. Several lines of information point to there having been some reduction in the rate of subsidence and some narrowing of the troughs during the Jurassic Period. For example, parts of the Jurassic sequence in eastern Southland become thinner, not thicker, in the direction of the Haast Schist belt, suggesting that parts of the axial zone were rising. The importance of this bit of evidence could be overrated, however, for it will be recalled that indications of a rising median ridge are detected even in the Permian Period in Nelson, and very probably there were rising strips at different places and times throughout the history of the geosynclinal phase. The most telling signs of when the geosyncline as a whole ceased to subside at an overall rate sufficient to keep pace with sediment supply are the change to different kinds of sediment suggesting shoal conditions and nearness to land, the final disappearance of marine fossils and the incoming of abundant plant debris. These changes become a strong trend from mid-Jurassic time onwards. A central, anticlinal up-arching developed, though it may not have risen high above sea level at this stage. Marginal belts continued to sink, to allow the accumulation of hundreds of metres of Jurassic mudstones in the Auckland region and to account for scattered occurrences of late Jurassic marine fossils in the South Island. Fleming, indeed, doubts whether a median ridge existed at all in the Auckland region at this late stage.

Rock that had been buried tens of kilometres deep in the core of the geosyncline before being raised and laid bare by erosion developed a variety of fine-scale structural features from which the structural petrologist can work out the order of tectonic events. For instance, it has been determined that the strata had already been compressed and crumpled into tight, flat folds (rather like a rucked-up mat), and the whole then refolded, while the mass was being carried down into the metamorphic realms of high temperature. Moreover, it is known that deforming of the schists and gneisses thus produced continued after the peak of metamorphism was passed, and again during the eversion phase while the whole geosyncline mass was being uplifted and eroded deeply enough to uncover schist at the surface. Each phase of tectonism left its imprint in the detailed structure of the rocks now exposed. Also, we know that granitic plutons came into existence during the orogeny.

Let us now look at what may be called the 'time-constraints'. The youngest strata of the New Zealand Geosyncline, though not as hard as some of the older unmetamorphosed sedimentary beds or the gneisses, are indurated enough to require that several hundred metres at least rested on top of them at some stage. In some places at least it is known that this cover could not have been Cretaceous or Tertiary strata, leaving as the only alternative the conclusion that rapid sediment accumulation went on there until the very end of the Jurassic Period.

The timing is tight. The final crisis of upheaval and erosion must have been dramatic and rapid indeed by geological time standards. Radioactive isotope datings tell us approximately when the rocks ceased to be at the high temperatures which prevailed when metamorphic schists and granitic plutons were forming. When applied to dating the metamorphic climax of the New Zealand Geosyncline they are not without anomalies, and the spread of dates ranges from approximately the Jurassic-Cretaceous boundary at about 135 million years ago to mid-Cretaceous, say, 100 million years ago. The oldest strata deposited after the upheaved geosyncline contents had been eroded to a land of low relief and again submerged are also of mid-Cretaceous age, pehaps between 105 and 110 million years old. At most, therefore, 20 million years is available for the orogenic crisis, and probably less. This seems too little, until we compare the Rangitata events with what happened later during the Kaikoura Orogeny and connected events over the last 20 million years. The Rangitata orogenic tempo does not seem to have been exceptional.

C. A. Fleming in his review of New Zealand Mesozoic geology given in London in 1968 told of how a simple pattern of ancient foreland, volcanic belt, coastal hingeline and geosynclinal trough with a NW-SE elongation gave way before the end of the Rangitata event to a much more complex arrangement of numerous tight folds crowded together near the Alpine Fault, and then how these structures as a whole were bent or dragged round into a back-to-front 'S' form by the great shear of that fault. Opinions have differed as to when the main part of the 450km lateral displacement along the Alpine Fault occurred. Fleming supported R. P. Suggate's view that most of it happened in the later phases of the orogeny.

Things were never the same again. The entire subsequent history appears to have been more complex and disjointed than in the times of the New Zealand Geosyncline and before. Perhaps it is merely because we have more complete and detailed knowledge of subsequent episodes, but probably not. Increased complexity might be expected to result from the later tectonic events which impressed a strong, new NE-SW structural grain upon the older trends. The newer trends are expressed today in much of our geographical relief.

Fig. 4.10
A simplified picture of the evolution of the New Zealand Geosyncline, as envisaged by C. A. Fleming (1970). In a series of paleogeographic sketches he shows an inferred sequence of events from the initiation of the trough in early Mesozoic times by underthrusting of the Pacific Ocean floor 'south-westwards' (in terms of present-day compass directions) beneath the foreland of Buller Geosyncline and older rocks; then its later warping and final dislocation in response to new stress directions, when an ancient continental nucleus ('Gondwana') fragmented in Cretaceous times and the New Zealand segment swung away from Australia.
A—Auckland Islands
C—Campbell Island
H—Lord Howe Island
NC—New Caledonia
B—Bounty Islands
CH—Chatham Islands
N—Norfolk Island

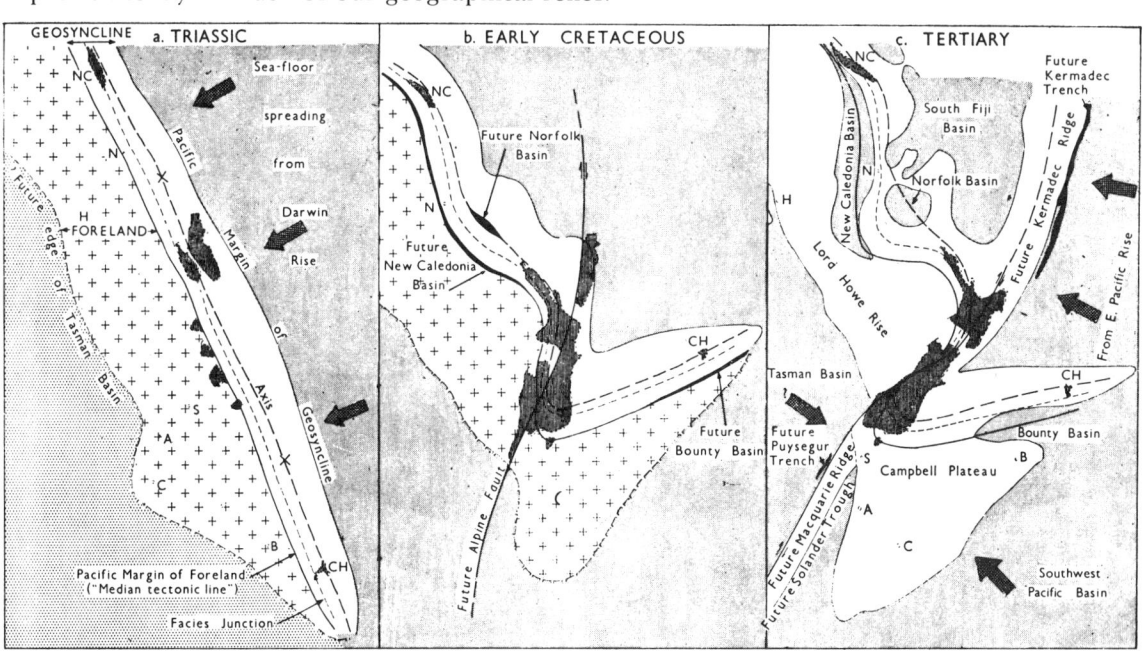

Chapter Five

METAMORPHISM AND METASOMATISM

Rock in the Cooking Pot

There is a distinctive quality about landscape underlain by metamorphic rock. In New Zealand it is associated with bold and striking scenery like the precipitous fiord country of western Otago and Southland, the broad, bare mountains of Central Otago, the craggy marble summits of western Nelson and, at the other end of the scale, the finely modelled ridges and gullies of the Marlborough Sounds. Metamorphic rock itself looks old, and in many first reconnaissances of newly-explored regions has often been regarded as the most ancient terrain and made out to be older than it subsequently proves to be. This may be partly because metamorphic rocks, being the result of physico-chemical alteration, are usually of an obviously crystalline texture, but chiefly because any fossils that might once have been present are now obliterated. It takes more than a little imagination to accept that coarsely crystalline gneiss such as one sees in the plunging walls of Milford Sound, here and there with garnet crystals the size of a cherry or larger, previously had the form of some kind of greywacke sandstone, and earlier still, that of newly-deposited, uncompacted sediment. A hint of such an ancestry is sometimes given by faint surviving traces of distinctively sedimentary features—not to suggest, of course, that all Fiordland gneisses were originally sedimentary rocks.

This is a convenient point at which to look more closely at what happens to the sedimentary contents of a geosyncline while becoming buried kilometres deep, and at the relationship between metamorphics and that other supposed 'rock of ages', granite (Chapter Six).

The Environmental Approach
It has been found helpful to approach this subject by noting that deeply buried sedimentary materials in the depths of a geosynclinal furrow have been moved into a very different environment from that in which the sediment was deposited.

The environmental concept is now well known in connection with plants and animals and the conditions under which they live. For living creatures at the earth's surface the biological environment is the sum total of many factors, none of which, if a given species is to thrive or even survive at all, must vary beyond certain limits. Important limiting factors for survival of life include temperature, availability of moisture, food sources and, in some instances, light. For deep-sea organisms the factor of pressure is important, but it is constancy of pressure rather than the intensity of pressure.

Finding themselves in a changed, less-favourable environment, organisms may have the ability to adjust to it up to a point, as for example the colour changes some creatures can assume so as to remain inconspicuous in all seasons, or they may be lucky enough to have the option of removing themselves from environmental stress. Adjustments may be possible for individuals in their lifetime, or it may take place over many generations through natural selection of resistant or tolerant strains. It would be easy to stretch the biological comparison too far, but it is valid enough to be helpful in the environmental approach to understanding metamorphism, for it is true that mineral species, rock structures, etc. have their limits of tolerance of terrestrial environmental factors, and that in some circumstances rock material may change form or even move to another place in response to environmental contrasts.

Particles of sediment being deposited on the floor of a comparatively shallow sea are under pressures, at most, a few times greater than that of the atmosphere at the surface. The pressure is a few *more* times greater on the bottom of a deep ocean. Much of the sediment supply to the ocean, as we have seen, is the product of weathering of rock under atmospheric pressure and moderate temperature, so as far as pressure goes there has been no significant change. This is true even for the very small proportion of marine sediment that finds its way into the great ocean deeps, because the pressure on each particle when first deposited is uniform all round it, that is, 'hydrostatic' as experienced by any object standing freely in a column of still liquid. Sea-bottom temperatures, too, are likely to range from around zero Celsius to twenty degrees above—still very moderate except in a few situations—while the chemical environment is governed by the gases, salt and other soluble substances dissolved in sea water, none of them particularly powerful chemical reagents. Up to this point, the environmental 'shock' experienced by sedimentary materials has not been too severe.

As a sediment particle becomes buried ever more deeply under a load of later deposited sediment its environment progressively changes. The water film which at first (thanks to 'surface tension') separated the particle from its neighbours and with which it may have been reacting chemically in a mild way is squeezed out in the direction of lower pressure—usually upwards, if there is still room for it to move between particles above. At the same time, the whole assemblage of particles settle themselves into a more compact mass, often moving about in the process and producing curious structures which may modify the original sediment layering. At this stage the pressure ceases to be quite as uniform as when the water-envelopes still remained, but there is as yet little coherence or internal friction. Thus, if the sea bed is on a slope, the accumulating sediment will be unable to resist the pull of the down-slope component of gravity. It will tend to slide or flow down the gentlest of gradients. All of the factors affecting the sediment at this stage, physical or chemical, during and after deposition, are summed up by the word 'diagenesis'. In many ways it can be looked upon as the very first step on the long road to metamorphism. To begin with, however, its main effect is to turn incoherent sediment particles into a continuous, coherent (i.e. 'lithified') sedimentary rock.

Sedimentary material buried beneath kilometres of later deposits in a subsiding geosyncline has entered a very different environmental realm. Temperatures are now measured in hundreds of degrees and the confining pressure is thousands of times greater than that of the atmosphere at the surface. In addition to confining pressure, other stresses are applied which differ in intensity in different directions, so the sediment mass (we can now surely say rock mass) is subject to non-uniform pressure or 'shear-stress' tending to deform it. The greater the depth below the surface, the more intense these deforming tendencies may be, for it must be remembered that horizontally-applied compressive stress at any depth cannot for long exceed the vertical component of stress acting downwards at that depth due to gravity—the load of overlying rock—without relief occurring in the easiest direction, which in this case is likely to be upwards. Fluids and dissolved matter entombed with the sediment when deposited may still be a factor in the chemical environment, but are likely now to be augmented by other chemical agents dissolved in circulating waters brought in from deeper levels of the crust. We are accustomed to thinking of water as a mild dissolving agent with limited ability to corrode carbonate rocks and stratified salt beds, but under high pressures and temperatures its physico-chemical activity is greatly increased.

An alternative definition of metamorphism could be 'the physical and chemical adaptation of rock materials to new environments upon entering hot, pressurised zones'. Strictly, it should involve no more than reconstitution of the same initial materials, with the replacement of the original minerals with new ones better adapted to the pressure/temperature conditions at depth, and of the original sedimentary structures (such as bedding) with new ones suited to the stress conditions. The chemical factors should have chiefly the functions of catalysts in the metamorphic reactions. The other term, 'metasomatism', would take care of the case where the bulk composition of the rock is altered as a result of substances being brought in and taken away by chemical agents moving through the intergranular and intra-crystalline spaces.

In fact, metamorphism and metasomatism go hand in hand, and it is difficult to maintain a sharp distinction when one does not know exactly, after metamorphism, what was the composition of the parent rock before alteration. It would be wrong to think of rock undergoing metamorphism as a completely closed system. Metasomatism occurs through atom-by-atom replacement, so that although there may apparently be no change of crystal form, the internal structure may be greatly altered. Metasomatic replacement thus may result in the new mineral becoming a pseudomorph of the one whose shape it has adopted. Metasomatism is by no means exclusively a high-temperature process. One of the commonest examples is the replacement of aragonite crystals by the other familiar form of calcium cabonate, calcite; another is the replacement of fluorspar by quartz. Neither of these replacements requires a high temperature. At moderate depths the chief agent involved in the conveyance of materials to and from the site is water, usually with carbon dioxide, the sulphur oxides, chlorine, fluorine, boron and other things in solution too. Movement through the rock is easier where there are fissures and bands of crushed rock due to faulting.

In the early stages of metamorphism, and at moderate depths of burial, the effects of metasomatism cannot easily be separated, but progressively with increasing depth the closed system of metamorphism applies to a greater degree.

The Geothermal Gradient

Anyone who has been down in a deep mine will have been uncomfortably aware that the temperature rises as the distance from the surface increases, despite powerful ventilation. The rate of increase over each 100 m of depth varies in different regions from less than one degree to about 4°C near the surface, and theoretically the rate of increase should diminish towards the earth's deep interior.

These temperature gradients reflect the necessity for the earth to keep on discharging its internally generated heat. They are governed essentially by the limited ability of rock to transmit heat by conduction, coupled (at very great depths) with the possibility of bodily transfer of heat contained in rock which is taking part in very slow convection flow. Rock that is being depressed to deeper crustal levels in an active geosyncline rises in temperature as fast as heat can flow into it, but thermal conductivity of rock is so poor that the passage of heat into the geosynclinal mass cannot keep pace. The geothermal gradient will therefore be below average while sinking is going on. The reverse is true in the orogenic phase, because material is being brought up and stripped by erosion at a rate too fast for its heat to be discharged to the surface, and the gradient should be steeper. The gradient beneath stable, continental areas represents an equilibrium and is more uniform. Heat may also be brought upwards by uprising molten igneous rock (magma), with effects upon the invaded rock confined usually to the area near the surface of contact. For this reason, and because raised temperature is the dominant environmental influence, the resulting alteration of the invaded rock is called 'contact' or 'thermal' metamorphism.

Geosynclinal accumulations of sedimentary rock, while they are still subsiding and afterwards in the orogenic phase, are entering realms not only of higher temperature (due to the geothermal gradient) but also of high pressure and deforming shear-stress. The resulting alteration is therefore called 'dynamo-thermal' metamorphism, and because the effects are not localised, as with thermal metamorphism, also 'regional' metamorphism. Without going too far into the problems connected with the origin of the stresses (apart from those due simply to depth of burial), it may be noted here that while some theories

Fig. 5.1
Block-diagram and cross-section to illustrate a simple form of 'pluton'. It could represent, for example, a batholith of granite which invaded the roots of a mountain chain at depths of several kilometres, and which has since been exposed at the surface by uplift and erosion. Such a pluton would be surrounded by a zone of thermal metamorphism sometimes called a 'contact aureole' in which the country-rock has been altered to schist, spotted slate, hornfels, etc.

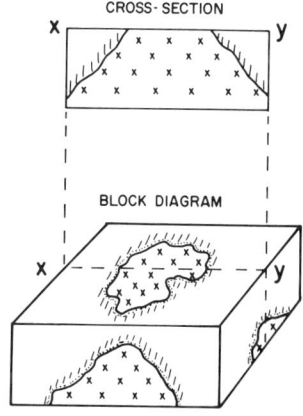

CROSS-SECTION

BLOCK DIAGRAM

Fig. 5.2
Metamorphism, especially of fine-grained sedimentary rocks, is likely to result in a parallel arrangement of crystals of flaky minerals like mica and chlorite which will give a sheen to the rock when light falls on it from certain directions, and may cause a tendency to split easily (cleavage) in directions governed by the parallel orientation of mineral flakes (a). Although typically a product of metamorphism, as in slate and schist, a splitting tendency is also present in fine-grained mica-bearing sedimentary rocks deposited from still water; it is then called 'fissility' and the rock will be called 'fissile shale'. Parallel alignment of needle-shaped crystals (e.g. types of amphibole) in metamorphic rock is one way in which rock can acquire a linear grain (lineation) (b).

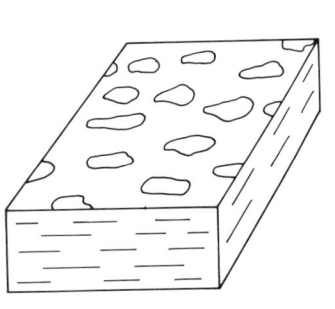

a

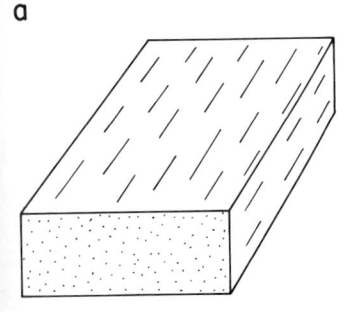

b

invoke nothing more than gravitational flowage or 'creep' of deep rock layers down the flanks of upwardly-bulging 'hot-spots' of the mantle, more generally the stresses are explained in terms of frictional drag applied below the crust by horizontal movements of the mantle, which in turn are attributed to thermal convection movements on a vast scale in the earth's interior. Over the last ten years or so we have been hearing a lot about the origin of lateral, compressive stresses during both the geosynclinal and orogenic phases from the continual expansion of the main ocean floors, which thus are caused to underride the edges of continents. (See Chapter Ten.)

Mineral and Rock Transformations

Long established laws of physics require that metamorphic changes must always be of a kind, and in a direction, that will tend to relieve rather than intensify an existing state of stress. Thus rock minerals that were in equilibrium with conditions of minimum pressure when deposited as sediment have to be replaced by other forms which are stable under the more severe conditions at depth and which generally are more compact and of higher density. As the new, high-stress environment is generally a non-uniform one the new minerals tend to be of a flaky or lamellar habit (like mica), or needle-like and fibrous (like hornblende and actinolite), which forms assist the rock mass to accommodate to shear stress. The rock as a whole acquires not only a new mineral complement appropriate to the pressure/temperature environment (P/T for short), but also new textures and structures such as the cleavage of slate and the layered, foliated structure of schist.

Progressive Metamorphism

Deep erosion brought about by the rapid uplift at the climax of the orogenic phase lays bare rock which previously had been several kilometres below the surface; it brings its metamorphic features up with it, for many of the transformations are reversible only by the attack of weathering in the next cycle. Careful mapping over the eroded surface sometimes shows a progressive increase in the 'grade' or 'rank' of alteration across a belt of metamorphic rocks. Arbitrary scales have been worked out which recognise the presence of a particular key mineral species, or an association of several minerals, as indicating that a certain stage of advancement of metamorphic change had been reached by the rock containing them before the final uplift terminated the process. Belts over which the rocks yielding particular key minerals occur are mapped as metamorphic zones. Up to a certain grade the differences between them are ascribed to differences in their maximum depth of burial, but for the higher grades this amounts to an oversimplification of what must always have been a complex situation.

In the case of the greywacke type of sediment typical of most geosyncline deposits the first step on the way to metamorphism is recrystallisation of some components and minor mineral replacement, causing a textural change barely visible to the naked eye. The original clay matrix between sand grains is converted to fine muscovite (white mica) and to the finely-flaky, greenish-grey mica-like mineral chlorite, and the rock externally no longer looks quite like an unaltered sedimentary rock. Chlorite is later replaced by biotite in larger flakes, biotite by garnet, and so on. But while chlorite still remains other important changes to do with texture and structure are developing. The alignment of flaky or elongated crystals in preferred directions gives the rock as a whole first a tendency to split or 'cleave' in a direction distinct from those of the original stratification or joints and other fractures, and then an increasingly pronounced metamorphic layering, 'foliation' or 'schistosity'.

In New Zealand we have for many years used a system for classifying the metamorphosed greywacke members of the New Zealand Geosyncline which was adapted from classical studies early in this century by George Barrow in Britain. He designated zones of progressive regional metamorphism based on the appearance, successively, of chlorite, biotite, garnet, etc., as one passed into increasingly metamorphosed rocks. In the 1930s F. J. Turner and C. O. Hutton introduced four subdivisions within one of these, the Chlorite Zone in Otago, based on visible, textural differences which the field geologist could detect without special aids. The aluminosilicate minerals kyanite and sillimanite are known to require high temperatures for their formation, and presumably deep burial as well. They are useful 'geological thermometers', but do not lend themselves to zonal mapping. Kyanite has been found in South Westland in boulders, but not yet in place.

It is important to remember that unless the original rock contains the necessary constituents the zone index minerals cannot form even if the temperature and stress conditions are favourable. This means that while the absence of a particular metamorphic zone from its proper place in the sequence could be due to dislocation by later fault movements, it could also have resulted from the rocks at that place being deficient in some of the chemical ingredients of the index mineral. The zonal scheme as developed for New Zealand requirements is based on the metamorphic path followed by the typical sandy greywacke.

Rock in the heart of an active orogenic system is maintained at the maximum temperature for longer than it can sustain strong deforming pressures, so bodily transport or 'solid flowage' of the material takes place. The coarse crystalline texture of gneiss, commonly sorted out into layers of alternately lighter and darker coloured minerals, will there replace the cleavage and finer schistose structures more typically the product of non-uniform stress.

This brings out an important point about metamorphism in general. Although the whole mass of rock undergoing alteration is sheared and distorted, stretched, squeezed out and rolled up—as evidenced sometimes by distorted pebbles, fossils, sedimentary structures still discernible in low-grade metamorphic rocks—the entire process goes on strictly in the solid state without the rock having to become even partially molten. It may be found hard to visualise 'solid' rock distorting and bending, yet laboratory experiments using slowly-applied high stresses show how this can happen, as also does the sagging of unsupported stone lintels occasionally seen above doorways in ancient buildings. Let it be remembered always that the time-span of the metamorphic process is not likely to have been of a lesser order than a million years.

Having stressed the importance of distortion in the solid condition, I should now go on to say that the climax of every orogenic event seems to have been accompanied by some actual fusion within the core of the geosyncline forming an igneous magma commonly of granite or granodiorite composition (Chapter Six).

The typical New Zealand metamorphic zones have to take account of the differences of original composition in the Hokonui Facies and Alpine Facies sediments, both of which grade into the metamorphic belt now usually called the Haast Schist. This has led to complications, some lack of sympathy between the petrologist in the laboratory and the geologist in the field, and a plethora of special terms none of which has been adopted universally. Regional metamorphism of what were originally sandstones and siltstones composed of detrital quartz, feldspar and clay minerals derived from a previous cycle of weathering of granitic or greywacke rocks produces the suite of quartzo-feldspathic schists divisible into the several progressive zones and sub-zones. The Hokonui Facies included large amounts of volcanically-derived tuffs and lavas of a different bulk composition, so naturally the index minerals cannot be expected to appear in the same way as they do in progressive metamorphism of the Alpine Facies. Being richer in iron-magnesium silicate minerals tending to be green in colour, notably varieties of hornblende, yellowish-green epidote, and particularly abundant chlorite (in the Chlorite Zone), they are informally called 'greenschists' (which is a bit ambiguous because 'Green Schist Facies' has a rather special petrographic meaning, which we cannot go into here). A number of other terms have been suggested, some of which eventually will replace the 'zones' of Hutton and Turner, and the nomenclature is in a state of flux at present.

Rocks of originally simple composition suffer little mineralogical change in metamorphism, though they will be recrystallised or crushed and shattered by severe mechanical stress under low temperatures. Thus a relatively pure calcite limestone may merely recrystallise, changing from a granular texture to that of a crystalline marble; a pure quartz sandstone will recrystallise and weld into a quartzite. Such

Fig. 5.3
A greatly simplified, hypothetical cross-section across the Southern Alps between Haast (left) and Lake Ohau (right). It gives one interpretation of the succession of belts of rock showing various grades of metamorphism. A.F.–Alpine Fault; I, II, III, IV–subzones of the Chlorite Zone of progressive, regional metamorphism; B–Biotite Zone; G–Garnet Zone; O– Oligoclase Zone. (After B. Mason, 1962)

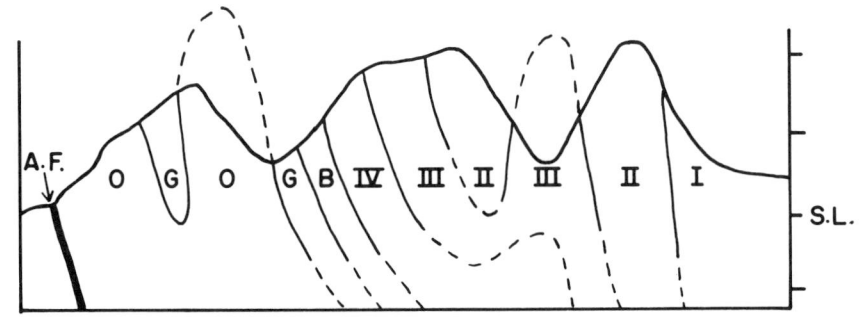

transformations do not lend themselves to the separation of progressive stages of regional metamorphism.

Repeated Metamorphism and Rock Folding

One of the most important areas of advance in New Zealand geology in the years since World War II has been in unravelling the history of metamorphism and tectonic deformation of the geosynclinal rocks. Very detailed observation of structural features, attitude of foliations, cleavage, etc., at the outcrop and refined methods of laboratory study combine to tell us when the schists of an area have suffered two or more phases of metamorphism and from which directions the stresses were acting at different times, and also when there have been alternate episodes of metamorphism and folding. In many cases it has been found that the foliated schist structure developed during one period of dynamothermal metamorphism has had the effects of a later one overprinted upon it, with folding of the earlier foliation pattern in between. It is also found that an older dynamothermal metamorphism has been obliterated in a later period of high temperature during which strong shear stress was lacking. This is one form of repeated metamorphism.

Recognition of repeated metamorphic episodes is an old problem in New Zealand. For a very long time it has been appreciated well enough that the age of the original rock and the date of its most recent metamorphism are different things, but it is also known that sedimentary rock formed in the geosynclinal phase will, if it descends deeply enough, become metamorphosed during the subsequent orogenic part of the same mountain-building cycle. The early phases of metamorphism cannot be separated from the crumpling and piling-up ('tectonic thickening') of strata at the bottom of the geosynclinal mass. Folding and metamorphism must go hand in hand at this stage. The folding involves much distortion and rotation of the material, and this alone must mean that the stresses of later metamorphic spasms will be applied effectively as from different directions, producing differently-oriented schist foliations and mineral or structural alignments ('lineations') superimposed upon the earlier ones, and upon any stratigraphic features that have not yet been blotted out. Further deformation may then be superimposed during a later orogenic cycle.

Confusing indeed! Sceptics are not unknown, but despite much critical scrutiny, some remarkable syntheses of metamorphic, folding and plutonic history have gained very general credence in the last decade or so.

The Mapping of Metamorphic Rocks in New Zealand

There has been occasion already in the earlier chapters to mention metamorphism, but we are now in a better position to take an overall look at New Zealand metamorphic rocks.

From the time of pioneering geological explorations in the mid-nineteenth century onwards, various schemes were brought forward to classify, name and correlate the metamorphic rocks in various parts of the South Island. Some authors took the trouble to

Fig. 5.4
Contorted Haast Schist in the Biotite Zone of progressive regional metamorphism, seen on the Otago coast near Brighton, south of Dunedin. Layering due to the separation of quartz and feldspar from the micaceous components of the schist ('metamorphic segregation') has been crumpled and ruptured in the course of severe internal deformation of the rock in later phases of its metamorphic history. (Photo: A. H. Dowling)

explain how and why their proposals differed from others, but commonly the alternative schemes of earlier or contemporary workers were just ignored.* The naming of metamorphic rocks in the early days was done chiefly for use in restricted areas. The units were valid enough for this limited purpose, and confusion arose when their application was extended by the original author or by others to imply correlations with similar rock in other areas. There was plenty of scope for guesswork and futile argument in the earlier correlations and classifications, and it is indeed only within the last few decades that the stratigraphy of our older sedimentary rocks has advanced to the stage where it could provide the essential background for a proper understanding of the metamorphic rocks derived from them.

Our present understanding of progressive metamorphism in New Zealand stems mainly from petrological work from about 1929 onwards by F. J. Turner and C. O. Hutton at the University of Otago. The relation between metamorphism and the structural evolution of New Zealand's framework has been advanced continually since then by J. J. Reed, G. W. Grindley and others connected with the Geological Survey, and also by D. S. Coombs and C. A. Landis and their associates at Otago. Radiometric means of determining the dates of metamorphic events were not applied in New Zealand before about 1960.

To understand the nature and origin of the metamorphism of New Zealand rocks is one thing; the mapping of our metamorphic rocks is another task, in which the earlier mappers were knowingly or unknowingly handicapped without that understanding. Neither objective could be achieved satisfactorily in isolation from the other, and I think it fair to comment that liaison between petrologist and field geologist could at times have been closer. As it was, some metamorphic petrologists were chiefly interested in studying mineralogy, mineral chemistry and microstructure in great detail, while the field mappers felt that they had to set up their own simplified units of rock which could be distinguished in the field but which did not make use of all the available knowledge from laboratory studies. It was petrologist C. O. Hutton, however, who was largely responsible for the introduction of the first practical mapping scheme. Collaboration all round is now excellent.

Once again it was the practical necessities and limitations of the 1:250,000 mapping programme which made it imperative to develop a classification of the metamorphic rocks which was simple and direct enough to be used in rapid reconnaissance of unmapped areas with a minimum of sampling for later laboratory study, yet soundly and adequately based on the findings of the petrologists. The compromise which was achieved will be outlined in the following paragraphs.

Contrasting Metamorphic Realms
A clear separation can be made between rocks that were

---

* A critical and rather uncompromising appraisal of the growth of ideas on the subject of metamorphism in New Zealand has been made by J. B. Waterhouse (*N.Z. Journal of Geology and Geophysics*, vol. 10, pp. 923–49, 1967). His observations are no less cogent because Waterhouse is a stratigrapher-paleontologist rather than a petrologist.

metamorphosed (most recently, anyway) in an environment of high temperature but relatively low deforming stress, and those produced under strong deforming stress but comparatively low temperatures, which by geothermal standards would mean not more than a very few hundred degrees Celsius. The distinction was already established, largely on field evidence, at the outset of the 1:250,000 mapping programme and was used to classify the metamorphic rocks for the 1:2,000,000 Geological Map of New Zealand published in 1959 to mark the hundredth anniversary of Hochstetter's pioneering explorations.

As explained briefly by G. W. Grindley in Chapter II of *N.Z. Geological Survey Bulletin 66* (essentially a greatly expanded legend for the 1:2,000,000 map) the first group embodies the products of essentially thermal metamorphism, not only localised contact metamorphism by the heat introduced by invading granitic plutons, but also the more extensive and uniform thermal metamorphism ascribed to direct and prolonged access by geothermal heat in the environment of the deeper zones of an orogenic belt. It also included gneisses in Fiordland which were suspected to have encountered very high temperatures at more moderate depths because the geothermal gradient had at some stage been exceptionally steep. The parent rocks of this group turn out to be mainly the early Paleozoic sediments of north-west Nelson and Fiordland, but included sediments of volcanic origin in southern Nelson and western Otago which could be as young as late Paleozoic.

The second group accommodated the schists and gneisses into which the New Zealand Geosyncline strata progressively grade, and which now constitute the Haast Schist belt. Their metamorphism was essentially dynamothermal, and there are no large plutonic intrusions. The parent rocks were the geosynclinal sediments ranging in age from late Carboniferous to Jurassic and the associated volcanic products.

C. A. Landis and D. S. Coombs in 1967 formalised this distinction in another way. Similar contrasts are known elsewhere, in Japan for example, separated by a major tectonic boundary, and the expression 'paired metamorphic belts' was brought in to sum up the situation. In New Zealand the western belt of high temperature/low stress metamorphism of Buller Geosyncline material constitutes essentially the Tasman Metamorphic Belt (that overworked prefix Tasman again), while the high stress/low temperature metamorphism of the New Zealand Geosyncline material essentially constitutes the Wakatipu Metamorphic Belt. But it should be noted that these divisions are tectonic and structural, not stratigraphic.

The stratigraphic legends on Geological Survey maps treat the rocks of the two belts very differently. Those of the Tasman Metamorphic Belt are named as formations and groups, i.e. as lithostratigraphic units, and outlined as such where sufficiently extensive to show on the mapping scale; localised contact metamorphic rocks adjacent to intrusive sills and dykes are not separately recognised. No attempt is made to portray zones of progressive metamorphism in these rocks. Grindley, however, records that in Nelson the highest metamorphism is attained in the Pikikiruna Schist which underlies Mount Arthur Marble

Fig. 5.5
Not many years before this photograph was taken the Haast Schist rock in the foreground was still being abraded and smoothed by the Franz Josef Glacier. Darker bands are rich in the micaceous minerals biotite and chlorite and other dark silicate minerals.
(Photo: R. Warburton)

on the west flank of the Pikikiruna Range, and that the grade generally decreases westwards. Rock units formerly known in their respective areas as 'Marlborough', 'Alpine' and 'Otago' Schists and since 1961 as 'Haast Schist Group', the metamorphic rocks in the Wakatipu Metamorphic Belt are subdivided not into lithostratigraphic formations, but into zones of progressive metamorphism; four sub-zones of the Chlorite Zone; Biotite Zone, Garnet-Oligoclase Zone.

The 1959 map and *Bulletin 66* also included in the high temperature group the southwards-narrowing band of dark gneissic 'metavolcanics' between Glenhope and Maruia now considered part of the Rotoroa Igneous Complex (see later in this chapter). Under the name Rotoroa Gneiss, they were described as probably the equivalent, across the Alpine Fault dislocation, of the Darran Diorite of the Darran Mountains between Lake McKerrow and the northern end of Lake Te Anau. Landis and Coombs appear to leave them in the Tasman Belt too, although in age they may range through the whole of the Paleozoic Era, overlapping into New Zealand Geosyncline times.

What does it all mean? How did it come about that belts of altered rocks suggestive of metamorphism under very different physical environments lie side by side? With the Alpine Fault in mind, and also the knowledge that the parent rocks of the two metamorphic belts were almost entirely different in age, it is no surprise that some writers suggested that the metamorphism took place at different places and perhaps different times, and that the boundary between them, known as the 'median tectonic line' and apparently now offset by the Alpine Fault, was itself originally a similar great lateral dislocation. This, however, does not appeal to C. A. Landis and D. S. Coombs, who took the cue from Japanese investigators of similar situations in the west and north-west Pacific and applied the 'paired metamorphic belts' concept in New Zealand. On the contrary, they believe it is necessary to explain the contrasting metamorphic realms in terms of different histories on opposite sides of the 'line', and an abrupt difference in the geothermal gradient (much steeper in the west) during the last 'heating up', or 'thermal event'.

Dates of Metamorphism

Opinions have varied widely as to when the various groups of schists and gneisses became metamorphosed. In the earlier days the question was pursued along what we now realise were rather oversimplified lines, essentially stratigraphic. Thus, the age limits were sought by considering: (i) the youngest sedimentary rocks that appear to merge into schist (metamorphism must be more recent); (ii) the oldest sedimentary beds containing recognisable pebbles of underlying schist (metamorphism must be earlier to allow for uplift and deep erosion needed to expose schist formed at depths, presumably, of at least a few kilometres); (iii) relationships with igneous dykes and sills; etc.

The answer was never clear-cut and seldom conclusive. Next came the era of structural petrology, introduced to New Zealand chiefly by F. J. Turner, which enabled the metamorphic schists to speak for themselves at least about the number of times they had suffered stress, and thus indicated the number of episodes of metamorphism that had to be inserted into the stratigraphic geological history of New Zealand. Once radiometric dating was available here it was possible to establish when the radioactive 'clocks' in a large number of rocks sampled began to 'tick'. Two main periods are notable since the beginning of the Paleozoic Era, one around late Devonian or early Carboniferous (370—320 million years) and the other early in the Cretacous Period (140—95 million years). These times, of course, agree with the geological datings of Tuhua and Rangitata Orogenies, not surprisingly because of what we believe about the physical environment in an active orogenic belt.

Radioactivity dates, as given by the methods appropriate for the time-ranges involved here, actually tell us when rock substance that previously had been very hot (solid or molten) had cooled down to the temperature where crystallising (or recrystallising) minerals were thereafter able to retain the products of radioactive decay. One important source of error, giving anomalously young ages, is leakage of these products, especially the gases, helium and argon. This retention-point of temperature is the starting point of the 'clock', after which the ratio of parent radioactive isotope to decay product can be made to measure the lapse of time since the sampled rock cooled.

Metamorphic rock datings may be interpreted as meaning that crustal movements rapidly brought up into the realms of erosion at those particular times rocks which had previously spent some time down in the deep, high-temperature zones—in the case of the low-stress, high temperature group, long enough for the effects of their earlier passage down through cooler, high-stress realms of folding, shearing and low geothermal gradients to be eliminated. Along these same lines one can explain the surprisingly recent radiometric ages obtained by the potassium/argon method from schist and pegmatite in the Haast Schist along the western flank of the Southern Alps, and without claiming errors due to loss of the argon gas product. Ages ranging from 75 million to as little as 4 million years were at first calculated by this and other erroneous methods, but it was quickly realised that the dates indicated *not* the time when New Zealand Geosyncline rocks were altered to Haast Schist, but rather the time when, as a result of a strong vertical component coming into movements east of the Alpine Fault, schist metamorphosed in the Cretaceous Period or earlier was rapidly brought up and exposed by erosion. The maximum amount of uplift on the Alpine Fault during the Kaikoura Orogeny (see below, page 29) was estimated by R. P. Suggate at more than 14 kilometres.

As a postscript to the above, let us note that the environments of metamorphism undoubtedly exist today, and that schists are currently being formed at appropriate depths in mobile belts of the crust where thick sediments are accumulating; deeper still, where shearing stresses cannot be sustained, high temperature thermal metamorphism and melting are going on. The radioactive clocks of the material concerned will not start until orogenic uplift in the future brings them up to cool off.

Fig. 5.6
Haast Schist is exposed at many places alongside Grove Road between Picton and Havelock. Having developed a very distinct cleavage (dipping to the right) but without the quartz and other minerals being segregated into separate layers (foliae), it has been mapped here as belonging to the Chlorite-III Subzone of progressive metamorphism.

Regional Metamorphism on Parade

Few countries in the world can provide more accessible, less complicated and more geographically compressed illustrations of progressive regional metamorphism than those available for any geological tourist travelling the trans-Alpine highways of the South Island. All three of these have long sections cutting fairly directly across the trend of the metamorphic zones of the Wakatipu Metamorphic Belt (or, if you prefer, the Haast Schist Zone). In the case of the Lewis Pass and Arthur's Pass highways the sequences of metamorphic zones on opposite sides respectively of the Maruia and Taramakau rivers are offset along major faults branching from the Alpine Fault. Within a few minutes, representative samples of the Chlorite Subzones, the Biotite Zone and the Garnet-Oligoclase Zones can be picked up from the beds of the many rivers crossing the Haast Schist belt west of the main divide. If searching for specimens in the small streams crossing the trans-Alpine roads east of the Alpine Fault, it is necessary to allow for the fact that past glacier and river transport have brought debris of lower-grade schists down-valley from outcrops nearer the alpine divide. One should therefore look for the highest rank pieces one can find at each locality.

Most of the main valleys of South Westland also give excellent sections, including spectacular and handsome examples of hornblende- and epidote-bearing schists presumably derived from sediments containing a fair amount of igneous material. The Chlorite-IV Subzone is missing north of Fox Glacier but it extends over vast areas in Central and western Otago. Marlborough schists are mostly in Chlorite-II and Chlorite-III Subzones.

The 1:250,000 geological maps are an adequate guide to examples of the different types of metamorphic rocks, while the maps and descriptions in J. J. Reed's 'Regional Metamorphism in South East Nelson' (*N.Z. Geological Survey Bulletin 60*, 1958) give more details about the Lewis Pass Highway section and surroundings. I do not propose to describe the individual units named in the maps. In the case of metamorphic rocks, to go into matters more deeply than is done in the published map legends and accompanying summaries soon brings in matters beyond the scope of lens-and-pocket knife studies in the field. Before concluding this section, however, it seems advisable to say something about the more or less metamorphosed igneous rocks in south-east Nelson which have in the past been distinguished by the impressive name 'Rotoroa Igneous Complex', though it will be necessary to bring them into the discussion again in later chapters.

The name 'Rotoroa Igneous Complex' was first used for this assemblage of rocks by J. Henderson and H. E. Fyfe when reporting on regional mapping from Motueka south to Maruia in the 1920s. Originally they were igneous rocks ranging in composition from diorite to gabbro and in texture from granite to that of shallow intrusive bodies, perhaps even lavas. The metamorphosed products include varieties of hornblende gneiss intruded by sheets and veins of granodiorite and pegmatite. Rotoroa rocks compose the basement over much of the upper Buller Valley, and there is geophysical evidence that they underlie the late Tertiary sandstones and conglomerates covering the Moutere Depression and extend seawards into Cook Strait. There are good exposures of Rotoroa rocks alongside Highway 63 between Kawatiri and Howard River.

It is likely that the parent igneous rocks were emplaced at different times during the Paleozoic Era. The unmetamorphosed intrusive igneous rocks could perhaps have been part of the feeding system for volcanoes along the western border of the New Zealand Geosyncline (see also Tasman and MacKay Intrusives and Fiordland Complex, below, pp. 144–5). Northern readers may ask 'Did the North Island regions escape metamorphism altogether?' The answer is certainly that they did not. Although exposure is limited to a belt of low-grade (Chlorite-II Subzone) schist in the central Kaimanawa Mountains east of the Tongariro volcanoes, and another narrow strip along the east coast of Kapiti Island, pebbles of gneiss, marble and quartzite in Tertiary conglomerates make it very likely that more extensive metamorphic terrains underlie the covering strata elsewhere.

17. *Two higher positions of the shoreline during the Pleistocene Period are indicated in this view of Te Miko Point on the South Island west coast north of Punakaiki. The lower, gently sloping platform was carved by the sea across Tertiary strata during one of the later interglacial times of relatively high sea level. A steep cliff at far left exposes beach gravel and sand deposits marking an earlier, still higher, interglacial shoreline. Both raised shorelines are now higher than they would have been without the continual uplift suffered by this region through the late Pleistocene (p. 245).*

Chapter Six

GRANITE

Rock of Ages?

Geologically Unfortunate Metaphors

Marble symbolises coldness and immobility probably because for so
long it has been used by sculptors to memorialise the dead and to still
the motion of living things. Geologically the metaphor is not very apt
because marble was formed from limestone under confining pressure
and raised temperature, because it often shows internal signs of having
flowed like an extremely viscous liquid, and because it can still be made
to deform under long-applied stresses. Granite, too, is associated with
an unfortunate metaphor. It is represented as a standard of stability,
permanence and great antiquity, whereas granitic rocks decay relatively
quickly in humid conditions, and although granite is plentiful among
the most ancient Precambrian rocks, it is no longer identified
necessarily with relics of a supposed primordial earth crust. In fact,
granite now exposed by vigorous erosion in the Himalaya region almost
certainly came into being in the Tertiary Period.

Granites and Granites

As students, geologists of the author's generation learnt about igneous
rocks according to a genetic classification developed in Europe a
century earlier. There seemed to be no question about the place of
granite in the scheme of things. It was produced at considerable depths
by slow cooling and therefore coarsely-textured crystallisation of
molten magma of the appropriate composition, i.e. rich in silica,
relatively rich in sodium and potassium, relatively poor in lime, iron
and magnesium. The excess of silica meant that besides feldspar, mica
and perhaps hornblende, quartz was visibly present and the rock
therefore belonged to the 'Acidic' family on the basis of its composition.
At the same time, being wholly, evenly and rather coarsely crystalline,
granite was in the 'Plutonic' class. Well towards the other end of the
line as regards composition, we learnt that the silica-poor, lime-, iron-
and magnesium-rich rocks belonged to the 'Basic' family in which the
Plutonic Class was represented by gabbro and the fine-textured
Volcanic Class by basalt. The erupted lava rock of the granite family
was rhyolite; and so on. Today we use 'silicic' or 'felsic' for acidic,
'mafic' for basic.

Within the deceptively simple framework of the above scheme there
was a limited number of pigeonholes, but as more advanced students
we were to discover that a host of special categories had had to be
provided (and we were supposed to know them) for all the
intermediate compositional and textural variations. At times it seemed
as though the igneous petrographers forgot that the class subdivisions

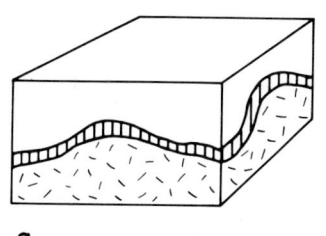 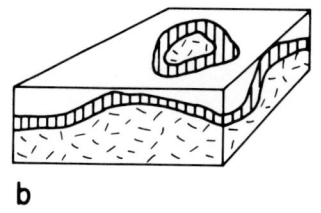 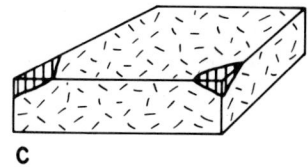

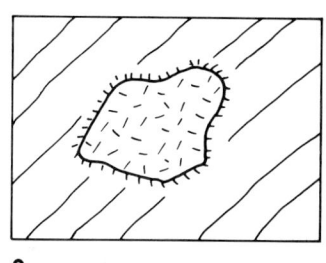

which they had invented were artificial, whereas the igneous rocks in nature grade continuously between family and family, class and class, within broad limits set by laws of physical chemistry. It did seem odd that some of the most abundant types of plutonic rock exposed in the eroding cores of younger mountain chains around the world (granodiorite or quartz-diorite and their volcanic equivalents, dacite, etc.) did not fit readily into pigeonholes of the simple scheme. I am glad to say that the subject is now approached from the beginning in a way that emphasises the continuities.

Granite in its many varieties with more or less orthoclase feldspar, more or less white or black mica, hornblende, etc., whether it occurred as relatively small dykes and sills or vast masses continuous for hundreds of kilometres ('batholiths' or 'plutons'), was an igneous rock. Its origin by crystallisation of a rock-melt was not in question. Nevertheless it was recognised that granite could be found in two quite different settings. First, as 'discordant' plutons it cut across the boundaries and the structural 'grain' of the invaded 'country rock', causing thermal metamorphism along the margin. Secondly, granite occurred as 'concordant' plutons, lens-shaped masses running as it were *with* the structural grain of the country rock, lacking sharp boundaries with it or contact metamorphic effects. Moreover, in the second case sedimentary stratification, conglomerate pebbles and other textural contrasts typical of sedimentary rock may fade, in an apparently continuous transition, into what can only be described petrographically as granite. But how could such a rock have crystallised from a molten magma? And if not, should it still be called 'granite'? Other old questions, unresolved but pushed aside, cropped up again. For example, how to account for the volumes of granite and similar rocks in enormous plutons around the world by the orthodox method of fractional crystallisation from a supposedly world-encircling basaltic layer at the base of the crust, remembering that from a given volume of basalt only a small fraction of that amount of granite can be obtained directly.

In Europe generally, and especially in Scandinavia and Scotland, it was widely believed that some ancient granites had been derived in place from sedimentary rocks or from their metamorphic products without ever having been liquefied. The origin of vast volumes of granodiorite in the mountain belts was thus no problem because the composition of that rock is similar to the bulk composition of geosynclinal greywacke. The argument between conservative 'magmatists' and the rebel 'granitisers' was prolonged and fierce, almost

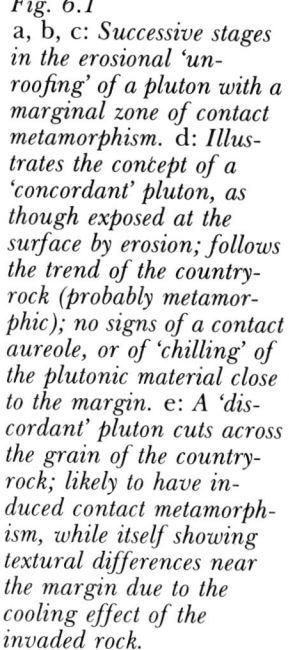

Fig. 6.1
a, b, c: *Successive stages in the erosional 'unroofing' of a pluton with a marginal zone of contact metamorphism.* d: *Illustrates the concept of a 'concordant' pluton, as though exposed at the surface by erosion; follows the trend of the country-rock (probably metamorphic); no signs of a contact aureole, or of 'chilling' of the plutonic material close to the margin.* e: *A 'discordant' pluton cuts across the grain of the country-rock; likely to have induced contact metamorphism, while itself showing textural differences near the margin due to the cooling effect of the invaded rock.*

18. *All kinds of glacier-margin sedimentary deposits are exposed from time to time along the highway between Hokitika and Fox Glacier after slips and road works, but they quickly become overgrown. Virtually all were laid down during the Otira Glaciation by the valley glaciers and piedmont ice sheets of Westland. These laminated silts were laid down in an ice-margin lake and then buried under coarse, bouldery till when the lake was overwhelmed by a new ice advance. The large block would have fallen off the glacier or from a floating iceberg. Bending of the laminae around the block show that the silts were later compressed by the loading of till and ice (p. 237).*

19. *Oligocene limestone beds near Kaikoura showing complex folding, which may have developed through sea-bottom slumping and sliding soon after the sediment was deposited and before it had become indurated into a hard rock. These are merely minor folds superimposed upon a larger fold structure known as the Puhipuhi Syncline which originated during the Kaikoura Orogeny.*

20. *Layers of angular greywacke chips on the roadside near Porters Pass, Canterbury, were formed when frost-riven rock debris was streaming down the hillside during the intense cold of the Otira Glaciation. The soil has developed since the end of the glaciation (p. 260).*

21. *In this bank of the Kowai River in Canterbury, alongside Highway 73, two sets of gravels of different colour are easily distinguished. (Actually, there are three, but the uppermost bluish-grey set is hard to separate in the photograph.) The lower, yellowish set represents aggradation by the Kowai during the cold of the Woodstock Advance by the Waimakariri Glacier (p. 256), though the Kowai catchment itself was not glaciated. Erosion and deep weathering occurred under mild, interglacial conditions before aggradation occurred again during the Otarama Advance at the beginning of the Otira Glaciation.*

22. *A fault exposed in Hanmer River, Canterbury, displacing Tertiary mudstone and Pleistocene gravels. Although mudstone shows on both sides, only a minimum amount of dip-slip (up-and-down) displacement can be estimated because on the left side the older brown gravels had been stripped off before the younger grey gravels were deposited. 'Drag' on the gravel layers can be seen close to the fault-plane. A long history of intermittent movement in the same direction is shown by decreasing amounts of displacement at successively higher levels, but none since the surface at the top of the bank was formed. It is a reverse fault because the fault-plane dips towards the upthrown side. (Photo: Lee Clayton)*

as hot as magma itself, and came to a head after World War II. Today the reality and ubiquity of the granitisation process is accepted by many, although the geochemistry is obscure. Some petrologists emphasised the importance of metasomatism involving migrating inter-grain fluids; others invoked migrations of elements in the ionic state; all envisaged recrystallisation under raised temperatures, and most agreed that once sedimentary materials had been chemically reorganised into the mineral groupings of granite ('granitised') they could later become molten and invade higher zones as an intrusive, truly igneous granite magma. The largest plutons, however, are more happily explained by partial melting of the roots of the sedimentary contents of a geosyncline in the climax of its subsidence.

By the middle of this century the debate had collapsed. We now see the real importance of the unity of all these phenomena: metamorphism; granitisation; intruded granites, granodiorites, diorites and gabbros; andesite and rhyolite volcanic lavas; and the evolution of great mountain chains. Varying composition of granitised rocks is seen as reflecting both the original composition and the changes due to differential migration of components while in the solid state, while in intrusive rocks the old fractional crystallisation scheme is adequate to explain the differentiation of more acidic from more basic portions within the same body.

The grand synthesis of granite origins helped to clarify other issues, including relationships between granites and banded granite-gneisses in the same region, and the long-debated origin of gold-bearing quartz veins. It also justifies the use of 'granite' in a widened sense beyond the originally restricted petrographic meaning, in fact to denote a terrain of granitic-textured plutonic rocks including not only true granites but also the more mafic types which nearly always are present in some amount as well.

How did it go in New Zealand? As long ago as 1890 James Park had discussed granitic rocks in north-west Nelson as both igneous and metamorphic without comment, but the granites were largely neglected until recent times. The great controversy was heard here as distant thunder only, and I recall little objection when J. J. Reed in 1948 suggested that the granitisation hypothesis could explain very well the distribution of mafic plutonic rock types in Southland.

Essentially South Island Rocks
A granite terrain hidden beneath younger rocks in the Auckland region would account for the few granite pebbles in conglomerates of Jurassic and Tertiary age, and for granite fragments brought up with volcanic lava ('xenoliths' = 'stranger rocks'), but no true granite is exposed at the surface in the North Island. Classed with the plutonic rocks, though not in large masses and probably not of very deep origin, gabbro and pyroxenite at North Cape and a light-coloured relative of granodiorite called tonalite near the tip of Coromandel Peninsula (western side) provide North Islanders with their best chances of seeing what plutonic rocks look like.

In the South Island a wide variety of granites (in the strict

petrographic sense) occur in larger and smaller masses from Nelson all
the way down the western side to Preservation Inlet and again on
Stewart Island. Sizable bodies of the feldspar-hornblende rock syenite
('granite without quartz') intrude the Torlesse Group greywackes of the
Kaikoura Ranges and in North Canterbury but there are no true
granites on the eastern side. Farther to the south and east, granite
composes the Bounty Islands, the Snares, and the basement under the
basaltic pile of Auckland Island.

The Tuhua Granites (Tuhua Intrusive Group)

The older Geological Survey under Hector did not make a practice of
naming the plutonic rocks as formations, but the first geological
bulletin issued by J. M. Bell's reorganised Survey in 1906 described
granite and associated rocks in the Hokitika district as 'Tuhua
Formation'. The granites were rather neglected in New Zealand for the
next few decades, but the name 'Tuhua' was revived and extended to
cover all granitic rocks west of the Alpine Fault for the 1:2,000,000
Geological Map of New Zealand in 1959. At that time also the belt of
granites east of the line of the Alpine Fault in the far south, though
thought of as a probable continuation of the Tuhua Granites, was
referred to as 'Fiord Granites'.

In Nelson and Westland the Tuhua Granites (or 'Tuhua Intrusive
Group' as they have recently been called) occur in three well-defined
belts trending roughly north and south. The westernmost belt has been
called 'Paparoa Granite' since 1959. It extends from Kongahu Point on
the coast north of Mokihinui southwards across the lower Buller Gorge
into the Paparoa Range, thence tapering to a narrow strip in the
northern catchment of Moonlight Creek. An offshoot from what is
probably the same body of granite at depth outcrops alongside the
coastal highway at Barrytown. As in the other Tuhua areas, a rapid
inspection of river-bed boulders is sufficient to show a considerable
range of granite types, though predominantly they are normal
biotite-granite. The most conspicuous variants are darker types, some
of which are partially-digested xenolith inclusions while others are local
concentrations of the biotite and hornblende crystals similar to those
normally dispersed through the granite. Attractive-looking porphyritic
types (meaning that notably larger phenocrysts are present in a matrix
of normal granitic texture) are not uncommon, the phenocrysts being
well-formed pink or cream-coloured potassium feldspar crystals.

Clear contacts between Paparoa Granite and host rocks have been
hard to find, but it has been shown north of the Buller River that
granite invades Greenland Group greywacke, converting it near the
boundary into spotty-looking contact-metamorphic hornfels. In detail,
the boundaries are irregular because wedges and fingers of country
rock protrude into the granite. Within the Paparoa Range itself this
granite intrudes banded granitic gneiss formerly called 'Charleston
Gneiss' and later 'Constant Gneiss' (after Constant Bay, the incredibly
tiny rock-girt harbour of Charleston in its goldmining heyday). The
relations between the granite, the Constant Gneiss and the Greenland
Group have been rather perplexing, and will be discussed later.

23. *The gap through which the Opihi River escapes from the Fairlie basin in South Canterbury is an excellent example of a 'superposed gorge'. The bold, straight front of the hills illustrates what is meant by 'fault-line scarp'; it has gained height relatively as a result of the softer Tertiary strata underlying the Fairlie basin having been removed by erosion more rapidly than the harder greywacke rock on the upthrown side of an old fault (pp. 303, 309).*

24. *The Victoria Range, west of the Maruia valley, Nelson, a granite mountain landscape broadly modelled by glaciers and dissected by widely spaced streams and gullies since the ice departed (p. 317).*

Detailed mapping by M. G. Laird, D. Shelley and S. Nathan has clarified them, and several distinct formations have been proposed to subdivide the Paparoa Granite.

The largest continuous tract of granitic rocks we have is the central belt of Tuhua Granites, known as 'Karamea Granite'. Covered locally by strips of younger sedimentary rocks, it stretches from Kahurangi Point north of Karamea southwards through central Nelson and into north Westland as far as Ahaura River. Southwards from there the gaps between granite exposures are wider, but in part at least this is because of deep erosion by glaciers during the Pleistocene Period. Very likely the granite of Bell Hill, Te Kinga, Hohonu and the other isolated mountains down through South Westland to Mount McLean at Jacksons Bay is more or less continuous at depth.

The range of types is much the same as in the Paparoa belt, so the distinction is purely a geographic separation. The first New Zealand discoveries of orbicular granite with its concentric shells of different mineral composition—a rare and handsome rock—were made near Karamea. Ample opportunity to examine Karamea Granite is afforded by roadside exposures in the upper Buller Gorge between Murchison and Inangahua Junction. The western boundary of the belt crosses Highway 6 at 1.3 km on the Murchison side of Lyell. Thermal metamorphism of adjacent Greenland Group rocks can be seen in a rather complicated contact zone crossed by Highway 7 between Springs Junction and Rahu Saddle.

Veins of very coarsely-crystalline, granitic pegmatite (see below, page 152) cut through both Paparoa and Karamea granites; they may be of later origin than the main granite mass. At Mount Radiant, south of Karamea, pegmatites contain ores of copper, lead, zinc and molybdenum as well as of gold and silver. Needless to say, these have attracted prospectors from time to time but without any marked success. A similar kind of mineralisation is found in some Fiordland granites which could well be a continuation of the Tuhua Group. Other granites in Fiordland and Stewart Island are younger and apparently unrelated.

The third granite belt is petrographically distinct from the other two. Under the name 'Separation Point Granite' (taken from the cape between Tasman Bay and Golden Bay), it skirts the western side of the Waimea Depression as far as Mount Murchison, south of the Buller River at Gowan Bridge. The bulk of the rock is lighter in colour than the other belts. Instead of orthoclase, the dominant feldspar components are albite or oligoclase, these being sodium-rich members of the plagioclase series, rather than potassium-rich orthoclase or microcline, while the darker minerals are hornblende instead of biotite mica.

Compared with the other Tuhua granite belts, Separation Point Granite appears to be more readily and more deeply affected by weathering. Kaolin produced by decomposition of the feldspar has been concentrated to produce deposits of potentially useful ceramic clays in the Motueka Valley, while the quartz grains and mica flakes thus released are the source of the golden sands of Kaiteriteri,

Totaranui and Tatas beaches. Vacationers at Kaiteriteri can be pardoned for failing to recognise the soft, gritty rock in the sea cliffs as granite.

Interesting effects of contact-metamorphism are to be found where Separation Point Granite has invaded marble of the Mount Arthur Group and other Paleozoic rocks. Revision of the earlier mappings of the Takaka district, besides resulting in the creation of several new formation names, has thrown more light on the relationship between Separation Point Granite and Rotoroa Diorite (part of the 'Rotoroa Igneous Complex'; see below, page 157) which adjoins it on the eastern side. It seems likely that still further detailed mapping and petrographic studies will produce a story more complicated than one of mere intergradation between them.

Fig. 6.2
The layering in this block of Karamea Granite from Cape Foulwind, near Westport, is emphasised by alignment of large white phenocrysts of potash feldspar. It has been ascribed to flow movements within the granite magma while it was being intruded into ancient metamorphic gneiss.

Is 'Tuhua Group' still a Valid Unit?

Rock-stratigraphic names become obsolete, or at least are replaced for reasons that may be adequate in the eyes of some though not of others. But even in good cases the maps with older names live on, and the older uses of them live on as well, so there is sometimes confusion. Since one of the objectives of this book is to help people to use our geological maps, a short digression will be made to look at this problem.

It may have been noticed already that the 1:250,000 mapping programme had marked effects in the progress of geology in New Zealand. It was provoking because many arbitrary decisions had to be made in the absence of decisive evidence if 'gaps of ignorance' were not to appear on these maps—and that was against policy. It was stimulating in that the applying of arbitrary solutions showed up further weaknesses and pointed up the need for intensive studies. Over all, though, the programme resulted in a great step forward.

25. The Hooker Glacier, with Mount Cook above it, photographed in 1957. A row of cirques occupied by ice or snow overhang the glacier trough on the far side, their rear walls shaping the ridge crest into a sharp, uneven arête. The glacier is heavily encumbered with ablation moraine, and the obvious signs of collapse show that the lower part is no longer moving, but is wasting away. Alpine vegetation is well established upon a series of stranded lateral moraines on the lower slopes opposite, each ledge of moraine marking earlier levels of the glacier margin over the last few centuries (pp. 264, 320–1).

But what about afterwards? The tempo of regional exploration has slowed down but certainly has not stopped, and many detailed studies have continued. Meanwhile the classification schemes, the namings and the ages adopted for it will continue in use at least until this very popular map series is replaced by new editions, notwithstanding that some groupings and namings have since been found incorrect or made obsolete for other reasons. The reasons, however, seldom affect the bulk of users of the original map edition.

'Tuhua Group' provides an example. The name was revived as a convenient mapping unit to include the granites of Nelson and Westland first for the 1:2,000,000 map and then for the 1:250,000 series. It is now known that the Separation Point Granite, besides being petrographically different from the other components of the group, largely belongs to a different chapter in our geological history. All three components have now been subdivided petrographically into several named formations. There is little need now to retain the Tuhua Granites as a descriptive grouping or a mapping unit, but it will certainly remain in use while there are still copies around of the present edition of 1:250,000 scale maps. This was the main reason for retaining it as a main heading in this chapter. The same applies to Paparoa and Karamea Granites, which are indistinguishable petrographically as groups, and likely to be replaced by finer subdivisions. In this case, fortunately, confusion is not likely to result from their continued use, though map users may be puzzled when they come across new names attached to familiar formations on later maps, with no reference to the changes or to the terms that have been dropped.

Tasman Intrusives—the 'Boulder Bank' Rock

Granite-like rocks along the eastern shore of Tasman Bay north of Wakapuaka came in for attention in the earliest days of Nelson geology (which were very early for the colony as a whole) because they were obviously the source of material for the Boulder Bank, a remarkable 15 km-long barrier beach of cobbles and boulders enclosing the lagoon and harbour of Nelson Haven. Hochstetter saw it in 1859 and referred to it as 'syenite', a description which stuck for a long time afterwards in the name 'Mackays Bluff Syenite'. W. R. Lauder made a close study of this rock from about 1960 onwards and found that the Mackays Bluff rock is part of a complex of granitic rocks, dykes and gneiss extending from Wakapuaka north to Pepin Island and small islands immediately beyond. It includes granite, diorite, granodiorite—but not syenite. For the 1:2,000,000 Geological Map of New Zealand this group of granitic rocks was associated with the Permian-age Brook Street Volcanics under the name 'Tasman Intrusives'. It was an unhappy association because the Brook Street rocks are reliably connected with the early phase of the New Zealand Geosyncline, while Lauder's work showed the gneiss and some of the granitic rocks to be older, and the granodiorite intrusive rocks to be more recent than the Brook Street volcanic episode. To make matters worse, Sheet 14 (Marlborough Sounds) of the 1:250,000 maps restricts the application of 'Tasman

Intrusives' to the younger dykes which Lauder named 'Cable Granodiorite'. As a prefix, 'Tasman' is also used in other, distinctly different contexts.

This collection of igneous rocks has become important again in recent years since it has been seen to occupy a critical position between the older and newer geosynclines. It is suspected that there have been sources of magma available here from time to time both before and after the Rangitata Orogeny, the later ones being perhaps due to partial re-fusion of intrusive rocks emplaced during the Tuhua Orogeny. They may share in the history of both the 'New Zealand Geanticline' (as some would call the upheaved mass of Buller Geosyncline rocks) and the New Zealand Geosyncline in the interval between the two orogenies. Such a view makes all things possible. It can explain, for example, how the granodiorites of the Mackays Bluff group intrude volcanic rocks of Permian age in some places, are faulted against them in others, and are intruded by younger granitic rocks.

Geologists are rather prone to invent new and sometimes cumbersome terms to identify special situations and relationships. The need being felt strongly to attach a name to the whole, long succession of deep-seated magmatic intrusions, metamorphism, metasomatic alteration and volcanic outbreaks along this belt, and to include the Rotoroa Diorite complex—not in a stratigraphic or even a petrographic sense, but rather in a tectonic and historical one—the hard-worked prefix 'Tasman' has appeared once more. Thus we now have the 'Tasman Igneous Complex and Metamorphic Belt'. I doubt whether the term will be found very useful.

Granites in the Far South

Contrasts between the bold relief of Fiordland and the alpine scenery in other parts of the South Island have a lot to do with the fact that even the precipitous fiord walls which less than 13,000 years ago were still being eroded by Pleistocene glacier ice did not afterwards collapse back to gentler slopes. The main reason for this is to be found in the great predominance of crystalline granites and gneisses in Fiordland. Not so much that the rock substance is harder—some of it is marble, a comparatively soft yet tough rock—but rather that tightly interlocked crystalline texture and structural massiveness, with relatively few joints or discontinuities of any kind, do not make it easy for water to soak in from the surface and promote either chemical decay or the disruptive action of frost.

Throughout Fiordland granite rocks are to be seen in practically any extensive exposure, from thin veins and dykes a few centimetres thick up to continuous masses vanishing beyond visible limits. On the eastern side of Fiordland granite tends to occur as discontinuous bodies in various forms, including sheet-like masses 30 m or more in thickness cutting across the grain of the schist and gneiss host-rocks, and as swarms of regularly orientated smaller dykes. Sills of granite are injected along the planes of foliation of gneiss. In the south-west more continuous granite masses invade the metamorphic rocks and unmetamorphosed Paleozoic strata of Preservation Inlet and Dusky

26. *A non-glacial cold-climate landscape near Burkes Pass, inland South Canterbury. The lower slopes on the right are underlain by a mantle of coarse, angular waste brought out from the foot of the hills at the right by solifluction processes (p. 260) during late Pleistocene times. Even today, this region can experience very severe overnight frosts.*

Sound. These could well be connected with a large, continuous pluton taking in most of the southern part of Stewart Island. Pink, porphyritic granite with large phenocrysts of potassium feldspar is the dominant type, but there are also granodiorites and darker, more mafic plutonic varieties, as for example at Wilmot Pass.

The history of Fiordland granites is complex. It has been shown that some were intruded before metamorphism of the schist was complete, more during later deformation of the schist, and more at a still later date after the tectonic phase was over. An interesting feature of the Fiord Intrusives as a whole is the regional variation in the effects of their emplacement on the intruded rocks. In the south-west, where the rocks were relatively low-grade schist or sedimentary strata, strong thermal metamorphism resulted, producing prominent contact aureoles, that is, zones of decreasing alteration with increasing distance from the intrusive contact. Close to the granite, schists now contain the very high-temperature minerals andalusite and sillimanite; farther away, the contact rocks grade to hornfels and to almost unaltered sedimentary rock. Where the host-rocks are higher-grade schist and gneiss the invading granite had little effect, suggesting that the invaded rocks were hot at the time of intrusion, or that the temperature of the molten granite was never as high as that to which the enclosing rocks had already been subjected during metamorphism.

Fiord Intrusives have been mapped as several different formations, of which the most extensive and continuous are Kakapo Granite in the south-west of Fiordland and Pomona Granite in the south-east. Both are white or pink porphyritic types. Similar rocks making up the southern part of Stewart Island have recently been separately named Rakeahua Granite.

During compilation of the 1:2,000,000 geological map an attempt was made to distinguish systematically between discordantly-intrusive Fiordland granites and those granitic rocks which occur in belts running concordantly with the structural grain of the metamorphics, though the distinction was not easily made everywhere. Concordant intrusions were linked in origin with the main metamorphic event. First described in detail by Grindley in the Eglinton Valley, they were then identified with the Tasman Intrusives of Nelson, and like the latter may span a range of ages from late Paleozoic to early Cretaceous.

Layered Granites, Granite-gneisses and Gneissic Granites
This merry-go-round of words does have a geological message to convey. The distinctions are real, though in combination the expressions remind us once more of the basic unity of plutonic and metamorphic rocks, both being products of deep-seated environments with many factors in common.

It was observed long ago in Scandinavia, in the north-east of Britain and in northern Canada, where ancient crystalline terrains were still fresh from the abrasive effects of Pleistocene ice-scour, that layers of granitic rocks from a metre or less up to a hundred metres or more thick occur widely as sheet-like injections along the foliation planes of metamorphic rocks. Separate sheets connected here and there by cross-cutting veins or dykes make up a complicated network or boxwork. Frequently the whole mass, schists, granites and all, has been folded and distorted in a complex way and has had new metamorphic cleavages superimposed upon the old, distorted structures. Contact metamorphism is present, too, at the boundaries, to confirm that granitic magma was injected as a hot liquid. Finally, some of the thicker layers are divided internally into subsidiary sheets or zones of different composition. Layered granites produced by the latter process, that is, by separate injections along sub-parallel planes, are called 'lit-par-lit' intrusions (French: 'bed-by-bed').

In the same regions one can find also layered crystalline rocks, in composition resembling granite, diorite, granodiorite, etc., but with crystal forms and textures of gneiss. Being at least finally the product of metamorphic processes, such layers are called granite-gneiss, diorite-gneiss, etc., according to composition. If supposedly derived directly by metamorphism of sedimentary materials, they would be put in the 'paragneiss' class. And lastly, still in the same regions, there are extensive bodies of granites for which a magmatic origin is inferred (from contact metamorphic effects, etc.) but in which later stresses have imposed a reorientation of minerals, giving the rock an internal 'grain' or 'fabric', in other words, turning it into a metamorphic rock which would be called a gneissic granite, or a gneissic diorite, etc. The parent material having been igneous rock, the product is classed as 'orthogneiss'.

The above terminology was developed overseas, but all the words will be encountered in descriptions of New Zealand granitic rocks, and especially in Fiordland, which is our great showcase for all kinds of products of the plutonic environments. Bare rock surfaces abound, still freshly ice-worn, but access is not easy except by boat around the steep fiord and lake shores and along the few roads and tracks. Given helicopter transport, it is now more feasible to explore the summits and ridge crests, but the sudden moods of Fiordland weather may cause serious problems.

Much of the strongly layered gneiss in Fiordland is composed of quartz-feldspar sheets alternating with layers ('foliae') rich in hornblende or other dark silicate mineral. Relict sedimentary features and occasional layers of marble indicate that some of it is paragneiss.

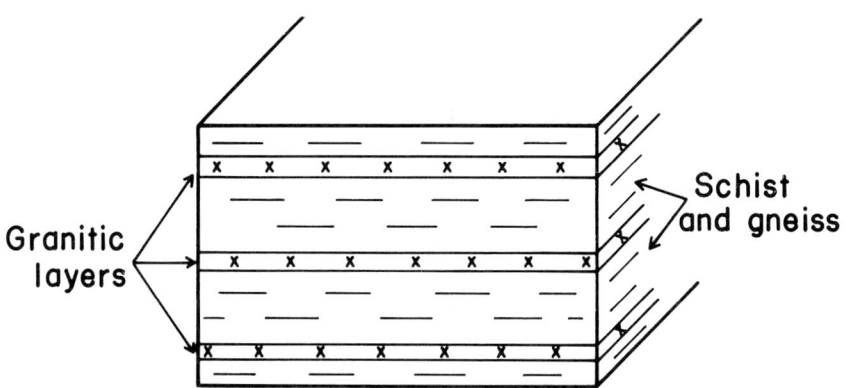

Fig. 6.3
Lit-par-lit injections of igneous rock between bands of layered metamorphic rocks; commonly, the injections are of granite or pegmatite and the invaded rock is gneiss or coarse-textured, foliated schist.

27. A series of three valley-plain terraces carved by the Awatere River, Marlborough, in soft, Tertiary strata. Each terrace has resulted from lateral cutting by the river during a halt in the progressive deepening of the valley. Note that each terrace is underlain merely by a thin capping of gravel; this makes it unlikely that the pulses of downcutting and terrace formation are simply related to climate changes. (Compare with Plate 28; see also pp. 323–5.)

28. The uppermost of these prominent terraces of the Poulter River, Canterbury, was mainly built up of glacial outwash gravels during the last invasion of this valley system by ice (that is, neglecting a few metres of fan gravel subsequently deposited from the side valley behind). Its origin is thus linked with a climatic event. The lower terraces were formed in the course of post-glacial erosion of the former valley-filling of glacial gravels. (Compare with Plate 27.)

Elsewhere large masses have the composition of granite, diorite or granodiorite, but with sufficient lining-up or parallelism of hornblende needles and prisms or mica flakes to justify calling the rock, for example, 'gneissic granite'. In many cases, though, the ancestry of the rock remains obscure. For even with such a wide-spread and important formation as the Bradshaw Gneiss in northern and central Fiordland it is difficult to decide which parts were originally intruded igneous rock and which parts were already metamorphic gneiss prior to the latest phase of metamorphism.

We will now leave Fiordland and return to consider the problems already mentioned in connection with the granite-gneisses and gneissic granites of Charleston and the northern Paparoa Range, and their relationships with the non-gneissic Paparoa Granite and the Greenland Group sedimentary rocks.

Earlier mapping by the Geological Survey did not distinguish granite from gneiss in this area, though J. Henderson, the author of *Bulletin 18* (Reefton Subdivision, 1917), thought the gneiss could be the basement under the Greenland Group (he called the latter 'Aorere Series') whereas the non-gneissic granites were introduced into Greenland rocks. Detailed structural mapping and petrological studies combined have thrown more light on the problems. One suggestion was that the granite-gneiss resulted from partial melting of Greenland Group greywacke, at a time and in a situation where the geothermal gradient was unusually steep—it had to be, because the transition from greywacke to gneiss takes place over a few metres. Later, granite magma invaded it all from below, and introduced the pegmatite dykes as at Charleston. Recently, however, B. M. Hume re-examined contacts in the central Paparoa Range, and is convinced of the truth of the old hypothesis that the Greenland rocks were deposited unconformably on ancient gneiss. It seems that the last word has not yet been pronounced on this question.

D. Shelley places great importance upon structural features to be seen in the Constant Gneiss at Charleston which, in his opinion, show that the later uprise of granitic magma was accompanied by an up-arching of this area, so that the gneiss, in its re-heated, mobile condition—weak, but not quite molten—slumped by gravity down the gently sloping flanks of the arch. This may sound far-fetched, but I can assure you that the notion is in accord with modern petrological views. We will be discussing ages later, but it may be mentioned here that interpretation of a few radiometric dates suggests a time no later than Carboniferous for metamorphism of the granite-gneiss, and no later than Cretaceous for the later phase of tectonic deformation and granite intrusions.

Pegmatite—Prime Source of Mica and Gemstones
Uses of the silicate mineral muscovite ('white mica') are many. One of the most essential is to provide either light-transparency or electrical insulation in situations where high temperatures have to be withstood. Though partly replaced by silica glass and synthetic substances, it is still required as the base on which to wind heating elements in some

household appliances. Besides its excellent insulating properties, even when in very thin sheets, mica also has eminent natural cleavage which enables sheets of any desired thinness to be produced easily, so long as the original piece is wide enough for the job. The maximum width is limited by the size of the natural crystals; in crystallographic language the direction of the cleavage is 'basal'. Though on the way to being giants as natural crystals go, they seldom exceed 20 cm across.

This is where pegmatite comes into the mica story. Although many kinds of rock contain one or more of the mica minerals, the best known being muscovite and biotite, only in this excessively coarse-textured form of granitic rock does one find mica crystals large enough and perfect enough for splitting into commercial sheet mica. Pegmatite occurs as veins, sills or dykes, from a few centimetres to a few metres thick, cutting across a host-rock which may be either the 'country' invaded by a granite pluton, or within the granite itself. It is also found in sill-like layers parallel with the foliation of schist. As regards origin, pegmatite was formerly regarded as a late product of progressive, fractional crystallisation in a deep magma body, solidified from a residual liquid squeezed out from an already partially-frozen melt. Water and other volatile constituents of magmas were seen as playing an important role in the late stages, acting as vehicles for the enrichment of the end-products in the components of orthoclase feldspar, mica, quartz and a variety of minerals containing uncommon elements such as beryllium and lithium. In this day of dualism, it is accepted that while some pegmatite veins were generated in the way described above, others have been derived by metasomatic and granitisation processes, and especially by partial liquefying ('mobilisation') of geosynclinal sediments or metamorphic rocks that have been depressed into high-temperature realms but not wholly melted.

Mica became critically important in New Zealand during World War II when our electronics factories were supplying radio communications equipment to the Allied forces in the Pacific. Normal sources of mica from which capacitors were then made had been cut off, so the search for domestic supplies was stepped up. Hopes of finding pegmatite veins with mica crystals large enough for the purpose were limited to Stewart Island, remote parts of Fiordland, the Mataketake Range in South Westland, and Charleston. Eventually effort was concentrated on the last two areas, where large pegmatite bodies were contained respectively in the Haast Schist (as concordant sills) and the Constant Gneiss. The Mataketake Range occurrence was worked for a year or two, the mica and some high-grade feldspar needed for ceramic industry being packed out on horseback (there being no Haast Highway then). Mica-bearing pegmatite had been located by prospectors at uncomfortably high altitudes, 2000 m up on the range summit, but H. W. Wellman and W. F. Heinz predicted from the trend of the pegmatite sills above the bush-line that outcrops would be found in the bush above the Moeraki River, and that is where the small mica mine was established.

The Mataketake Range pegmatite sills raised some questions because

29. *Slump and earthflow landforms in the Motunau valley, North Canterbury. Blocks detached from the Tertiary limestone outcrops on the ridge are being rafted down-slope by superficial mass movement, and water is ponded in an undrained hollow on the slump surface. As in many other eastern districts of both the North and South Islands, the instability here is due to the presence of weak bentonitic clays in strata of early Tertiary age underneath the limestone (p. 325).*

they were on the opposite side of the Alpine Fault from the rest of the granite occurrences in Westland. Before the full significance of the fault was appreciated, ideas were entertained that the pegmatites might have been offshoots from Tuhua Granite, but it is now seen as far more likely that they were derived by partial-liquefying and mobilisation of granitised New Zealand Geosyncline material, since uplifted by the vertical component of movement on the Alpine Fault.

Dark-coloured Granitic Rocks

The word 'mafic' was introduced earlier as an informal term that has largely replaced 'basic' to denote igneous rocks composed mainly of lime-rich feldspars, the dark iron-, calcium- and magnesium-rich silicate minerals of the amphibole and pyroxene groups, but no obvious

Fig. 6.4
Dun Mountain, in the ranges east of Nelson City, was named in the early days of settlement because of its drab bareness compared with the neighbouring forested slopes. It is now known that the ultramafic rock of which it is composed yields a deficient soil, unfavourable for forest growth (p. 101). (Photo: N.Z. Geological Survey)

quartz. Where there is little feldspar, but substantial amounts of darker minerals, including olivine, the adjective is 'ultramafic'. Of the common mafic plutonic rocks, 'gabbro' and 'norite' will be the most familiar names; of the ultramafics, 'dunite' (olivine rock named from Dun Mountain, Nelson) and 'pyroxenite'. There are many more.

As noted earlier too, finding an explanation for the range of variation in igneous rock composition was the chief quest of petrology in the first part of this century. It now appears that although the granitisation approach to the problem has made futile most of the earlier controversies, many of the diverse physico-chemical mechanisms then suggested probably do have some application somewhere, some time. In the modern liberal mood of petrology there is no longer any difficulty in accepting that a mafic, intrusive rock of granitic texture can have been produced in several possible ways, e.g. by metamorphism followed by mobilisation of the constituents of older mafic igneous rocks or of sediments derived from them; or by one of several modes of crystal fractionation which concentrate early-formed (and usually denser and darker) minerals and separate these crystals from the light residual fraction while it is still molten; or in special cases by contamination of a granite or granodiorite magma through engulfment of older rock material.

Large mafic intrusive masses are most common in New Zealand in areas where geosynclinal rocks contain much mafic igneous material, suggesting an origin by partial remelting. One of the better known examples of mafic plutonic rock of this origin is the gabbro forming the hills near Bluff—a handsome, durable rock which holds a good polish and has long been popular for ornamental and memorial stonework. In the stonemasons' trade it is 'Bluff Granite', but to the petrographer that is perhaps taking liberalism a little too far. Gabbro and norite occur widely also in the mountains of southern and western Otago, where they have been mapped under several formation names. In west Nelson they occur, though less widely, where Haupiri Group volcanic rocks were metamorphosed and plutonised in the Tuhua Orogeny (e.g. Cobb Intrusive Group). Gabbro is present with the diorites of the Rotoroa Igneous Complex, and gabbro and norite turn up again in the Auckland region, notably near North Cape, though much altered by metasomatism.

The ultramafic igneous rocks and serpentinites have been mentioned previously in connection with the New Zealand Geosyncline (Dun Mountain Ophiolite Belt) and with the Buller Geosyncline. Much has been written about the origin of these rocks, especially as regards the serpentinised ultramafics and how they were emplaced in the associated rocks. It cannot be said that we have the complete answer.

It is not only in New Zealand, however, that the origin, date, manner of emplacement and tectonic significance of such rocks is obscure and a subject of controversy. I would not be justified in taking up the large amount of space that would be needed for a useful survey of these questions as they apply in this country. Those who would like to pursue them may read 'The New Zealand Serpentinites and Associated Metasomatic Rocks' by R. G. Coleman (*N.Z. Geological Survey Bulletin 76,*

30. *This view northwards along the gravel spit enclosing the Okarito Lagoon, Westland, shows clearly how the building of these spits can eliminate a former embayment of the coast, trap the sediments brought in by streams and thus contribute to coastal straightening (p. 331) by progradation. In profile in the distance we can see the crests of the high lateral moraine ridges between which a large glacier from the Whataroa valley flowed to beyond the present coastline during the Otira Glaciation.*

31. *Lake Pearson lies in the trough of a former ice stream which split off from the Waimakariri Glacier near Cass. It is there because drainage down Winding Creek towards the valley of Broken River (in the background) is blocked by a broad gravel fan built by the Craigieburn Stream, which comes in from the right (p. 329).*

32. Deep erosion of the Lyttelton Volcano has exposed pre-volcanic basement rocks, which in this view form the peninsula towards the right. Looking eastwards from Dyers Pass, on the western rim of the great erosional caldera which now forms Lyttelton Harbour, we can see in the distant hills outcrops of the lava flows which built up a broad andesite-basalt dome in late Miocene times, and also the smooth surface of younger basaltic lava which in the Pliocene Period flowed from vents high up on the flank of the adjoining Akaroa Volcano, after the centre of the Lyttelton Volcano had already been hollowed out by erosion. The erosional depression was finally drowned when sea level rose in post-glacial times (p. 216).

1966) and an article by J. B. Waterhouse in *N.Z. Joural of Geology and Geophysics*, vol. 10, pp. 949–60, 1967, the latter bringing in some of the 'human aspect' of New Zealand studies covering more than a century.*

Granite, Metasomatism and Mineralised Veins

Mention has been made more than once in earlier paragraphs of the processes summed up in the words 'metasomatic alteration'. One meets it both in the contexts of plutonic activity and the late stages of cooling of magmas, as well as finding it an associated process in some kinds of metamorphism. A major role has always been assigned to what were once loosely known (and indeed vaguely conceived) as the 'volatiles'. Principles of physical chemistry taught that water as vapour (or, under high confining pressure, as liquid) in company with sulphur, chlorine, fluorine, boron and other reactive agents, had a great capacity to perform many operations involving the transfer of rock substance from place to place underground, to aid recrystallisation, and to effect metasomatic replacement of one mineral by another. A useful result of these activities was the filling of rock fissures with vein deposits of metalliferous ores, and even the bodily replacement of rock minerals by ore minerals.

The immediate source of the 'volatiles' was once held to be the end-products of crystallisation of granitic magmas at depth. Nowadays it is more fashionable to stress the 'closed-system' aspect of metasomatism, wherein the changes are more of the nature of a reworking of materials already present, even the water agent itself being mostly from that which was entombed with sediment particles during the preceding sedimentation phase of a geological cycle. It is not ruled out, however, that some material enters such systems directly, through magmatic processes, from the earth's deeper resources. Movements of the volatile 'messengers' and the materials transported took place through inter-particle or intra-crystal spaces of microscopic or sub-microscopic size, driven by pressure-differences and other environmental contrasts. Today, however, the physico-chemical view of 'volatiles', etc. is far more precise.

This brings us to an old, long-debated question as to the mode of formation of quartz veins containing sulphide ores of copper, lead, zinc, etc., native silver and gold. An influential school of European mining geologists maintained that the silica and metallic compounds of quartz veins had all been leached or otherwise chemically extracted from the country rock by percolating solutions and then redeposited (with some concentration and enrichment) in rock fissures or else as whole-rock metasomatic replacement. This was mineral formation by lateral secretion. In our metal-mining heyday, however, most New Zealand geologists subscribed to the older and more orthodox magmatic theory, according to which the constituents of mineralised veins were all concentrated in the course of crystallisation of a parent magma and introduced into the country rock in the final stage.

---

* A more general, though less easily available review of the problem by W. R. Lauder appeared under the title (abbrev.): 'The Geology of the New Zealand Ultramafic Rocks . . .' in *Pacific Geology*, vol. 7, pp. 97–130, 1974.

There were exceptions, notably J. Henderson, who in his latter days as Director of the Geological Survey became a supporter of the lateral secretion origin. Earlier he had been impressed by spatial relationships between quartz veins (with or without other minerals), pegmatite veins and offshoots from the Karamea Granite in the Reefton district, which seemed to support a magmatic origin for the gold and sulphide ores of that region. Again, we are looking at a former distinction that seems to have vanished. In terms of varying fields of temperature and stress the environmental approach suggests an essential continuity between vein-formation by magmatic, metamorphic and metasomatic processes. It also helps to predict with more confidence the geological circumstances in which useful mineral concentrations can optimistically be sought.

Ages of New Zealand Granites

It is fitting to conclude this chapter by taking up again the question implied in the subtitle—how old are our granitic rocks?

To expand a little on the opening remarks of Chapter Five, it was in countries of Western Europe and North America, where modern geology chiefly developed, that the stratigraphical record of events seemed to imply a great age for crystalline granites and gneisses. They tended to be the lowest formations present in any area, and whatever sedimentary aspects they might once have had were now destroyed or almost obliterated. The widespread notion of great antiquity was conveyed by these rocks being named 'Archaean'. For a long time there was little reason for even considering the possibility that some crystalline rocks might be younger than some unaltered sedimentary formations.

It was not surprising, therefore, that in the course of nineteenth-century geological exploration of the rest of the world the granites and gneisses were at first *assumed* to be ancient—until, first, detailed study of their relationships with other rocks in the field and later, radiometric dating compelled a rather reluctant acceptance of the fact that late Paleozoic, Mesozoic and even Tertiary granites exist at the surface today.

The influence of the older philosophy can be detected in New Zealand in some writings lasting well into the present century. Yet even before 1900 it was being suggested that some of our granites were intruded in the Jurassic and, in an extreme case, in the Eocene Period. In those days everything depended upon finding the correct geological age of the rocks intruded by granite and of the rocks with conglomerates containing pebbles of the same granite, and we have already seen the wide range of opinions in some such matters.

Current views as to the ages of our granites are based upon generally very good agreement between the dates suggested by different lines of evidence. The sources include firstly the straightforward stratigraphical-geological relationships as noted above; secondly, modern refined studies of the microstructure and petrology of metamorphic schists and gneisses into which granites have been intruded—giving evidence of the history of tectonic deformations in

Fig. 6.5
Crests of granite mountains, typically rugged where glaciers have been active, near the southern end of the Victoria Range in south-west Nelson. The view is south-eastwards from Mount Gore, and the rock is Karamea Granite.

relation to plutonic events; and thirdly, within the last decade or so, radiometric datings by uranium/thorium, rubidium/strontium and potassium/argon isotope ratios, applied to the whole rock of the granites, to particular constituent minerals and to younger igneous dykes intruding the granites. Datings are now coming forward rapidly, and by the time this book is in print some of the conclusions summarised below may be affected.

Taken altogether, the indications from these lines of evidence, though not without anomalies, suggest the following:

(1) Granitic rocks were already present when the ancient foreland to the west was supplying sediment to the early Paleozoic Buller Geosyncline. Radiometric dates imply at least two occasions of high temperature in late Precambrian times.

(2) Granite may have been intruded in the core of the Buller Geosyncline during the mid-Cambrian 'Haupiri Disturbance', followed by erosion deep enough to bring it to the surface and yield granite pebbles for the Lockett Conglomerate. More likely, perhaps, these pebbles came from the ancient foreland granites, as there is no direct radiometric support for such an event.

(3) A considerable number of datings between 300 and 380 million years reflect the cooling times of plutonic granites intruded before and during the Tuhua Orogeny, in late Devonian and Carboniferous times. The main period of emplacement of the Karamea and Paparoa granites belongs here.

(4) There is also plenty of support from radiometric datings for emplacement of granite plutons during the early-mid-Cretaceous climax of the Rangitata Orogeny. These are the youngest New Zealand granites exposed at the surface today; they include the Separation Point Granite and the younger post-gneiss granites in the Buller region.

(5) Rapid uplift of the block on the east side of the Alpine Fault near the highest regions of the Southern Alps has brought to the surface metamorphic rocks that cooled down during the Tertiary Era. Pegmatite sills ascribed to partial melting of the Haast Schists have come up with them, but no granite plutons of Tertiary origin have been exposed by erosion yet.

Obviously, 'Rock of Ages' is strictly a relative term.

We are reminded incidentally by paragraph (5) above that the rapid uplift and erosion necessary to lay bare granitic plutons emplaced during an orogenic climax really makes only modest demands upon available geological time. The erosional tempo during such events must be greatly accelerated. Thus, for example, conglomerates in the mid-Cretaceous Hawks Crag and Ohika Formations in the Buller area (see pp. 165–6) include pebbles derived from the Paparoa Granite which could have been intruded only a few million years beforehand.

Postscript—Granite Study Areas

The foregoing paragraphs contain many clues to where to find granitic rocks in New Zealand. Perhaps one of the best, most accessible and most varied areas is in the Buller catchment, from Lake Rotoiti to Westport and southwards along the coast to Maybelle Bay. The Bluff-Greenhills District, Invercargill and North Cape display the mafic plutonic types, and Nelson has the most accessible ultramafics. Gabbro and syenite are well seen at the head of Akaroa Harbour, on the shores of Onawe Peninsula, though unfortunately their relationships with the basal lava flows of the Akaroa Volcano are not very clear. For easily accessible exposures of granite-gneiss, gneissic granite and pegmatite, the coastal area between Cape Foulwind and Charleston is magnificent.

Chapter Seven

THE LATEST CYCLE

Beginning and Ending with Mountains

Bold Mid-Cretaceous Landscapes

We return to the New Zealand geological story in the middle of the Cretaceous Period at a stage when the climax of the Rangitata Orogeny was past but there was still a substantial topographic relief. Part at least of the land which had replaced most of the New Zealand Geosyncline was mountainous. Vigorous erosion of the mountains produced quantities of coarse rock debris which mantled the slopes and spread across the valleys between elevated blocks, after the fashion of the rows of shingle fans seen in such situations at present in inland Canterbury and Marlborough. These breccia and conglomerate accumulations rest on eroded surfaces of schist, granite or older sedimentary rock, making them the basal members of our youngest rock succession, the products of the last of three major cycles in our geological history.

It is often convenient to use a simplification suggested by C. A. Cotton more than sixty years ago, which was to call the deposits of this latest cycle the 'covering strata', distinguishing them as a whole from the 'undermass' of older rock* upon which in most areas they rest with obvious unconformity between them.

Fig. 7.1
This early example of the late Sir Charles Cotton's flair for conveying three-dimensional geological concepts by means of serial block-diagrams was first published in 1913. It represents a cross-section of the Seaward and In-land Kaikoura ranges and middle Clarence valley, and from it we gather that such downfaulted wedges of younger strata are remnants of formerly extensive cover which overstretched the older rocks where mountain ranges now stand; following vigorous crustal movements in late Tertiary and early Pleistocene times, during which mountain blocks rose along major faults like the Clarence Fault, the covering strata were stripped away by erosion, laying bare the undermass rocks except in protected situations such as the Clarence depression.

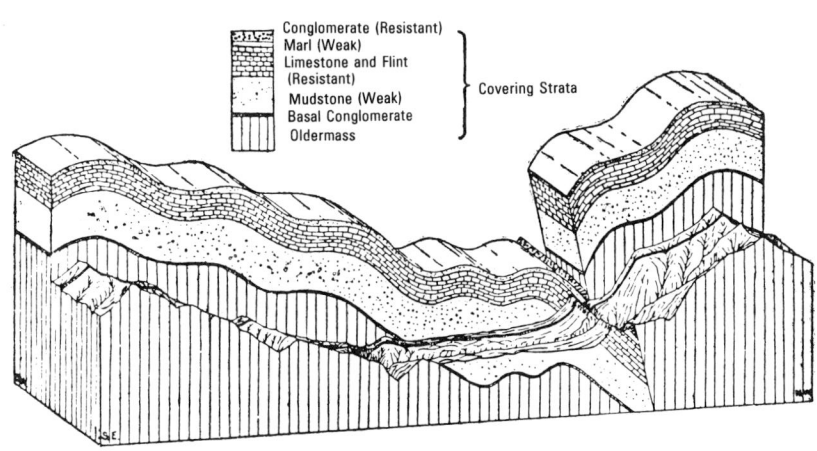

Conglomerate (Resistant)
Marl (Weak)
Limestone and Flint (Resistant)
Mudstone (Weak)
Basal Conglomerate
Oldermass

Covering Strata

* The chief objective in making this distinction was to note the very different course followed by landscape evolution in the two types of terrain and the contrasting landforms that have resulted. Cotton had adopted terms proposed by his great inspirer, the American geographer W. M. Davis. At first he used 'oldermass' but later changed it to 'undermass' when it was found that the youngest 'undermass' in one area could be younger than the oldest 'covering strata' in another. It was not the age difference that was important; rather, that the younger rock formations in all but a few areas are softer, more easily eroded and structurally very different from the undermass.

The distinction is well marked except in a few areas where the strata of the latest cycle are exceptionally thick and therefore in their lower parts almost as hard as the undermass. One of these areas is the eastern belt where the New Zealand Geosyncline sedimentation persisted longest and was followed in the Cretaceous Period by a rather similar kind of marine sedimentation. Another is in western South Island districts where, as we shall see later in this chapter, the thicknesses of late Cretaceous and early Tertiary strata locally amount to several kilometres in thickness. For this reason it has been unclear whether the basal breccias and conglomerates we are about to describe in detail belong to the undermass or to the covering strata. Nevertheless these expressions are often useful and are still widely applied in an informal way.

The poorly stratified lowest members of the covering strata in a few widely scattered South Island areas are made up largely of fragments of angular shape in all sizes, from a few centimetres up to several metres across, with a sandy or gritty matrix between them. Apart from the fact that they are now compacted into a tough, coherent rock, these deposits do very much resemble the scree, or talus, mantling mountain sides today. Parts are more obviously stratified and contain a higher proportion of rounded pebbles, resembling the gravels seen where roads now cut through alluvial fans along the South Island mountain highways. A strong indication that the basal breccias were made up of material eroded near by is given by the fact that their composition closely reflects that of the underlying basement rocks. Thus where basement is granite, the constituent pebbles will be mostly if not all of granite, and so on. Among the various theories advanced to explain these deposits was the suggestion that they represent ancient glacial moraine deposits. It is no longer doubted, however, that they originated as scree and piedmont gravel fan deposits laid down in the setting of a former mountain landscape.

These distinctive breccia and conglomerate formations occur in a number of separate areas between Pakawau (north-west Nelson) and Kaitangata (South Otago) as discontinuous masses varying in thickness abruptly from a few metres up to several hundred metres. The best known examples are the Hawks Crag Breccia of the Buller, Inangahua and North Westland regions, the Kyeburn Formation of the Naseby area, Central Otago, and the Henley Breccia of the hills between the Taieri Depression and the Otago east coast.

Hawks Crag Breccia is particularly well exposed alongside the Buller Gorge road (Highway 6) for about 5 km west from the famous half tunnel through a craggy bluff which furnished a name for the formation as long ago as 1883. It attracted special attention in 1955 when parts were found to contain enough uranium/thorium mineral to encourage a burst of vigorous prospecting, though the results did not warrant mining.

A feature common to all these breccia-conglomerate formations is a strong, red colouring of both pebbles and matrix which comes through present-day weathering into the modern soils. Due to a film or rind of hematite, the colour does not always penetrate to the cores of pebbles

Fig. 7.2
Chief occurrences of coarse, basal breccia formations. 1: Buller Gorge (Hawks Crag Breccia); 2: coast at Fox River mouth (H.C.B.); 3: central Paparoa Range (H.C.B.); 4: Big River, Paparoa Range (H.C.B.); 5: Reefton and Waitahu valley (H.C.B.); 6: Roa, southern Paparoa Range (H.C.B. or basal part of Jay Formation); 7: Kowhitirangi Hill, near Hokitika (H.C.B.); 8: Paringa area (H.C.B.); 9: Naseby district (Kyeburn Breccia); 10: east of Taieri lowlands (Henley Breccia).

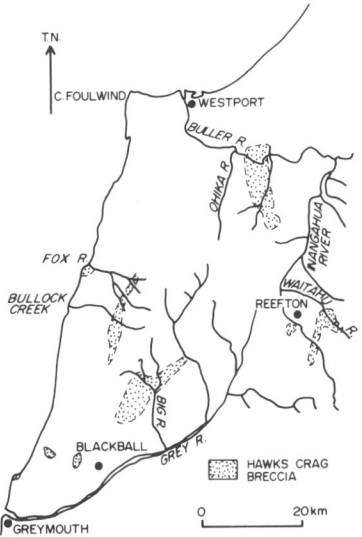

Fig. 7.3
Main occurrences of
Hawks Crag Breccia in
North Westland and
Buller districts.

Fig. 7.4 (right)
Typical roadside exposure
of the Hawks Crag Brec-
cia in the lower Buller
Gorge. The formation is
not always as distinctly
stratified as it is here.

and does not occur at all where carbonaceous matter is present in the rock. Once thought to require either a hot-dry or hot-humid climate during deposition, the redness is now seen as possibly the result of subsequent alteration of originally normal, yellow-brown hydrous iron oxide weathering products after the breccia had become deeply buried under younger sediments.

Widely different ages, from Triassic to Pleistocene, have been proposed for the several occurrences of these breccias. Modern plant-microfossil studies and a better appreciation of regional geological histories now suggest that all are about mid-Cretaceous in age. This does not necessarily mean that all were formed in precisely the same interval. The Cretaceous was a very long period.

At the type locality in the Buller Gorge the Hawks Crag Breccia rests, not as usual on pre-Cretaceous basement rocks but conformably upon the Ohika Formation. This unit was first recognised as a separate entity in 1950 by H. W. Wellman. It is made up of freshwater stream gravel, sandstone and impure coal beds once thought to be of Jurassic age, but now known also to be mid-Cretaceous though a little older than Hawks Crag Breccia. It is the only significant case where the coarse breccias are not actually at the very base of the younger rock succession, and it emphasises that fault-bounded mountain blocks continued to rise well into the Cretaceous Period, presumably as a late phase of the Rangitata Orogeny. It should be noted, too, that block-faulting movements were occurring outside the boundaries of the New Zealand Geosyncline, and within the borders of its former foreland.

For reasons given earlier, it is now realised that emergence of land due to the Rangitata Orogeny had begun before the end of the Jurassic Period. When land first became extensive and continuous is uncertain. The Ohika Formation and the red breccia beds with their associated sandstones and rare carbonaceous layers must have accumulated in a setting that was entirely terrestrial, not marine. While this does not mean that dry land need have been continuous between the various

areas in which the breccias now occur, such an assumption would not really be in conflict with our knowledge that marine sediments were being deposited at the same time in eastern areas in the North Island east coast and coastal Marlborough. Subsiding troughs in those areas are regarded by some as continuing the existence of the New Zealand Geosyncline and by others as initiating the final cycle of sedimentation.

The final, abrupt upheaval of fault-bounded mountains coming after the initiation of the new sedimentation cycle in some areas shows that the Rangitata Orogeny ended raggedly and had no distinct climax. Nor were its effects entirely confined within the bounds of the New Zealand Geosyncline. Such a conclusion helps to explain the range of datings of our Cretaceous granites, from 105 to 130 million years ago. It shows also how the ensuing long period of erosion, in which the mountainous relief was reduced and eventually destroyed, did not begin everywhere at the same time.

Mountains Worn Down

'How long does it take for erosion to wear mountains away?' A simple question to ask, one that is crucial when piecing together a regional geological history and determining the time-spacing of events, yet it cannot be given a direct answer. The rate at which erosion can destroy relief depends upon many factors, some of them interrelated, and no one yet has found out how to apply the results of short-term measurements of modern erosional processes or the results of experiments to the above question, which is a long-term one.

Ideas have clarified considerably, however, and the time-scale difficulties are more widely appreciated. Although we may never be able to produce an accurate, quantitative answer to fit each of the multitudinous combinations of different kinds of rock, different structures, different climates, different rates of orogenic uplift, etc., a few generalisations are possible. For instance:

(1) The overall tempo of erosion must increase as the mountains

Fig. 7.5 (left)
Closer view of the breccia at the same locality as in Fig. 7.4. Note the uniformity in composition, here entirely of Greenland Group greywacke, and the lack of rounding of the fragments.

Fig. 7.6 (right)
Basal conglomerate of the Tapuwaeroa Formation (Cretaceous) exposed on the roadside near Porangahau, southern Hawkes Bay. It includes pebbles of a variety of rock types and fossil fragments which were derived from the erosion of older Cretaceous and pre-Cretaceous rocks.

become higher, because of increased exposure, severe physical weathering and steepened gradients.

(2) Once orogenic uplift has ceased, the early stages of destruction of mountainous relief must be relatively rapid, and it will slow down as relief is reduced.

While we cannot give a precise estimate of how long it must have taken to wear away the relief of the ranges raised in the Rangitata Orogeny, it is a reasonable guess (supported by the timing of events of a similar kind in the Pleistocene Period) that the initially steep, lofty alpine relief would perhaps have faded a few hundred thousand years after uplift ceased, but that several millions of years were probably needed for the completion of the task of reducing the region to low relief.

The question was a crucial one when it seemed that the Rangitata Orogeny and the subsequent destruction of its mountains ran to an incredibly tight schedule. To a degree, though, the difficulty is exaggerated by time perspective. The Cretaceous Period, to repeat, was a long one—as long, in fact, as the whole of post-Cretaceous time. The tempo of erosion therefore did not have to be greater than that which applied during the Kaikoura Orogeny (about which we have more direct knowledge) to have destroyed the mountains of the Rangitata Orogeny well before the end of the period. It is helpful also to know that the timing of the end of the orogeny was a bit 'ragged', which means that the wearing down of mountains could have been well advanced in some areas even before the late block-faulting which gave rise to the Hawks Crag Breccia.

Still later, though before the Cretaceous Period ended, a new trough began to subside in North Westland and continued doing so until well into the Tertiary Era. Eventually it was 80 km long and 16 km wide and contained up to something in excess of 3 km of sediments, which, incidentally, include the Paparoa Coal Measures, our only important

Fig. 7.7
These steeply-dipping Cretaceous rocks in the Ruatoria district are in an area of complex fold structures. An example of 'drag-folding' is visible near the centre of the picture. When a sequence of sandstone and finer beds is being sharply folded, some slippage between the layers along the bedding. planes is a mechanical necessity, just as when a leaf spring or a pack of playing cards is bent. As a result, weaker mudstone and siltstone layers may be puckered into small, superimposed drag-folds. The direction of 'drag' thus indicated gives valuable information when complex folding is under study. (Photo: R. D. Black)

Fig. 7.8
Rugged terrain in the
heart of the Paparoa Coal
Measures country in the
upper reaches of Seven
Mile Stream and its
branches; the summit of
Mount Davy, capped by
Brunner Formation con-
glomerate, is in the centre
background. Well-forested
slopes to the right (middle
distance) show the higher
fertility of soils developed
on shaly lake sediments
(Goldlight Formation)
compared with the poor
soils on the coal measures
formations above and
below. Landslips on the
near valley-side are caused
by subsidences following
extraction of coal in the
long-abandoned Liverpool
No. 1 Colliery. (Photo
taken in 1946)

source of valuable, low-sulphur, low-phosphorus bituminous coals. The
Paparoa Trough, however, does not conform with the direction of the
earlier movements, but seems to have cut across the fault-bounded
depressions which accommodated the red breccias and the lowest coal
measures. This would suggest that it had nothing directly to do with
the Rangitata Orogeny.

The marine Cretaceous strata of the Raukumara district, eastern
North Island, show that a sharp uplift occurred late in that period.
According to I. G. Speden, as much as 1600 m of mid-Cretaceous
marine strata were stripped away and the harder Torlesse greywacke
exposed and trimmed. This erosion might have resulted wholly
through wave and current action below sea level with little or no
emergence of land, but it requires that the undermass surface came up
by that amount, or more, at the very time when relief produced by the
Rangitata Orogeny had been greatly reduced elsewhere. Speden
assigns about five million years of late Cretaceous time for this
interruption.

The Late Cretaceous Peneplain

Despite the local interruptions mentioned above (and perhaps others
we don't know about), reduction of topographic relief continued
through the first part of the late Cretaceous* until a landscape of
limited relief and easy slopes existed over virtually all of a 'proto-New
Zealand' which must have extended well beyond the present coastline
(see above, pp. 56, 113). In 1935 W. N. Benson brought forward one of his
many challenging generalisations by giving this former landscape an
identity under the name 'Late Cretaceous Peneplain'. Young covering
strata still hide this old land surface over wide areas, but where the

---

* Early Cretaceous and late Cretaceous are internationally recognised subdivisions of the
period, but 'middle Cretaceous' has no formal status. One can still speak informally of
'mid-Cretaceous times' as I have done in many places in this book.

Fig. 7.9
Goldlight Formation,
uppermost of three
distinctive, thick, brown
mudstone units within the
Paparoa Group, is well
exposed on the Greymouth-
Rewanui railway.
Uniformly fine-grained
and evenly bedded,
they had already been
interpreted as lake deposits
before the discovery of
fossil shells of fresh-water
mollusca gave confirma-
tion. Well-preserved fossil
leaves can also be found.

Fig. 7.10
At Ten Mile Bluff, on the
coastal highway north of
Greymouth, the strati-
graphic position of the
Late Cretaceous Peneplain
is shown up by the light
colour of leached and de-
composed conglomerate in
the uppermost few metres
of the Dunollie Formation
(non-marine; late Cretace-
ous). The Brunner For-
mation, representing the
Quartzose Coal Measures
in this region, is less than
1 m thick here and hidden
by vegetation on the ledge
at the base of the Island
Sandstone (marine;
Eocene) which tops off the
cliff. Only a few kilometres
inland the Brunner For-
mation reaches a thickness
of more than 100 m.

covering strata are markedly softer than the undermass, much of it has been re-exposed by erosion during the Kaikoura Orogeny. Either as an 'exhumed fossil land surface' or just as an unconformable contact separating strata of the last two cycles, the old peneplain can be identified at many places throughout the South Island and in some North Island areas as well, still retaining its original soil and weathered zone in protected situations.

The prime region for viewing remnants of the exhumed fossil landscape is Central Otago. The broad, up-arched surfaces of Rock and Pillar, Rough Ridge, Dunstan Range and others were once all continuous parts of it. Eroded across the schist belts during late Cretaceous times, it was buried under Tertiary coal measures and marine strata, remnants of which survive under the floors of the Maniototo, Manuherikia and other depressions. Trampers on the Heaphy Track in north-west Nelson see another stretch of the old surface when they look across Gouland Downs.

The reality of the Late Cretaceous Peneplain is not in doubt, but the question has been whether it existed everywhere at the same time. We will shortly see that the ages of the oldest strata overlying it vary systematically from region to region, and it is impossible to establish for certain that the peneplain had in fact been fully developed over its full extent before some parts of it began to founder. Also, as noted above, some relief may have survived, or may have been regenerated by local uplift, so perhaps the whole Late Cretaceous Peneplain did not exist continuously as a single surface of vast extent at any one time. On the other hand, there is little basis for arguing a case against that simpler and generally more attractive view.

Rubbish on the Peneplain—The Quartz Conglomerates

It would not be practicable to try to assess the volume of older rock that was worn away during the Rangitata uplift and we can only say, rather weakly, that it must have been enormous. Some of the erosional products undoubtedly would have been trapped along the eastern belt where sediment accumulation continued in subsiding marine trenches while the Orogeny was going on and also during the early stages of the next (and final) cycle of sedimentation. However, it seems probable that the bulk of the waste produced when erosion was vigorous in the early phases of Rangitata uplift has failed to be preserved in the New Zealand geological record. Retained in greatest amounts were the more stable components released by erosion in later less vigorous phases—materials resistant to further weathering because they were already products of chemical decay. Mantling the deepening zones of weathering beneath the wasting highlands, these residual accumulations thickened because as relief faded away the streams became less and less competent to export them.

Debris lagging on the peneplain as it developed consisted mainly of quartz and clay minerals. The quartz came from pebbles and sand grains in the older sedimentary rocks, from foliae in schist and gneiss, from quartz veins in all the older rocks and from crystals in the granites of the former foreland belt. The clay came from decay of

Fig. 7.11
Earliest Tertiary (Paleocene) marine sands with a basal lens of pebbles, resting unconformably upon a wave-planed surface over Haast Schist in Raupo Stream, a small tributary of the Kakanui River, North Otago. In this locality currents and waves of the transgressing sea had swept away the quartz-rich debris, carbonaceous clays and old soil which elsewhere lie between the schist and the overlying marine beds.

Fig. 7.12
Thick beds of conglomerate with layers of sandstone, mudstone and coal, ranging in age from late Cretaceous to early Tertiary, overlie the Paleozoic rocks in far north-west Nelson. They are mapped as Pakawau Group. These cliffs near Cape Farewell expose some of the upper part, which here is rich in quartz pebbles and locally represents the Quartzose Coal Measures. Bituminous coal formerly mined near by at Puponga came from lower in the sequence.

Fig. 7.13
Hard Eocene sandstone beds containing coaly layers at Puysegur Point seem able to resist the stormy seas which continually batter the south-west corner of the South Island. The western, landwards-dipping flank of a syncline is showing here. (Photo: N.Z. Geological Survey)

aluminosilicate minerals, especially feldspars. All these sources contributed the essential components of sandy, leached soils and pebbly alluvium on the peneplain which were to become one of the highly distinctive and most widespread facies in the lower part of the younger rock succession. Whether as friable, quartz sand, as cemented quartzite or as the ubiquitous quartz conglomerate formations of Southland, Otago, Buller and Nelson, these sediments appear consistently just above the stratigraphic level of a major unconformity which in many areas separates New Zealand Geosyncline or older rocks from accumulations of late Cretaceous or Tertiary age.

Subsequently the peneplain was destined to founder beneath the sea. In most places marine erosion failed to scour the old soil away altogether, exposing fresh rock, but waves and currents did work over the upper layers of quartz-rich debris, indeed all of it where the mantle was thin, redepositing the material with new types of stratification and with carbonaceous layers. Where thick, however, the deposit retained much of the appearance of stream gravels and sands.

The concentration of decay products derived from a very large volume of basement rock has had important economic consequences. In addition to the quartz conglomerates and sands, which are a source of industrial silica and glass, other valuable minerals were retained and concentrated at the same time. Gold, for example, initially concentrated by the processes of vein-formation in the parent rock, is an insoluble, stable mineral of high density and it therefore lagged behind along with the resistant quartz debris. This is the origin of the 'Cements' quartz conglomerate once mined and crushed for gold near Reefton, the 'older Auriferous gravels' of Central Otago, and, incidentally, the 'Banket reefs' of the Rand mining field in South Africa. Under various local names, the quartz conglomerates were worked at some of the most famous goldfields in Otago, Southland and Nelson, including that of Gabriels Gully where the first Otago discovery was made in 1861.

Besides gold, quartz conglomerates in the Collingwood district of Nelson locally contain concentrations of limonite iron-ore, both as replacement of the original clay matrix between pebbles and as irregular masses within the underlying deeply decayed schist and corroded marble. The ore was worked for a time at Onekaka and Parapara and smelted to make pig-iron and pipes in the depression years, and prospected vigorously again in 1938–9.

Major Fuel Source—The Quartzose Coal Measures

Nowadays the greatest economic importance attaching to the quartz debris accumulations on or just above the Late Cretaceous Peneplain is that in several districts they contain valuable coal deposits and associated refractory clays. For this reason they also go by the general name 'Quartzose Coal Measures'. It should be mentioned here that there is a common practice of referring to all sedimentary strata associated with beds of coal as 'coal measures' even where no workable amounts of coal are present. 'Measures' is an old-fashioned synonym for 'strata', now obsolete except in connection with coal.

The name 'Quartzose Coal Measures' is hardly a formal stratigraphic term, but rather an informal collective name for all the coal beds, carbonaceous clays, quartz-rich sandstones and conglomerates found on or not far above the Late Cretaceous Peneplain. As noted above, the lower parts of the conglomerates often appear to have been freshwater accumulations, but the higher parts have been reworked by sea currents and waves and sometimes contain fossil marine shells. Coal beds in the upper parts may occur in alternate layers with claystone beds containing shells, as for example in the Waikato and Ohai coalfields. We are in fact about to deal with the origin of the most important source of New Zealand coal. From this stratigraphic horizon come the sub-bituminous coals mined in North Auckland, Waikato, Buller Gorge, Kaitangata and Ohai; bituminous coals at Westport, Reefton and Greymouth; lignites at Charleston, inland Canterbury, North and South Otago and Southland; in fact, all the significant coal sources except those of a small field near Ohura in inland Taranaki, of late Tertiary age, and the lower ('Paparoa') bituminous coal measures

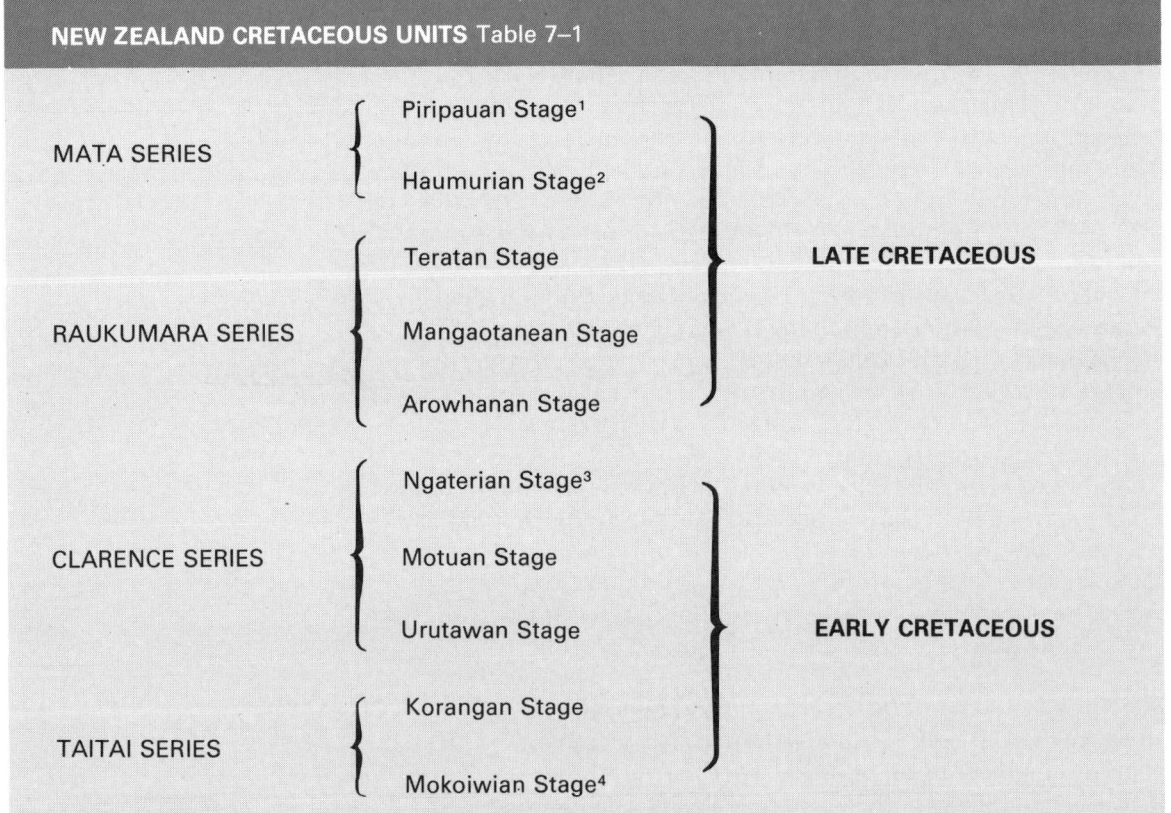

NEW ZEALAND CRETACEOUS UNITS Table 7–1

MATA SERIES	Piripauan Stage[1]	
	Haumurian Stage[2]	
RAUKUMARA SERIES	Teratan Stage	**LATE CRETACEOUS**
	Mangaotanean Stage	
	Arowhanan Stage	
CLARENCE SERIES	Ngaterian Stage[3]	
	Motuan Stage	
	Urutawan Stage	**EARLY CRETACEOUS**
TAITAI SERIES	Korangan Stage	
	Mokoiwian Stage[4]	

Notes:
[1,2] *The Haumurian and Piripauan stages have been placed by some in the Paleocene, but they are usually regarded as late Cretaceous units.*
[3] *In the original stage classification of the Cretaceous of New Zealand by Wellman in 1959 a 'Coverian Stage' was recognised; it has since been found contemporary with Ngaterian, and abandoned.*
[4] *The Mokoiwian Stage has recently been found by I. G. Speden to be equivalent to at least part of the Motuan Stage, and it is therefore likely to be dropped from the classification. It will of course be found on maps issued in recent years.*

NEW ZEALAND TERTIARY AND QUATERNARY UNITS (CENOZOIC) Table 7–2

Series	Stage	Epoch
HAWERA SERIES	8 climatic stages[1]	PLEISTOCENE-HOLOCENE
WANGANUI SERIES	Castlecliffian Stage[2]	
	Nukumaruan Stage[3]	PLIOCENE
	Waitotaran Stage[4]	
	Opoitian Stage	
TARANAKI SERIES	Kapitean Stage	
	Tongaporutan Stage	
SOUTHLAND SERIES	Waiauan Stage	
	Lillburnian Stage	
	Clifdenian Stage	MIOCENE
	Altonian Stage	
PAREORA SERIES	Awamoan Stage[5]	
	Hutchinsonian Stage[6]	
	Otaian Stage	
LANDON SERIES	Waitakian Stage	
	Duntroonian Stage	OLIGOCENE
	Whaingaroan Stage	
ARNOLD SERIES	Runangan Stage	
	Kaiatan Stage	EOCENE
	Bortonian Stage	
DANNEVIRKE SERIES	Mangaorapan Stage	
	Waipawan Stage	PALEOCENE
	Teurian Stage	

Notes:
[1] Refer to Chapter Nine
[2] Substages 'Okehuan' and 'Putikian' given stage status by some.
[3] Substages 'Hautawan' and 'Marahauan' given stage status by some.
[4] Substages 'Waipipian' and 'Mangapanian' given stage status by some.
[5,6] No longer recognised by most stratigraphers, but still appear on older maps still in print.
   Most of the stage names in the middle column were in use during the Geological Survey's 1 : 250 000 mapping programme. With revisions and advances going on all the time, further changes in this classification can be expected.

of Greymouth. They have been mapped under different formation names in the various mining districts, including 'Kamo Coal Measures' in North Auckland, 'Waikato Coal Measures' in the South Auckland coalfields, 'Brunner Coal Measures' (or just 'Brunner Beds') in the South Island West Coast fields, and 'Taratu Coal Measures' in south-eastern Otago.

The coal-bearing sequences amount to anything from a few metres up to hundreds of metres in thickness. There may be just one, but more usually a number of separate coal beds of varying thicknesses which in several coal fields exceed 20 m. The total thickness of coal-bed components of the coal measures in one area can amount to 100 m or more. Since it requires from 100 to 160 m of peat to make a 20 m bed of bituminous coal, an interesting question arises. How, on an old peneplain surface, could such great thicknesses of peat have accumulated?

The interlayering of coal with other strata containing fossils of marine or estuary-dwelling shellfish provides one clue. Others come from observing that the coals we are discussing now contain relatively large amounts of sulphur, some of organic origin but mostly as pyrite or marcasite, and further that coal beds at the top of coal measures and close beneath overlying marine sediments have the highest content of sulphur. The only likely origin of the inorganic sulphur compounds is from reduction of sodium and magnesium sulphates present in sea water and introduced into the coal-forming peat by downwards percolation before it was too deeply buried and compacted. It is inferred that progressively rising water level in extensive coastal swamps with abundant vegetation growing in and around them allowed thick peat to accumulate, while occasional incursions of sea water increased salinity to levels acceptable by marine shellfish but not so well by the vegetation, and brought in silt, sand and gravel at the same time. With many oscillations of the coastline to and fro, the sea crept progressively over the old land and covered the peat swamps with normal marine sediments.

Fig. 7.14
Block diagram to illustrate a fringing swamp–a favourable situation for the accumulation of coal-forming peat–varying in width from time to time but persisting behind a sandy shore while the landmass slowly submerges beneath the sea. The shore has been fluctuating sea-wards and shorewards depending upon a balance between supply of sediment, rate of submergence and strength of waves and currents. Note also that even if the coal-forming peaty beds, beach sands, etc. form continuous rock units, they must vary in age from place to place within the span of geological time taken up by the submergence; hence, they are said to be 'time-transgressive' strata.

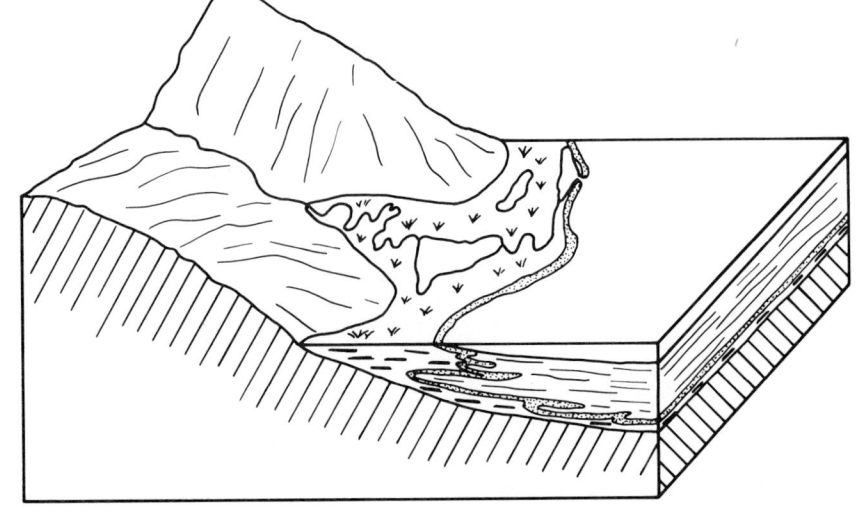

So consistently does the above story apply in all areas where the quartzose coal measures occur that we recognise it as unmistakable evidence of the beginning of a flooding of the ocean across former land areas. Known as the 'Late Cretaceous-Early Tertiary Marine Transgression', it initiated the last major episode of marine sedimentation in New Zealand's geological history. Its progress will be examined in the following section.

Going Under the Sea Again

Before going on, we must pause and consider more closely some of the obscurities which for a long time enshrouded our early Cretaceous history.

It has already been mentioned that the break between New Zealand Geosyncline deposits and the younger covering strata is least well marked in the Raukumara region and in Marlborough. It had been suspected by some people that marine sediments continued to be deposited in the eastern trough of the geosyncline uninterruptedly during the upheavals going on elsewhere as the Rangitata Orogeny progressed towards its climax. A corollary to this view was that *no* New Zealand-wide unconformity existed, and that the popular and convenient distinction between 'undermass' or 'basement' and 'covering strata' was invalid, or at least meant such different things in different districts that it had better be abandoned.

This was the state of affairs in the late 1950s when H. W. Wellman undertook a thorough re-examination, which in 1959 resulted in the first complete classification of the New Zealand Cretaceous rocks, based on modern paleontology and painstaking stratigraphic field work in critical areas of both islands. It seemed that little, if any, of the lengthy span of Cretaceous time, as understood overseas, was *not* represented by sedimentary strata in this country. The geology of the eastern North Island and Marlborough, however, is exceedingly complex, perhaps the most difficult of all to resolve. Following Wellman's great and welcome contribution it was some time before anyone else was able, or tenacious enough, to make much advance upon his results.

There are always doubters and constructive (mostly, anyway) critics among geologists. A few had always been sceptical of the fossil basis for some of the proposed New Zealand Cretaceous time-rock subdivisions, and especially about the supposedly unbroken stratigraphic continuity from New Zealand Geosyncline times through into the Cretaceous Period. Over the last decade a great deal more detailed work has been put into the Cretaceous problems, refining the paleontology and bringing in additional groups of fossils besides those on which Wellman chiefly based the 1959 classification. It has been done by teams and individuals from the oil exploration companies, the Geological Survey and the universities.

The outcome has been a strong case for regional unconformity in the critical Raukumara district between Torlesse basement and covering strata. Revision of old fossil collections combined with expert comparative study of much new fossil material indicates that four of the international divisions of the early Cretaceous (out of twelve covering the whole of the period) are in fact missing in New Zealand. It is nevertheless an unusually full record of the Cretaceous Period.

It is now generally accepted that the eastern geosynclinal trough was obliterated and everted late in the Jurassic Period and then sedimentation was renewed, either along the original line of depression or a new one similarly aligned, well before the end of the early Cretaceous. This was the curtain-raiser for the late Cretaceous-early Tertiary marine transgression. The real beginning of a new sedimentation cycle got under way when the sea overspilled from the trough on to the eastern margin of the Late Cretaceous Peneplain, which probably had never had an opportunity to develop over the persistently orogenic belt between Raukumara Peninsula and eastern Marlborough.

To sum up the situation in the early Cretaceous, the New Zealand Geosyncline had been everting since late in the Jurassic Period, that is, its contents had been upheaved and eroded energetically and no sedimentary record of this phase survived within the present land perimeter of New Zealand. By the end of early Cretaceous time, introduction of granite plutons in the metamorphosed roots of the geosyncline may have been largely over, and erosion was beginning to get the better of orogenic uplift in most parts of the country. Along the eastern border of 'proto-New Zealand', however, greywacke ridges were still rising and adjoining belts subsiding to receive sediment at such a fast rate as to promote a good deal of instability and re-deposition. Eventually, through a combination of continuing subsidence and marine erosion along the flanks of the rising ridges, the sea had spread generally westwards across a 20 km-wide zone by the end of the Cretaceous, thereafter overflowing ever farther in that direction across the more stable regions which by that time had been reduced to land of low relief. These events, thus compressed in words, sound rather hectic. It was a dramatic part of our history, but it did take forty million years to happen.

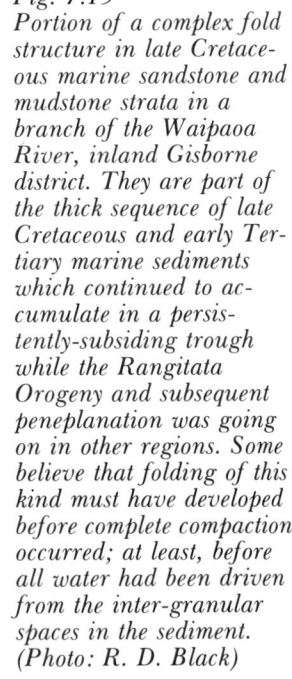

Fig. 7.15
Portion of a complex fold structure in late Cretaceous marine sandstone and mudstone strata in a branch of the Waipaoa River, inland Gisborne district. They are part of the thick sequence of late Cretaceous and early Tertiary marine sediments which continued to accumulate in a persistently-subsiding trough while the Rangitata Orogeny and subsequent peneplanation was going on in other regions. Some believe that folding of this kind must have developed before complete compaction occurred; at least, before all water had been driven from the inter-granular spaces in the sediment. (Photo: R. D. Black)

It is time to get back to the theme of the marine transgression. We should note that although the main flooding by the sea over the old land began some time before the end of the Cretaceous Period and continued to spread until half way through the Tertiary Era, its whole duration was less than that of the overlap period, the lingering on of Rangitata movements described in earlier paragraphs. There is an understandable tendency to minimise the time value of episodes recorded not by direct material evidence in the form of sedimentary accumulations, but by indirect evidence of sediment removal and tectonism.

It had not escaped the notice of earlier generations of geologists that the 'Younger Rock Series' (an alternative version of Cotton's 'covering strata') began with coal and terrestrial sediments at the base and passed up through sandstones and mudstones into widespread mid-Tertiary limestone strata. Some fifty years ago P. Marshall made much of the spread of limestones of this age from end to end of the country. He saw these as marking the maximum *depth* of submergence, which of course is not the same thing as the maximum extent of flooding by the sea. The limestones themselves actually contain evidence that they were not deposited in very deep water. More significantly, the amount of land-derived debris mixed with calcium carbonate of organic or chemical origin is small, showing how little land then remained above sea level to suffer ordinary erosion. The chief non-organic components are of volcanic origin, from eruptions which could have been entirely submarine.

Travellers between Dunedin and Oamaru are encouraged by a signpost on Highway 1 a little south of Hampden to make a side trip down to Hampden Beach to see the 'Moeraki Boulders'. These are not strictly boulders in the usual sense of large rock fragments rounded by the wear and tear of river or beach transportation, for in fact they are large, spherical concretions up to more than 2 m in diameter, lying on the beach or still in the process of weathering-out from the early Tertiary grey siltstone of the sea cliff. Some are smooth and entire, but many have been broken by marine erosion on the beach, exposing a softer core, or are reduced to small segments. Others are divided into innumerable, regular small segments by veins of the mineral calcite, which also sometimes forms a crystalline crust inside the shell where the core has been hollowed out. The veined variety of concretions are known as 'septaria'. Similar but less regular concretions litter the tidal rock platform of Katiki Beach about 10 km farther south, where the rocks are of late Cretaceous age, and they are not uncommon in strata of late Cretaceous to early Tertiary age in other districts, as for example in the Kakanui Valley, North Otago, near Five Forks bridge; in the middle gorge of the Waipara River, North Canterbury; in the lower reaches of the Conway River, Marlborough; and in East Coast districts of the North Island.

A few words in general about concretions would be appropriate here. The term is applied to a body of material within a rock which differs from its surroundings by being harder and tougher, and often paler in colour. The difference has resulted from the chemical precipitation of calcite, quartz, limonite or another binding mineral between the detrital grains. Thus toughened, concretions may survive more or less intact when the enclosing sediment has been washed or weathered away, whereupon they may become true 'boulders', in the usual sense, on beaches or river beds. In form, the concretionary mass may be anything from perfectly spherical to highly irregular, and fantastic shapes are not uncommon. One can also speak of 'concretionary beds', these having been made more resistant to weathering by the same precipitation process. A theoretical distinction is kept up between concretions roughly of the same age as (or at least very little older than) the containing rock, and those which have grown much later. Very probably, the great majority of concretions are in the first category, having grown after the sediment was compacted to the point of stability, but still permeable and saturated with fluids. Perfectly formed, spherical concretions like the Moeraki ones must have developed in conditions where the nature of the sediment allowed completely uniform growth outwards in all directions from some kind of nucleus, such as a small fossil shell, fish-scale, bone fragment or even a scrap of wood. It is important to realise that the growth of a concretion does not involve a pushing-apart of the surrounding sediment, but rather a transformation of it, in place.

Concretions in North Otago have been of unusual interest because of the light they have thrown on the tempo of sediment accumulation during early stages of marine flooding over the Late Cretaceous Peneplain. Some show, internally, clear evidence of an original stratification and well-preserved shells, while the enclosing sediment is barren of shells, at best faintly stratified and thoroughly modified by borings and castings of worms and other benthic animals. The hardening process must have taken place soon enough to protect the fossils and original sedimentary structures within a concretionary armour while the surrounding sediment remained wet and accessible to burrowers. Such a state of affairs is conceivable only in situations where the net rate of accumulation of sediment was slow, and the rate of build-up of the sea floor not too great to allow the activities of the organisms to continue at the depth of concretion-formation (perhaps a few metres below the sea bed) after the concretions had grown. It may be added that a lining-up of stratification within and outside of the concretions establishes without doubt that at least some of them grew in the sediment where now found, and were not washed in from somewhere else.

Only after Marshall's time did the means become available for tracing the progress of the transgression in detail. The first requirement was a better knowledge of the New Zealand Cretaceous and Tertiary fossil mollusca and brachiopods, which J. Marwick and R. S. Allan in particular studied intensively after World War I. Another was adequate methods for dealing with strata in which there were no fossils visible to the naked eye, or only types whose time range was too long to define the age within useful limits.

The study of minute fossils visible only under the microscope was not new, but the development of techniques for extracting microfossils from various kinds of sedimentary rock and dealing with large numbers of samples came about mainly in recent decades. Micropaleontology came to be applied in New Zealand in the mid-1930s during a resurgence of interest in petroleum exploration. Apparently unfossiliferous marine Cretaceous and Tertiary strata were made to yield determinable microfossils, mostly of the protozoan foraminifera. The name of H. J. Finlay is chiefly remembered in connection with the raising of New Zealand micropaleontology during that period up to a standard at which it became a useful stratigraphic tool. As a result, the classification of our Cretaceous and Tertiary rocks was greatly improved by Finlay, working together with J. Marwick, in 1940 and again in 1947. Further refinements have continued since then, bringing in other groups of fossils including the plants. Plant microfossils include the pollen of the higher plants and the spores of gymnosperms.

During the same period, attention was directed to the containing sediments, too, with the object of determining as closely as possible the environments in which they had been deposited—on land, under water, deep sea, shallow lake, and so on. This is 'sedimentary ecology'.

With the help of these new approaches it was discovered that the oldest marine strata overlying the Quartzose Coal Measures varied in age systematically from place to place, becoming generally younger from eastern Otago towards the interior of the South Island. This much was known by 1950, but a few more years were to pass before the study of plant fossil material in the coal measures themselves and of plant microfossils in the overlying marine beds reached a standard to confirm that the upper parts at least of the Quartzose Coal Measures were similar in age to the marine sediments that had buried them. In some cases it was clear that the entire thickness of the Quartzose Coal Measures consisted of debris on the Late Cretaceous Peneplain that had been reworked by waves and currents and embodied with the coal-forming peat deposits in coastal swamps and lagoons as the seas slowly advanced over the old land. R. A. Couper pioneered the plant fossil work which made this conclusion possible; it has been continued by D. B. McIntyre, D. Mildenhall and others.

Although the transgressing seas reached different areas at different times, the sequence of lithological types of marine sediment was similar: from coal measures to sandstones and mudstones, limey mudstones and limestones, the latter underlain sometimes by green sandy beds of glauconite ('greensands'). The main exceptions to the foregoing pattern are in eastern areas, where local orogenic impulses continued.

From the days of Hector, McKay and Hutton on into the twentieth century there were interminable disputes about the number and ages of the limestone formations, and whether the succession was complete and continuous from Cretaceous to mid-Tertiary times. Hector and his assistants believed that it was, bridging the gap in Europe when the dinosaurs vanished and modern land vegetation appeared. It was not then realised how different our fossil faunas are from the contemporary life of the northern hemisphere. Holders of the

Fig. 7.16
Hard white sandstone and grey mudstone (argillite) of the Whangai Formation (late Cretaceous or very early Tertiary age) in an old quarry near Poran gahau, Hawkes Bay. Towards the top of the quarry face, curving of the strata to the right is attributed to gravitational down-slope 'creep' of the soil and sub-soil.

Fig. 7.17
The culmination of marine
flooding over the North
Otago region in mid-
Oligocene time is marked
by glauconite-rich sands
(greensands), often richly
fossiliferous, and massive,
finely-granular hard
limestone. The picture
shows the Otekaike Lime-
stone overlying a softer
greensand bed which, be-
sides molluscs and
brachiopods, has yielded
bones of sea-birds and
whales. This locality, south
of the Waitaki River near
Duntroon, is called 'The
Earthquakes' because of
the many tumbled lime-
stone blocks, chasms and
cracks. The real origin of
these phenomena is not
seismic shocks, from which
the region is relatively immune, but rather the pre-
sence under the greensand
of a mudstone formation
that becomes very weak
and slippery once water
gains access to it.

alternative view that a Cretaceous-Tertiary break occurred in the New
Zealand succession were no better served by paleontology, but its
supporters were ever ready to see an important unconformity wherever
a visible sedimentary break, however slight, could be made out at the
appropriate stratigraphic level.

North Otago, North Canterbury and Marlborough were the classic
battlegrounds. Pilgrims may, with a little patience and at some peril
from traffic, find an unevenly corroded sedimentary contact between
pure white Amberley Limestone and greenish Weka Pass Limestone
east of Highway 7 on the southern approach to Weka Pass, and marvel
that it was once claimed as strong support for the theory of
Cretaceous-Tertiary unconformity in New Zealand. To others, it was
part of the conformable 'Cretaceo-Tertiary' transition beds, the white
limestone being considered identical with the Amuri Limestone of
Marlborough, then also believed to be Cretaceous. The two limestones
at Weka Pass, however, are now known both to be of Oligocene age, the
break between them representing only a trifling amount of time.

R. M. Carter and some of his associates have recently revived the
suggestion, not so much of an important time-interval unrepresented
by strata as of a sharp interruption to the tectonic calm at the
maximum of the transgression, a regional upwarping which elevated
the sea-bottom to within reach of wave action and caused numerous
minor local unconformities within the Oligocene beds in various areas.
Such an event is 'epeirogenic' as distinct from orogenic. Its timing
would coincide with an important Oligocene break in the Tertiary
succession of New Caledonia.

Over most of both the North and South Islands the sediments
deposited from the transgressing seas are nowhere thick, not by
geosynclinal standards at least. The 100 m order of magnitude applies
almost everywhere. The main controlling factor was relative rise of sea
level rather than subsidence of the crust. Events followed a different
course in a few areas, including that now occupied by the Paparoa
Range in North Westland. The Paparoa Trough was described briefly
on an earlier page. It has been likened to a miniature geosyncline
which began to subside in the middle of the Cretaceous Period. By the

Fig. 7.18 (centre) Wavy stratification in this limestone member of the Thomas Formation in Porter River, near Castle Hill, Canterbury, suggests the currents in the Oligocene seas were continually re-working the limey sediment while accumulation went on. The dip of the distant limestone outcrops is towards the camera, and in fact the strata here are folded into a syncline which plunges at a moderate angle towards the right of the picture. (Photo: D. J. Jones)

Fig. 7.19 (right) Looking up at the contact between Thomas Formation limestone of Oligocene age and the overlying Miocene sands and shellbeds (Enys Formation) above the west bank of Porter River. The upper surface of the limestone was scoured and corrugated by currents and solution and excavated by bottom-dwelling animals which left no clue as to their identity (several of their burrows are visible in the picture). Material from the basal shell bed of the Enys Formation now fills the cavities. Although some late Oligocene and early Miocene time is missing at the break, the beds above and below it are parallel with one another, which makes the contact a 'disconformity'. (Photo: D. J. Jones)

time the transgressing ocean had reached the area early in the Eocene Period, and before the adjoining areas had been reduced to low relief, up to 1000 m thickness of gravels, sands, carbonaceous silts and coal-peat beds had accumulated in the Paparoa trench. There must have been a pause while the Late Cretaceous Peneplain was carved across the area, but subsidence was resumed so that the Quartzose Coal Measures (Brunner Formation) are followed by a further 3000 m thickness of marine sandstone, siltstone, calcareous mudstone and limestone (Island, Kaiata and Omotumotu, Port Elizabeth and Cobden Formations), in other words the usual succession of lithological types associated with the marine transgression elsewhere, but here exceptionally thick. The trough ceased to subside at about the time of the maximum spread of seas.

Lignite and Bituminous Coal—Why the Difference?

This is a convenient point at which to bring in a short digression about the geology of coal, which will crop up again in Chapter Thirteen. First of all, it should be mentioned that the correct term to use when speaking of the different degrees of alteration from the original peat to lignite, towards bituminous coal and anthracite is 'rank'.

The geology and resources of West Coast coal areas were investigated in unusual detail during World War II. Some lignite beds, including those mined at Charleston and at Fletcher Creek, Inangahua Valley, which had been thought to be of low rank simply because they were supposedly younger than the bituminous coals, turned out to be of the same age as the Brunner Formation which yields high-rank coking coals in the Greymouth and Westport fields, that is Eocene. The surveys also disclosed that the strata deposited on top of the coal measures varied systematically in thickness, and in sympathy with variations in the rank of the coal below. In other words, the maximum depths of burial suffered by coal beds after deposition seemed to have a bearing on the rank attained by the coal. This is a corollary to an old principle of coal geology known as 'Hilt's Law', which had not been tested in New Zealand previously. Shortly afterwards there was an opportunity to try it out in the Reefton coalfield. Surveys here were

Fig. 7.20
Massive, grey-brown compacted muddy siltstone of the Kaiata Formation (Eocene) is one of the most easily recognisable sedimentary units in North Westland and in the Reefton and Buller districts. Crystals and nodules of pyrite and marcasite are often found in it, as well as odd-shaped calcite-cemented concretions. In this picture, taken in about 1939 at Nine Mile Beach north of Greymouth, a piece of abandoned road is evidence of coastal erosion, assisted by the seawards dip of the Kaiata beds and their rather weak resistance to attack by the waves. The buildings (long vanished) were connected with a small gold mining operation in a layer of uplifted beach gravel.

showing that the marine strata above the Brunner Formation thickened towards the east, and Hilt's Law predicted that coal discovered by prospecting in that direction should be of higher rank than that mined at Reefton. In due course the Garvey Creek bituminous field proved this to be true.

Clear Seas, Shoals and Volcanoes

The above heading sums up the situation over most of New Zealand and the surrounding continental platform in early Oligocene times when the ocean attained its maximum spread. Parts of the old land survived the inundation, though not necessarily as a single strip of continuous land, in west Otago and central North Island. There are signs in inland Canterbury that small areas were temporarily uplifted, most likely by renewed movements on faults that originated in the Rangitata Orogeny. Coarse, clastic sediment appears here and there in Oligocene strata in Auckland, East Coast and far southern areas but otherwise land-derived gravel and sand is generally absent. Minor sedimentary breaks within the limestone succession were caused either by a drop in ocean levels, or by mild, epeirogenic upheavals of the crust.

The seas, though shallow, were clear of sediment except when limey ooze was stirred up by currents or by the movements of bottom-dwelling animals, or where material erupted from oceanic volcanoes was raining down. Shoal areas built up of shells and skeletal fragments of crustaceans, sea-urchins, corals, coral-like bryozoans, foraminifera and lime-secreting algal plants probably resembled the Bahama Banks of today's Atlantic Ocean rather than coral reefs. However, calcium carbonate precipitated from sea water bonded the ooze into compact limestone, which at times suffered scouring by currents and corrosion.

Fossil vertebrates are not common in New Zealand rocks. Ocean-swimming saurian reptiles are represented by one example from the Torlesse rocks and by a few more in deposits from early stages of the marine transgression, but the greatest number of vertebrate fossils come from the Oligocene limestones and greensands. They include whales, penguins and other sea birds.

Before consolidation, the calcareous ooze was inhabited by burrowing creatures of many kinds—some known, others not—which left trails, casts and burrowings, all of which are called 'trace-fossils' because remains of the creatures causing them have not survived but merely the marks of where they have been. Among the known makers of trails and burrows are crustaceans, molluscs and worms. The green potassium-iron silicate mineral glauconite, formed by reactions between detrital muds and sea water, grows into pellets on (or inside) minute fossil particles, and these pellets may then be concentrated by the action of sea currents to form greensand beds. Abundant glauconite, as distinct beds or scattered particles in other Oligocene strata, indicates when the net rate of sediment-accumulation was low.

Our Chief Source of Lime

This limestone-forming episode turned out to be economically important for us. Much of our wealth still comes from pastoral industries, which demand large amounts of lime. The Oligocene limestones are the main source from Northland to Otago. Southland limestones are a little younger, formed, curiously enough, at a time when in most other areas an increasing input of land-derived sediment had put an end to pure limestone formation. The specifications of limestone suitable for use in cement manufacture are stringent, but the Oligocene limestones meet them in terms of both quality and quantity to supply the cement works at Whangarei, Tarakohe, Cape Foulwind and Burnside, in the latter case coming from Milburn in South Otago.

Fig. 7.21
Thick limestone formations of the Te Kuiti Group (Oligocene) are wide-spread in the south and south-west of Auckland Province and contain numerous caves, of which the Waitomo group are best known. This outcrop, in the Port Waikato district, is typical. Weathering often picks out the soft, less well cemented layers, leaving the harder ones to stand out as ridges. (Photo: N.Z. Geological Survey)

Re-emergence—the 'Kaikoura' Uplift begins

No suggestion of simplicity can be attached to the second half of the cycle. With minor, local complications, the previous transgressive phase was a progressive submergence of most of the country beneath the sea, whereas the re-emergence which commenced at the end of the Oligocene Period and early in the Miocene proceeded differently in each region and culminated at different times.

The onset of the Kaikoura episode of mountain-building dates from about the beginning of the Miocene Period, but that does not mean that marine sediments ceased to accumulate immediately. Some of our thickest Tertiary marine sequences, amounting to 4000 m and more, had yet to be deposited, in separate basins wherein subsidence began and ended at different times and proceeded at different rates. This was the main period of deposition of the widespread 'papa' formations of blue-grey sandstone and siltstone, including the 'Blue Bottom' which underlies the gold-bearing gravels of Westland. All of these thick, distinctively-marine strata continued to be laid down during the latter half of the Tertiary Era and early in the Pleistocene Period, despite a continuing net gain in both area and elevation of dry land.

Thicknesses (period for period) and sedimentary facies vary so much that it is sometimes impossible to match successions of late Tertiary strata in areas only a few kilometres apart unless fossils are both abundant and suitable for the task of correlation. This state of affairs implies that deposition was going on to the accompaniment of uplifts and depressions of ridges or blocks of basement, moving independently at different rates between trenches which at times were depressed relative to sea level. Ridges were anticlinal up-archings of basement or elevated blocks bounded on one or more sides by faults; likewise, some furrows were synclinal downwarps of basement, others fault-angle depressions. So varied in fact were the situations and so often uncertain that it is a common practice to speak simply of 'tectonic highs' and 'tectonic lows' or just 'depressions'.

The sediment which went into the depressions came from various sources. Some of it was derived from erosion of conglomerates, sandstones and mudstones deposited in the transgression phase, and now being stripped from the rising elements virtually as fast as they rose; some came directly from erosion of basement rocks exposed along the crests of the 'highs'. Depressions tended to be elongated in north-south or north-east/south-west directions, but by no means exclusively so.

Changes in sedimentary facies and thickness over short distances are often abrupt. Localised angular unconformities are common in the upper Tertiary successions, as, for example, in central Hawkes Bay where a sharp uplift is recorded between early and late Tertiary marine beds. In North Canterbury and Marlborough the successions of Miocene and Pliocene rocks to the east and to the west of Lowry Peaks Range are not the same, and there are even more striking differences east and west of the Hundalee Fault north of Parnassus. Very coarse breccia containing limestone blocks up to several metres across and associated lenses of both coarse and fine conglomerate can be seen

Fig. 7.22
An example of the many complexities of stratigraphy and structure in the Cretaceous and Tertiary rocks of Hawkes Bay. The attitude of the older, light-coloured, gently dipping beds in the foreground (Whangai Formation, Paleocene or early Eocene) is strikingly different from the adjoining, younger, steeply-dipping sandstones and mudstones (Weber Formation, Oligocene). The actual contact is not exposed, but they may be separated by a fault, now followed by the gully on the right. Alternatively, the Weber Beds may have been deposited unconformably upon an eroded surface cutting across the then steeply-dipping Whangai beds and the whole assemblage later rotated through about 90° to produce the present situation. The locality is in an eastern branch of Mangaorapa Stream, near Porangahau.

alongside Highway 1 north of Hundalee and again from Clarence Bridge to northwards of Kekerengu. It is evident that ridges were rising rapidly here in Miocene and Pliocene times, and delivering coarse debris to the sea across a faultline coast located not far from the present one.

Eventually the physical axes of the North and South Islands and important subsidiary ranges were rising fast enough to more than make up for erosion, after which they began to attain some appreciable elevation. Ever increasing quantities of detrital sediment of increasing coarseness augmented by widespread volcanic emissions in the Miocene Period brought to an end the Oligocene times of clear seas and pure limestone accumulation. By the end of the Miocene the Kaikoura uplift was properly under way. Floods of sediment displaced the sea from most of the depressions, even those still subsiding, and the marine fossils that are so common lower in the succession disappeared while debris of land plants appeared instead. As noted in Chapter Two the outline of New Zealand filled out to its present general form by a combination of orogenic uplift and the building out of aprons of erosional deposits along the bases of the growing mountain chains. The mid-Miocene coastal outline must have been long and irregular, but thereafter it tended to become simpler and shorter.

The sea did not give up easily, however. It continued to occupy practically all of the Auckland Peninsula, Gisborne, Taranaki, eastern Canterbury, the West Coast and Southland through Oligocene and early Miocene time. Some impressive thicknesses of sediment (Plate 13) accumulated where rapid, local subsidence persisted. In the Murchison Basin, for example, more than 3500 metres of marine sands and muds were deposited in the early part of the Miocene Period alone, followed by another 2000 metres of coarser beds in the later part. In North Westland the Paparoa trough began to be 'everted' and replaced by the ridge which later became the Paparoa Range, while beside it on the eastern side the Grey-Inangahua Depression remained a sea-way, and a receptacle for some 2000 metres of marine sediments, some no doubt worn from the rising Paparoa ridge, until the sea was finally driven out early in the Pleistocene Period. A final marine flooding in the Pliocene Period affected South Auckland, the southern half of the North Island, North Canterbury and a strip of Westland. The Pleistocene Period was well under way before the ocean finally yielded up Southern Hawkes Bay, Wairarapa and the Wanganui-Rangitikei district.

Younger Limestone Horizons

It is not clear why some of the areas still under the sea (or resubmerged) during later Tertiary times escaped the great floods of eroded material to the extent that more or less pure limestone could still form. Perhaps the detrital sediments were intercepted by submarine ridges or trenches nearer to the source. Economically, the most important example is the Miocene limestone of the Waiau Valley in Southland, which has provided nearly all the agricultural lime requirements of that district. In North Westland, though the main limestone horizon is that of the Cobden Limestone at Greymouth, well known also at the Pancake Rocks, Punakaiki, and of Oligocene age, there is also a Miocene limestone above it on the hill behind the southern residential part of Greymouth and farther inland in Stillwater

Fig. 7.23
The hills east of Blenheim are fringed by tilted late Miocene and early Pliocene gravels and sandstone, remnants of a former mantle of coarse debris from the rising Kaikoura mountains which rest unconformably on older rocks of various ages. For the most part they are freshwater deposits, the sea having been expelled from the emerging areas.

Creek. Limestone lenses formed of shell-bank debris ('coquina') appear in the late Tertiary marine transgression deposits of the North Island, as for example in the lower Wanganui valley (Wilkies Bluff Limestone) and in Southern Hawkes Bay (Te Aute Limestone) quarried for agricultural lime at Hatuma. Usually the coquina beds are composed of oyster and other bivalve mollusc shells, but there are some curious ones made up entirely of barnacle plates, large and small. An example of these is exposed in a long-abandoned lime-quarry east of Stoke, Nelson.

The Younger Coals

We saw how the advance of the sea over the Late Cretaceous Peneplain was heralded by a migrating belt of fringing swamps with peat. As the sea later retreated before advancing deltas and rising coasts, marginal swamps again provided favourable conditions for coal-forming peat beds to accumulate. Thus in southern Nelson, Westland, inland Canterbury and Otago the upwards transition from marine to non-marine sediments is marked by coal measures at a number of places. In inland Taranaki a temporary shoaling prior to the late Tertiary return of the sea in this region permitted the deposition of the coal which was worked in the Tatu State Mine near Ohura, and on a small scale also at Waitewhena. In Canterbury picnickers alongside Broken River at the Highway 73 bridge frequently discover a coal bed of this age which crops out just below the old bridge site. Still younger coal beds occur at Longford near Murchison and near Glenhope, Nelson, the latter marking the retreat of the sea after the temporary Pliocene resubmergence of central New Zealand. The younger coals make only a minor contribution to our fossil fuel reserves.

The presence of coal, or at least of highly carbonaceous sediment in the late Tertiary succession as well as at the base of the covering strata brings in a satisfying sense of symmetry to this story of 'going down, coming up' but does not alter the fact that the last sedimentation cycle was anything but a simple affair.

Great Conglomerate Formations and the 'Kaikoura' Climax

Conglomerate beds containing pebbles and cobbles of greywacke or other older rock type are common in the upper Tertiary sequences of both islands. Sometimes it seems that the pebbles came out of earlier conglomerate beds that were locally 'up for erosion' at the time, but perhaps in most cases their origin is from areas of undermass rocks that had been exposed to erosion following sharp uplift along faults. Upwards through practically every extended sequence of late Tertiary strata in the South Island, and some in the North Island as well, one finds that conglomerate beds become progressively more common, thicker and coarser. Appearing first in beds of different ages ranging from Miocene to early Pliocene, they continued to be deposited until well into the Pleistocene Period during the climax of the Kaikoura Orogeny.

In the North Island the conglomerates are regarded as marine deposits because of their stratification and texture and because they

Fig. 7.24
At the top of this cliff at Castle Point, eastern Wellington, a friable, shelly limestone (Te Aute Limestone, Pliocene) rests unconformably on an uneven surface eroded over steeply-dipping Miocene strata. These were tilted into their present attitude during vigorous crustal movements which had already blocked out the rough outline of the region before a final invasion of the sea spread late Pliocene and early Pleistocene marine deposits over parts of the Wairarapa region.

Fig. 7.25
Withdrawal of the sea from much of the South Island area in the Miocene Period is marked by a change from marine to non-marine sediments. These sands and clays near Avoca, inland Canterbury, were probably deposited on a gently sloping alluvial plain; they overlie thin coal beds formed in coastal lagoons which followed behind the receding sea. Soft and easily erodible, these beds of the Enys Formation show the effects of uncontrolled surface run-off, and have developed small-scale 'badlands' gullying in places.

grade up from fossiliferous marine strata in most cases with no obvious unconformity intervening. Travellers through the Manawatu Gorge see glimpses of Pliocene or early Pleistocene marine conglomerate and pebbly limestone at either end, and there are more of these beds to be seen on the Saddle Road north of the gorge. Very likely the faulted and eroded scraps of conglomerate seen on the uplands north of Wellington City, e.g. from the Haywards and Akatarawa roads and in the upper Hutt Valley near Kaitoke, are part of the same story.

The greatest thicknesses of conglomerates occur in the South Island, known by various formation names in different regions. In central Nelson they are the 'Moutere Gravels', well exposed in countless road cuttings over the Moutere Hills, Spooner Range, Hope Saddle and around the southern shores of Tasman Bay; in south-west Nelson and Westland, the 'Old Man Gravels' (or 'Group'), particularly obvious in the Grey Valley as high, brownish-yellow conglomerate cliffs east of the Mawheraiti River between Mawheraiti and Ikamatua on Highway 6; in Canterbury the 'Bourne Conglomerate' and 'Kowai Gravels'; in Otago, 'Maori Bottom Gravels'; in Southland, 'Gore Piedmont Gravels'; and some other names as well.

Certain characteristics are common to these conglomerates wherever they occur:
(1) They are thick, up to two or three thousand metres in some places.
(2) Larger and smaller pebbles are found together in some beds, i.e. they are not well sorted as to size.
(3) There are individual beds or lenses up to several metres thick towards the top of the conglomerates, but thinner below.
(4) The stratification is rough, after the style of stream gravels toward the top, but is usually more regular lower down, with sandstone, mudstone and lignite layers towards the base.

(5) The conglomerates are deeply weathered; many component pebbles are too decomposed to be extracted whole from the upper parts, but often merely brown-stained lower down.

(6) They are made up largely, but not exclusively, of undermass rock types derived from the neighbouring region.

(7) In most areas they are moderately or steeply dipping, and involved with underlying strata in the folding and faulting due to Kaikoura movements.

It is not always possible to recognise the place in the succession of strata which marks the transition from marine to non-marine conditions of deposition. Conglomerates had begun to accumulate before the sea had withdrawn from some areas, but throughout Otago the main period of gravel accumulation did not begin until after it had receded beyond the present coastline.

In central Westland the transition took place early in the Pleistocene Period. The dating of this important event was first established, along with that of the earliest indications of glaciation, in Jones Creek, a small stream behind the township of Ross. The glacial deposits occur within the Old Man Group, which, together with the underlying Pliocene marine sandstone, has been severely folded and faulted. This shows that the first New Zealand glaciation occurred before the climax of the Kaikoura movements, although the Alps must by that time have attained elevations on the order of 1000 m (see Ross Glaciation, below, page 242).

Parts of other conglomerate formations similar in age to the Old Man Group, though lacking in distinctively glacial features, could well have been outwash gravels formed in the same manner as was the great piedmont plain of Canterbury at a later time. It is not difficult to imagine a scene, very early in the Pleistocene Period, when the Alpine chain and other growing ranges were surrounded by broad, outwards-sloping, coalescing plains, presenting an aspect like that of the Andean 'pampas'. They would have been built up vigorously by rivers heavily laden with the products of severe, cold-climate weathering and erosion. During the later writhings of the Kaikoura Orogeny, the fringing gravel plains were broken up by faulting, warped and folded, and deeply eroded before the onset of the late Pleistocene series of advances by South Island glaciers, as described in Chapter Nine (pages 253–9).

Smooth, gently sloping profiles of the Moutere Hills, the Loburn Downs and other hilly areas underlain by these conglomerates raise a question which was vigorously debated in the past. Do these profiles still reflect the form of the original top surfaces of the conglomerates, or have the present summits been shaped entirely by erosion, their original form entirely obliterated? Where present surfaces cut across the bedding of strongly-dipping gravels the question does not arise, but it is less easy to decide in cases where present surfaces and ridge-profiles seem to lie almost parallel to stratification in the gravels below. The enormous amount of erosion during the Pleistocene Period, however, makes it rather unlikely that any of the original surfaces survive in most areas, even as profiles. The conglomerate areas, incidentally, show a very distinctive surface pattern of shallow, evenly spaced erosional gullies separated by distinct though obtuse-angled crests.

'Kaikoura Orogeny'—How does it Stand after Sixty Years?

There is no doubt at all that the mountainous parts of New Zealand, other than the high volcanoes, owe virtually all their present elevations to upward movements of the crust continuing from the late Tertiary through into Pleistocene times. In order to gain sufficiently over erosion, the actual amount of uplift of the elevated blocks was three or four times more than their present altitudes. But is that all that is meant by the expression 'Kaikoura Orogeny'? We have already seen that it caused much deformation of the younger strata in both islands, from mid-Tertiary times onwards, and much erosion and transfer of material from one place to another. It initiated major faults, some of which are still active, and played a vital role in the shaping of New Zealand as it now appears. Let us now look at how the concept evolved, and how its significance, even its definition, has changed since the term was first introduced.

It is almost a lifetime since C. A. Cotton coined a name for what he saw as a paroxysm of block-faulting, folding and mountain uplift terminating the sedimentation of Tertiary strata in the Pliocene Period and giving rise to the structural framework of our scenery. Obviously envisaged as a sharply-defined event, it was identified for the first time as 'The Orogenic Uplift' (thus, with initial capitals) in a paper on the landforms of the Clarence Valley, Marlborough, which appeared in the *Geographical Journal* in 1913, and renamed 'Kaikoura Orogeny' in 1916 in a classic paper called 'The Structure and Later Geological History of New Zealand' in the *Geological Magazine*. At that time Cotton was under the excellent influence of a sound and perceptive paleontologist-stratigrapher, J. Allan Thomson (then of the Geological Survey and later Director of the Dominion Museum), with whom he had been working in the field in Marlborough. This paper was to become practically an article of faith among New Zealand geologists for half a century. In the same publication, incidentally, Cotton recognised as 'Mesozoic Orogenic Movements' what was later called the 'Post-Hokonui Orogeny', and more recently the 'Rangitata Orogeny'.

In both name and concept the Kaikoura Orogeny found general acceptance among Cotton's contemporaries and continued to be used without serious question until fairly recently, when it became apparent that the event was not as sharply defined in time as originally thought. The youngest strata strongly affected are of early Pleistocene age, but it can be argued that the slight deformation shown by some late Pleistocene deposits and the seismic activity of the present day indicate that Kaikoura movements still continue. In different regions the length of time missing from stratigraphic sequences at unconformities between deformed Tertiary and relatively undisturbed late Pleistocene deposits varies considerably, and in some critical areas it is doubtful whether unconformity exists at all. One of these areas is Taranaki, which is important because the formal time-stratigraphic term embracing deposits laid down after the Kaikoura Orogeny, as defined by J. A. Thomson, is 'Hawera Series'. (See below, p. 196.)

It has been brought out recently that severe faulting of the undermass rocks coupled with complex folding of the covering strata is confined essentially within a belt crossing New Zealand obliquely, roughly parallel with the Alpine Fault, outside of which deformation of the covering strata is less intense. Hardly any part of the country entirely escaped the effects of the Kaikoura movements.

In the light of detailed structural mapping and more precise geological timing of events, and bearing in mind also the implications of modern tectonic theories, the whole 'Kaikoura' concept has come in for some reappraisals and criticisms. Such developments are normal in any healthy science. The timings and unity of events in particular have come under scrutiny, and also the question of their relationship with the Rangitata Orogeny. Is 'Kaikoura' merely a continuation or renewal of 'Rangitata'?

These inquiries have drawn attention to the mild disturbances which in a few areas upset the regular progress of the early Tertiary marine transgression, and perhaps broke the idyllic calm of its Oligocene culmination. Should these be counted as preliminaries to the Kaikoura Orogeny? Should the Rangitata and Kaikoura Orogenies be viewed as connected events? Concerning the latter question, it is pointed out that the sedimentary phase between late Cretaceous and early Pleistocene times cannot be compared with either the New Zealand Geosyncline or the Buller Geosyncline. And what about the supposed early mid Pleistocene orogenic climax? Did the tempo of events actually speed up then, and if so, has it since slowed down? Has there been a Kaikoura plutonic or metamorphic event deep in the roots of the Southern Alps? Firm answers to these questions are still lacking, but we will be looking at some of them again in Chapter Ten.

Structural geologists and geophysicists who prefer to retain a distinction between the Rangitata and Kaikoura Orogenies claim to see important changes in the stress-systems within the crust of the New Zealand region after the Rangitata Orogeny; others see no great change. Whatever the outcome, it will still be convenient to have some verbal expression to cover a clearly-enough defined set of events including block-faulting of the undermass, folding of the younger covering strata, uplift and deep erosion which in late Tertiary times replaced the comparative calm of the early Tertiary. Regional differences and ragged timings there may well be, but not yet adequate reasons for abandoning the concept of a Kaikoura Orogeny.

Classification and Naming of the Younger Strata

A few stratigraphic names appear in this chapter, but only incidentally and it is not proposed to define or even to list the numerous rock formation names in use for the Cretaceous, Tertiary and Pleistocene strata of New Zealand. The series and stages and some sub-stages and their approximate international geological time equivalents are listed in Tables 7-1 and 7-2 but the biostratigraphic 'zones' defined by the stratigraphic ranges of particular fossils are not dealt with.

A reference index of New Zealand stratigraphic names introduced prior to 1951 was published by G. L. Adkin in 1954. It included 700 entries relating to Cretaceous and younger beds. The New Zealand volume of the *International Stratigraphic Lexicon* published in France (in English) in 1959 contained well over 1000 terms, by far the greatest number referring to post-Jurassic units, and the new edition now being prepared will contain a large number of new names. It would be neither practicable nor useful in this kind of book to name and describe the multitude of younger rock-stratigraphic units, but there is some point in setting out the current sequence of time-stratigraphic divisions of the Cretaceous, Tertiary and early Pleistocene of New Zealand, these being the units used for mapping deposits of these ages on the 1:250,000 geological map series. They are used also to supplement rock-stratigraphic units in papers about local and regional geology.

The Beginning of the Pleistocene Period

The Pliocene Period is looked upon by some as the last major division of a Tertiary Era, followed by the Pleistocene and Recent or Holocene divisions of a Quaternary Era (the European way of doing it); or alternatively, as the last-but-one division of a Cenozoic Era, taking in all of post-Cretaceous time including a Pleistocene which continues up to the present time (the American way of doing it). In New Zealand we do not seem quite to have made up our minds as to which practice to follow. A question of more substance is that of deciding where, within our younger deposits, is the stratigraphic level corresponding in time with that of the international Pliocene-Pleistocene boundary. Fixing that boundary has been internationally a knotty problem, and locating the corresponding position in New Zealand no less so.

Pleistocene was separated from Pliocene originally, on the basis of fossils, in Sicily by Charles Lyell in 1833. In those days a fallacious numerical scheme based upon the percentage of fossils supposedly identical with still-living forms was used to distinguish between the Tertiary periods. We cannot go into the long story of how this definition became confused with the beginning of the 'Glacial Period' or 'Ice Age', but will merely note that a meeting of the International Geological Congress in London in 1948 recommended that the earliest signs of markedly cooler climate coming in above Pliocene strata should betoken the beginning of the Pleistocene everywhere, whether glacial evidence is present or not. It is still the most widely accepted criterion, though others have been proposed, and many problems have arisen while trying to apply it around the world. As an initial step in implementing the 1948 recommendation, a 'type-locality' for the base of the Pleistocene was chosen in a sequence of marine sedimentary strata in the Calabria region of southern Italy, but there have been some disagreements about this. In any case there is no easy way to find the corresponding time-level in non-marine strata and other kinds of deposits elsewhere.

Locating the Boundary in New Zealand

Until the last few years no consistent practice governed the way

'Pleistocene' was applied in New Zealand. It generally covered any deposits formed prior to the most recent shift of sea level, but not deformed to any extent by the Kaikoura Orogeny. In terms of the time-stratigraphic classification introduced by Thomson in 1916 (and revised by R. S. Allan in 1933), the Castlecliffian Stage was regarded as Pliocene, and our youngest Tertiary division. When in 1945 the author came upon glacial beds at Ross, in the middle of the Old Man Gravels which were then regarded as Pliocene, interest was aroused because of a coincidence. It happened that C. A. Fleming when studying the molluscan fossils in the late Tertiary coastal sections north of Wanganui about the same time came to the conclusion that the seas around New Zealand had cooled sharply during the Nukumaruan Stage, which was the next one below the Castlecliffian in the Wanganui sequence. Next below again was the Waitotaran Stage, and this was also the age of the fossiliferous marine sandstone underlying the Old Man Gravels at Ross. Although at the time it was not possible to establish a correlation on any kind of internal age evidence between the Old Man Gravels and the marine strata at Wanganui, it seemed a reasonable inference that the Ross glacial beds marked the same cool phase as the one detected by Fleming at the beginning of Nukumaruan time.

If the 1948 London decision referred to in the previous section was to be applied in New Zealand (and, indeed, it was soon after by Fleming), the Ross glacial beds together with equivalent marine strata at Wanganui indicating cooling could be used to fix the beginning of the Pleistocene Period here. Much water has passed under the bridges since this was first proposed in 1952, and as usual the situation has tended to become more complicated. No international meeting involving Pleistocene matters passes without more discussion of the Pliocene-Pleistocene boundary. Moreover, earlier times of cooling seas around New Zealand have been recognised, and there have been differences of opinion about the correlations and classification of the marine strata at the critical levels in the North Island. However, the linking of the peaty layers at the transition between marine and non-marine parts of the Old Man Gravels with the lower part of the cool Nukumaruan Stage ('Hautawan Substage', or 'Stage', according to some) of the marine time-stratigraphic succession has been confirmed on the evidence of plant microfossils.

From various lines of evidence it now seems likely that the accepted date, in years, of the international Pliocene-Pleistocene boundary is going to turn out to be about 1.8 million years, and the Hautawan Substage could be older than that. If so, the Ross Glaciation may have to be regarded as a late Pliocene event rather than an early Pleistocene one.

Deposits Younger than the Kaikoura Orogeny

Not long after the Kaikoura Orogeny had been given its name the need was felt for some formal stratigraphic pigeonhole to accommodate all kinds of superficial deposits that had not been folded or faulted or uplifted to any extent because they had come into being after the peak of Kaikoura movements. These include river-bed and terrace sands and gravels, the younger glacial moraines, stratified volcanic tephra layers, wind-blown sands and loess. We will be dealing with some of these in Chapters Eight and Nine.

In various inconsistent ways and more or less informally, the above kinds of deposits were classified and mapped as 'Recent', 'Holocene', 'Drift Formation', etc. Back in 1917 J. Allan Thomson, when setting up the embryo of our modern times-stratigraphic classification of the Cretaceous and Tertiary, gave the name 'Hawera Series' to all deposits younger than his Castlecliffian Stage (Table 7-2, p. 175) but older than 'Recent'. It was a useful suggestion, except that it has proved hard to define 'Recent' (or 'Holocene'). For a long time 'Recent' implied the period since the end of the 'Glacial Period' or 'Ice Age'. But when did the Ice Age end? It hasn't ended yet in Greenland or Antarctica.

R. P. Suggate in 1963 tried to cut through the problems by proposing to use Hawera Series for *all* the superficial deposits of younger than Castlecliffian age, including the post-glacial deposits. The Castlecliffian Stage is the youngest stage-division of the Wanganui Series, which meant that the beginning of the Pleistocene Period in New Zealand, if marked by the base of the Hautawan (or Lower Nukumaruan), did not coincide exactly with one of our series boundaries. There is no compelling reason why it should do so. Hawera Series is regularly used now in the sense suggested by Suggate.

However, the need to define 'Recent' or 'Holocene' consistently still remained. This need was filled by a well-received proposal at another international congress in 1969, this time in Paris, which would fix the Pleistocene/Holocene boundary arbitrarily at 10,000 years ago (strictly at 10,000 years before A.D.—the 'present' is of course a moving point in time). It is the only international geological time-boundary so far thus defined, by convention, in terms of years. It may not be the last.

Further discussion of how the deposits within the Hawera Series are classified and named will appear in Chapter Nine.

Cycle Completed—the 'Notocene'

This brings us through to the end of another cycle, not quite the same, though, as the cycles before and after the Tuhua Orogeny. The present chapter has followed events from the peak of the Rangitata mountain-building through a phase of wearing-down of relief, a phase of submergence and marine sedimentation, and a phase of re-emergence with local deformation leading on to the peak of the Kaikoura Orogeny (assuming this now to be passed), when the basis of our present mountainous topography was blocked out. Except for glacial moraines, their associated gravels, sands and silts and for the latest volcanic products, all to be discussed in the next two chapters, we have completed a review of the geological stuffs of which New Zealand is made. Before concluding this chapter, however, I would like to say a little about a proposal made many years ago which offered a convenient, word-saving way of summing up this episode and indicating its time-span. For various reasons the suggestion never caught on.

The time-span covered is actually from half way through the

Cretaceous onwards to half way through the Pleistocene, as we now understand those terms—an awkward mouthful of a definition. Sixty years ago J. Allan Thomson recognised the unity of the cycle we have been examining, a well-defined period in our geological history which did not fit nicely into the main international divisions of geological time. He proposed that the last sedimentation cycle between the Rangitata Orogeny (the 'Post-Hokonui' then) and the Kaikoura Orogeny be called the 'Notocene', joining 'noto', meaning southern, on to 'cene' of Cenozoic. Time since the Kaikoura climax would be the 'Notopleistocene' (which term, incidentally, I am not defending).

Certainly there were difficulties, some apparent from the discussion of the Kaikoura Orogeny itself earlier in this chapter. This event was by no means as distinct as Thomson believed. Also, the break between the youngest of the 'Notocene' divisions, which was the Castlecliffian Stage, and the Hawera Series is insignificant in the very type-area for these units on the Taranaki coast. The full impact of the deformation, even where its effects were minimal, was nowhere delayed until the end of Castlecliffian time. The Castlecliffian Stage (see Table 7-2) is now part of the early Pleistocene of New Zealand, not the end of the Pliocene as it was in Thomson's day. There were other catches as well, but the advantages of the concept, affording a neat, concise expression to sum up this chapter of history, warranted more patience in overcoming the snags than it has received. One serious attempt was made by C. A. Cotton in 1954 and 1955 to bring back 'Notocene', giving it era status as 'Notocenozoic' and bringing its end forward to the present day because of the inconclusive separation of Castlecliffian from Hawera in a few areas. This departure from Thomson's original definition may have been one of the reasons why Cotton's attempt failed. As amended, 'Notocene' no longer would enable us to refer simply to the time span of the last major cycle without having to mention the imprecise timing of its beginning and end, in terms of the standard geological time scale, and without having to stammer out 'late Cretaceous-to-early Pleistocene' or something of the kind. In fact, it rather missed the point of the original proposal. Unfortunately, therefore, it seems unlikely that 'Notocene' will come into common use.

Chapter Eight

NEW ZEALAND VOLCANOES

Ancient and Modern

Our Long Volcanic Record

This chapter is devoted to the topic of volcanic action in all its many forms and at all times in the recorded geological history of New Zealand. It brings together numerous references in earlier chapters and one that follows, and considers the phenomena as a whole.

For its size, New Zealand has had an unusually long history of volcanic activity continuing up to the present day. As can be seen from the legends accompanying the North Island and South Island sheets of the 1:1,000,000 geological map, rocks of volcanic origin are represented in every period of the geological column except Silurian and Devonian. Our wealth in volcanic phenomena, ancient and modern, was recognised when New Zealand was selected as the venue for an International Symposium on Volcanology in 1965.

Built during later Tertiary and Pleistocene times, the Tongariro group, Pirongia and Banks Peninsula, for example, are impressive physiographic features. By a rough estimate the bulk of each of these is between 10 and 50 cubic kilometres of volcanic material, large amounts, yet small compared with the quantities of erupted material from vanished volcanoes, preserved along with non-volcanic sediments during the geosynclinal episodes. Most of our volcanic history of course is read in the stratigraphy of these deposits, together with a less complete record of vanished land-based volcanoes and submarine volcanoes active during the final period of evolution of New Zealand as we now know it. The fact that there is much fine fragmental material in the geosynclinal accumulations as well as lava and volcanic breccia reminds us that a large proportion of the products from recent eruptions remains air-borne long enough to reach beyond the coast, coming to rest on the neighbouring ocean floor.

What Constitutes a Volcano?

Mark Twain's definition of a mine was 'a hole in the ground owned by a liar'. The schoolroom definition of a volcano was even briefer—'a burning mountain'. Thanks to the dissemination of news in pictures around the world few people today still have the limited notion that all volcanoes are conical hills with smoke issuing from the top. True, some kinds of volcanoes are topped off by symmetrical cones which emit heavy, dark clouds and plumes of vapour during eruptions, but their less conspicuous lower parts, built up of many successive flows of lava, are often more important in terms of volume of material. Mauna Loa on Hawaii, Rangitoto and other Auckland volcanoes, though inactive today, will illustrate the point, and it might be noted here that the bases of such volcanoes may be far below sea level.

It is best to avoid the word 'smoke' in connection with volcanoes because it implies 'something burning', whereas combustion in the familiar sense plays only a minor incidental role, a side effect from incineration of vegetation and other carbonaceous materials on the surface. 'Ash' and 'cinders', too, are in disfavour for the same reason, though still widely used informally for fragmental materials which compose the dark clouds and mantle the surrounding landscape. Various suggested alternative words were not universally adopted, mainly because they had rather special meanings, and a new, more acceptable general term for airborne volcanic products was wanting until the Icelandic geologists, only too familiar with volcanic phenomena in their homeland, came up with a Greek-derived word 'tephra'. This has been widely adopted as a companion term for the older word 'tuff', still applied to stratified deposits of similar origin. 'Lava' serves both for still-molten magma outpourings and for the resulting hard rock after cooling and solidification.

Far from always being builders of mountains (let alone of regular, symmetrical cones), volcanic eruptions may, after activity has subsided, leave large depressions, long open fissures, extensive flat plains, or sheets of lava and tuff spread out in the ocean floor. The essential thing to justify using the word 'volcano' is to be satisfied that igneous* magma rose to the surface, or close enough to it at least to have caused explosive contacts between hot magma and water-saturated rock or soil.

Types of Eruption Reviewed

So great is the diversity of volcanic phenomena that it is advisable to describe the main types of eruptive activity and suggest reasons for the differences between them before beginning the main task of this chapter, which is to give a historical review of volcanic activity in New Zealand.

To begin with, it should be obvious that the geological effects and products must vary enormously according to the kind of situation where eruption occurs—on dry land or under water; on flat ground or steep slopes; whether brief or long-continued, and so on. Next to be considered is the composition of the magma, i.e., whether of basalt, or rhyolite, or andesite, etc.

At surface pressures molten basalt begins to freeze at about 1200°C. It remains mobile and freely-flowing until close to the point of solidification. Silica-rich rhyolite or trachyte lava, though it does not become solid until the temperature has fallen to about 1000°C, is extremely viscous above that temperature and has a far less sharp solidification point. Andesite and dacite, among the more common

---

* The word 'igneous', though unlikely to be replaced, is itself open to criticism because of its derivation from the Latin word for fire. Similarly, the first part of 'pyroclastic', a general term for all kinds of solid, fragmental products from volcanic action, comes from the ancient Greek word for 'burning'.

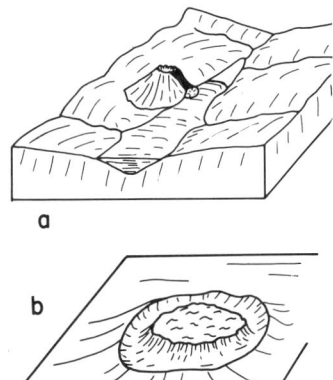

Fig. 8.1
a. *Fluid, basaltic lava flows have spread over the floor of the valley, forming a 'lava plain', and damming the headwaters of the stream. b. In marked contrast with a, a viscous mass of rhyolitic lava has been extruded from the pipe of the volcano within the ring of debris thrown out by an initial explosion at the beginning of the eruption.*

lavas of intermediate composition, are intermediate in these properties too.

The factor of viscosity of the lava during and before consolidation is of prime importance as regards the eruptive behaviour of a volcano and the kinds of surface features which result. For instance, basaltic lava is able to flow down much gentler slopes than rhyolite lava, and to keep on flowing until the moment of solidification, whereas rhyolite is so viscous when emitted that it needs a steep slope if it is to flow at all and, as lava, does not reach far from the point of emission. Also, the interior of the lava mass can remain molten long after an exterior crust has solidified.

Viscosity is important in another way as well. When rising up the feeding pipe or system of fissures beneath an erupting volcano, the magma is saturated with dissolved gases which are discharged under the reduced pressure as it nears the surface. From a highly fluid basaltic magma the gases coming out of solution are released freely and can go on escaping up to the last moment before solidification, after which any remaining gas is imprisoned in the pores and bubbles which give the resulting lava rock the property of 'vesicularity'. Basaltic eruptions, therefore, are not highly explosive once the initial outburst is over. On the other hand, viscous rhyolite lava restrains the release of gases, to the extent that disruptive pressures build up and cause the almost-solid lava to disintegrate explosively around the vent. Some of the most destructive kinds of eruption arise this way. Again, intermediate lava types generally are associated with intermediate kinds of behaviour.

Against this background it is easy enough to see why volcanoes on land can take such varied forms. Basalt welling up from fissures may spread out to form vast, flat lava plains burying a pre-volcanic topography, or, from a single vent, broad, gently sloping lava cones like the lower part of Rangitoto Island. From multiple vents, rounded domes like those on the island of Hawaii will be built. More viscous lava will build steep cones interlayered with fragmental material as in Egmont and Ngauruhoe. Rhyolite is likely to build irregular, humpy domes like Ngongotaha near Rotorua or craggy-topped plugs of stiffly-extruded obsidian glass. Essentially explosive eruptions may produce only small amounts of fresh magmatic material, but may leave lines of gaping fissures as did Tarawera in 1886. Besides lava features, a variety of forms are built almost entirely of fragmental material.

The term 'scoria', meaning vesicular, fragmental debris, and generally applied to basalt, is familiar to Auckland residents who live among a cluster of cones largely built of it. The cones grow, as may be seen from time to time in Hawaii, by vigorous fountaining of lava above the vent and may be more conspicuous features afterwards than the lava flows which account for the great bulk of material emitted. Sheets of fragmental rhyolitic debris, estimated to total 15 cubic kilometres in volume, were erupted in the central North Island during the Pleistocene Period. Emitted as floods of semi-solid particles, the material welded together, after coming to rest, to form a coherent, massive rock called 'ignimbrite'. In yet another form, aprons of volcanic rubble brought down as mudflows ('lahars') cover the lower slopes of many volcanoes, including Egmont and Ruapehu. (The nature and origin of ignimbrites and lahars are discussed again later in this chapter.)

Highly destructive, explosive eruptions are associated mostly with the emission of trachyte or andesite, but can occur also when magma of any type gains access to water-saturated ground near the surface or in the foundations of an earlier volcano. The amount of fresh magma erupted may not be large, especially in the initial stages, though later in the course of an initially destructive eruption the losses from the first blast may be more than made up by new accumulations. A classic example of this kind of behaviour is provided by the A.D. 79 eruption of Vesuvius which demolished most of an earlier cone and then in succeeding centuries built a new one partly on top of the ruins of the original. The explosive blast provided a name for this type of eruption, but it was taken not from the mountain or even from the city of Pompeii that was overwhelmed, but from the name of the contemporary author Pliny who described the event, hence, 'Plinian'. In New Zealand the Tarawera catastrophe of 1886 comes into a similar class.

The term 'crater' is widely known as meaning a conical or cup-shaped depression remaining at the top of a volcanic vent after an eruption; again, Aucklanders know them best. But holes and depressions can arise in other ways from volcanic action. Very large ones have been attributed to the collapse of the top of a volcano into a subterranean void left by the discharge of lava, etc.; these are called 'collapse caldera'. Similarly, a steep-walled vertical shaft or 'sink crater' or 'pit crater' may remain when the lava column falls in the open pipe of basaltic dome volcanoes like Kilauea on Hawaii Island.

How Does Magma Arise—and Rise?

How magma originates and how it travels to the surface are among the oldest of all geological speculations. The liquid lava emitted by some volcanoes was the very reasonable starting point for the now obsolete theory of a perpetually molten interior beneath a crust. This theory became untenable for a variety of reasons brought forward by astronomers and seismologists. Given, then, a substantially solid earth, the questions became 'Where below the surface does the material to make lava become liquefied?' 'Why does it liquefy?' and 'How does the liquid reach the surface?'

Increasing knowledge about the interior physical properties of the earth at different depths, measured from the surface by what is now called 'remote sensing methods', has narrowed down the possibilities as to what kinds of known rock substance best match the geophysical properties thus determined. There is no certain, unique solution, but one generally accepted model depicts a continental crust capable of melting to furnish, in different situations, andesitic or rhyolitic magma, sharply separated from a denser, lower shell of basaltic composition which in its turn overlies the 'mantle' with composition at the top akin to peridotite. There is much more detail than that, and several

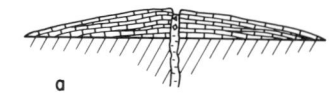
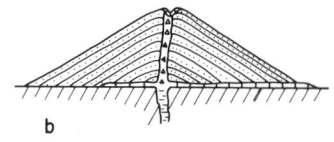
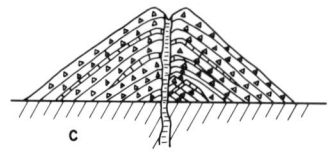

Fig. 8.2
a. Profile section through a basaltic lava dome. b. Fountaining of fluid, basaltic lava above the central vent has built a scoria cone on top of the lava flow emitted at an earlier stage of the eruption. c. Andesitic volcanic cones are usually composite, i.e., made up of alternating layers of steep, irregular lava flows (the lava is rather viscous) and angular, rubbly debris. d. An extruding plug of viscous, rhyolitic or trachytic lava is commonly surrounded by an apron of debris produced by the 'slag' from disintegration of lava as it cools to form a brittle, volcanic glass.

Fig. 8.3
Large topographic depressions which now occupy the central portions of extinct volcanoes are called 'calderas'. Sometimes these are ascribed to collapse following the discharge of vast volumes of material from an underlying magma chamber; but in many other cases, including that of the Akaroa caldera illustrated here in a pen sketch by C. A. Cotton, the depression is merely the work of erosion since volcanic action came to an end.

variations on the general theme with which we need not be concerned. The essential point is that melting at appropriate depths, transfers and mixing of melts, sometimes with further modification by 'crystal fractionation', seem competent together to account for the extreme and all the intermediate magma types.

According to seismologists and astronomers the earth is essentially a solid body, apart perhaps from the innermost core, so it now has to be explained why melting occurs to produce magma. The discovery of radioactivity provided the best answer to the question of a perpetual, internal heat-energy supply, and there are various versions of a hypothesis to explain how internally generated heat is continually brought up towards the base of the crust, involving slow, convectional flow and continual overturn of the 'solid' rock of the outer mantle. The physico-chemical properties of magma and the geothermal temperature gradient are such that rock materials at depth *would* be molten were it not for the influence of confining pressures (due to gravity) upon the temperature at which melting can occur. Liquefaction and the generation of magma can therefore happen at depths where the temperature is high enough—provided a reduction of pressure occurs. This is quite in accord with the laws of thermodynamics and experimental results, but the difficulty remains of finding a satisfactory, universal explanation why the pressure should change. The general answer must lie in the fields of stress involved in mountain building, but in detail the problem remains obscure.

Granted that magma has come into being within or below the crust, the question as to how it reaches the surface is not so difficult. A body of molten rocks will always be a little less dense than its solid counterpart so there will be a tendency for it to rise gravitationally, but that can hardly be the whole story. Another factor likely to be involved is the inability of magma under lessened pressure to retain components in solution which would escape as gases if they could. These are the 'volatiles' mentioned in Chapter Six, including water, carbon dioxide, the sulphur gases, etc. Liberation of the volatiles has for a long time been claimed to play a major role in mobilising magma, enabling it to find a way up through upper crustal zones of fractured rock, deeply penetrating faults and other fissures. A popular speculation, reasonable enough but difficult to establish as fact, is that considerable quantities of molten magma may accumulate beneath a volcano, probably in a network of fissures rather than in a single 'magma chamber' (though that expression is often used).

Once the magma reaches near enough to the surface to encounter 'cold' or water-saturated rock, and provided rising hot gases and thermal convection are still maintaining an adequate heat supply, the final break-through will be assisted by vapour pressures due to magma/water contacts. An eruption or cycle of eruptions will continue until the magma supply or the heat supply to maintain the molten condition has failed.

Considering the widespread distribution of volcanoes today, in several belts around the world, it may seem surprising that modern geophysical methods of study have not by now removed all speculation about how magma forms and how it reached the surface. However, it must be remembered that erupting volcanoes cannot be approached closely, so that even the eruptive phenomena at the surface have to be observed remotely or inferred from the results afterwards, while our only direct information about conditions deep beneath a volcano comes from geological studies of what has been exposed by erosion millions of years after it was last active.

Hot Springs, Geysers, 'Blowholes' and Silica Terraces

There is a natural tendency to associate our hot springs with our recent volcanic activity, until it is remembered that they are by no means unknown in the South Island, where no eruptions have occurred since early in the Pleistocene Period.

To ensure that a natural spring will discharge hot water rather than cold, it is only necessary that some of the water shall have been circulating through rock fissures and fractures to depths of a few hundred metres below the surface, in order to have picked up some of the normal outflow of terrestrial heat which establishes the geothermal gradient (Chapter Five). In regions where no volcanoes have been active in recent geological times it may be assumed that virtually all of the water percolated down from the surface, though not necessarily near the site of the spring. Where volcanoes have been active, however, water of magmatic origin may become mixed with it, and the heating then is more likely to be due to circulation near intruded magma that is still hot.

Geysers (the name comes from one in Iceland called 'Geysir') erupt hot water intermittently as the temperature in the water column feeding it rises to a point where boiling can begin at lower levels notwithstanding the static pressure, whereupon the discharge of a little water from the top lowers the pressure enough to trigger off boiling at all levels at once.

Some water vapour is expelled from crystallising magma in the final stages. It reaches the surface with other gaseous products through the volcanic vent during eruptions, and at other times contributes to the steam emitted from 'blowholes' or 'fumaroles'. Much of the steam, however, including that tapped when drilling for geothermal steam power, derives from ground-water of surface origin in areas where the outflow of heat is unusually high because heated magma is present at shallow depths.

Spring waters that are alkaline rather than acid dissolve silica from the country rock through which the waters are moving, but must

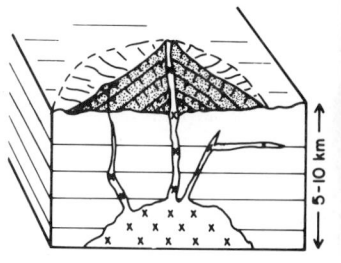

Fig. 8.4
From studies in many areas where ancient volcanoes have been deeply eroded (and especially the volcanoes of Tertiary age in the west of Scotland) it has been inferred that eruptions are fed directly from a body of molten rock that has appeared in upper levels of the crust (occupying a magma chamber). Magma reaches the surface through a system of dykes or a central pipe ('conduit'), while branching dykes may break through the outer slopes of a cone to produce flanking or 'parasitic' eruptions. This example would presumably have to be a basaltic or andesitic lava or composite cone. The vertical scale probably varies a good deal in different cases.

redeposit some of it upon cooling at the place of discharge, building up encrusted surface coatings called 'sinter'. Visitors to Rotorua see the sinter encrustations (in this case sometimes called 'geyserite') around Pohutu and other geyser vents at Whakarewarewa. Frequently these develop into a staircase of ledges or small terraces, reproducing in miniature the splendid Pink and White Terraces of Lake Rotomahana which were destroyed by the Tarawera eruption in 1886.

Volcanoes under Water

Unmistakable signs that a volcano is erupting on the sea floor have been observed many times out in the mid-Pacific and elsewhere. The sea surface is turbulent over a wide area, large volumes of gas bubbling up, and the water is heavily discoloured although the nearest land is far away. Some idea of what is going on at the bottom can be gained from watching the effects when lava from land-based volcanoes flows into the sea or a lake. Usually the results of contact between molten lava or hot fragmental material and the water are spectacular. The water boils furiously so that the operation is largely shrouded in steam clouds while the suddenly-chilled lava may go on disintegrating explosively as long as the flow continues. On the other hand, lava has been seen to enter the sea with remarkably little fuss and bother. A flexible, glassy shell or sheath forms instantly, within which molten lava continues to flow down the offshore slope until it congeals, volcanic rock being a poor conductor of heat. The elastic, glassy skin is liable to expand into blisters which fill with lava and expand like balloons until rupture occurs and another one begins to grow alongside. This is the origin of 'lava-pillows', piles of which can be seen on the coast south of Muriwai, Auckland west coast, and near Oamaru Harbour where volcanoes erupted into the sea in Tertiary times. They are sometimes seen in the process of forming in Samoa and Hawaii.

The fragmental products of underwater eruption, whether thrown above the surface or not, are likely to fall back and settle out on the sea floor to form sedimentary beds of volcanic tuff. Among the fragmental products from rhyolitic eruptions is pumice. Being frothed up and of low density, it may float thousands of kilometres away from the scene of the eruption before being washed up on beaches.

Undersea exploration since World War II has disclosed many more submarine volcanic piles than were known about previously. Rising from the ocean floor to within moderate depths below the surface, these 'sea-mounts' or 'guyots' are substantial mountains in their own right. However, unless eruptions are persistent and voluminous, the fate of a new mid-oceanic volcano that reaches the surface is usually rapid destruction, or at least reduction in a few months to a mere shoal below the limit of effective wave-action. Falcon Island in the Tonga Group is a well-known example of a volcanic island that has been built, worn down, and rebuilt several times in the past century.

The Evidence for Ancient Volcanoes

The original surface form and above-ground structure of volcanoes older than about middle Tertiary have usually been eliminated by erosion. Intruded 'dykes' and 'sills' as well as 'volcanic necks' or 'plugs' (stumps of lava-filled central feeding pipes) survive from the Eocene Period, as in North Otago, but the record of still more ancient volcanism is mainly to be read from stratigraphy. It is a matter of being able to identify igneous rock layers and masses which, by their internal texture and structure and their relationship with adjacent rocks, show clearly that magma actually reached the ground surface or sea floor of the time.

Basalt lava layers interbedded with other strata may have flowed over the top of the bed below, consolidated, and then have been buried under later deposits; or, alternatively, they may have been injected as sills at a later date, pushing apart the layers above and below at the same time. When trying to document the regional volcanic history, it will be important to know which is correct. Sills intruded at comparatively shallow depths do not differ very much in texture from surface lavas, although they are less likely to be vesicular. The best clues come from the boundaries with the strata above and below. The heat of a surface lava flow can affect the ground beneath it, perhaps baking it into a natural brick, but obviously not the next stratum above, which has yet to be deposited. The chilling effect of contact with moist ground underneath is often more obvious, as a glassy skin, than the effect of exposure to the air above. Sills, and for that matter dykes too, on the other hand, are likely to show chilling effects at both contacts about equally. There are other clues, and seldom is one line of evidence conclusive on its own.

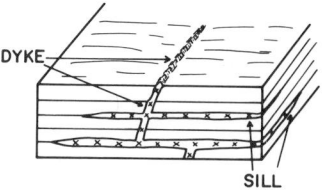

Fig. 8.5
Diagram of intruded sills and dykes. The latter term derives from an old name for stone walls, reflecting the fact that intrusive igneous rocks are often more resistant to erosion than adjacent materials, and thus may stand above the general level as a natural rock wall. (See also Fig. 8.9)

Fig. 8.6
A lava flow interbedded between sedimentary beds of sandstone and conglomerate. Most of the diagnostic features of an interbedded lava shown diagrammatically in Fig. 8.7 can be seen in this photograph. The locality is at Muriwai, Auckland west coast. (Photo: N.Z. Geological Survey)

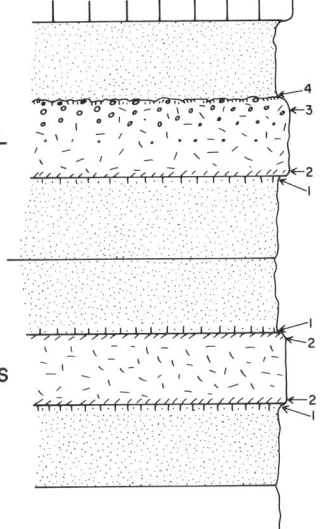

Fig. 8.7
Distinguishing interbedded sheets of lava from intruded sills. 1. Effects of heat from molten rock in contact with sedimentary rock include baking of clays, fritting or partial fusion of sediments; lava can affect only the lower contact unless buried while still very hot. 2. Effects of chilling of molten rock in contact with cold rock at margins, usually in form of a fine-grained or glassy skin ('selvedge'); present at top and bottom of a sill but usually more pronounced at base of a lava flow unless erupted under water. 3. Gas pores ('vesicles') are usually smaller or absent in sills; in lavas they are larger and more numerous towards top. 4. Upper surface of lava flow often uneven, porous ('scoriaceous') and may show weathering if exposed for some time before burial.

The presence of pillow structures is no guarantee that lava was emitted into water. In most such cases this process has taken place, but pillows have also formed where lava has squeezed into wet, unconsolidated sediments at shallow depth below the contemporary sea floor.

A compelling case for ancient volcanism has been based many times upon the occurrence of stratified deposits of tuff, or even upon minute proportions of unmistakably volcanic constituents, pieces of volcanic glass, for example, occurring with mainly non-volcanic sediments; also, upon inferences from the mineral and chemical composition of metamorphic schists. Again, seldom is a single line of evidence considered adequate.

Geosynclinal and Orogenic Volcanoes

The association of volcanic action with geosynclinal sedimentation was established on the above lines. The timing of events, too, was first deciphered as well as possible from stratigraphic and structural relationships, and in many cases has since been tested, clarified and confirmed by radiometric age determinations. Volcanic action seems always to have accompanied the subsidence and filling of geosynclines, and to have recurred in different modes during the orogenic phase of uplift and folding. The type of magma produced varied, presumably in relation to the depth at which it originated. Deep troughs tap a world-wide shell of basaltic composition or the deeper peridotite layer, so that basaltic tuffs and pillow lavas, peridotite lavas with serpentinite and other alteration products (the 'ophiolites') have made massive contributions to the geosyncline contents. Typically, as in the marginal belt of the New Zealand Geosyncline, the original plagioclase feldspar of these basalts, and the orthoclase of the associated but less abundant trachyte have been replaced by sodium-rich varieties; the dark minerals, too, have been affected by this so-called 'sodium metasomatism'. Some authors have suggested that the chemical environment of sea water, including that entombed with the sediment during deposition, is responsible. Others think that the changes are essentially metamorphic, resulting inevitably from deep burial beneath overlying sediment and volcanic rocks.

Before subsidence has ceased, and again afterwards during the orogenic phase, magma of intermediate compositions is generated beneath continental margins. Granodiorite plutons develop within the folded and metamorphosed mountain roots, and andesite and rhyolite is erupted to build chains of volcanic mountains on the crests of the upheaved and eroded geosynclinal rocks. Fine examples adorn the summits of the Andes and the coastal ranges of north-western North America.

The foregoing outline is, of course, a generalisation. There are differences in detail in different regions, and from one cycle to another in the same region, but the overall pattern has been consistent. Based upon observational facts of stratigraphy and structure, it holds up despite the revolutionary changes in thinking about the mechanism of crustal deformation that have come in over the past ten years.

Our Oldest Known Volcanoes

The volcanic history of the Buller Geosyncline in early Paleozoic times conforms to the pattern outlined above. Volcanic activity was important in the early stages. Early Cambrian strata of the Haupiri Group in north-west Nelson are dominated by sediments of volcanic origin together with flow rocks, sills, dykes and agglomerate of andesite and basalt composition. Peridotite intrusions date from the same period. Pebbles of the Lockett Formation (late Cambrian) are predominantly volcanic, and while it cannot be certain that none of these came from erosion of Precambrian rocks, there is every likelihood that the ancient foreland bore many volcanoes near the margin which continued to erupt throughout the Cambrian Period. The Ordovician Period saw a reduction of activity, after which there is no sign of volcanism until the initial stages of the New Zealand Geosyncline. Radiometric datings and field evidence together require that the Karamea Granite, and perhaps parts of the Rotoroa Igneous Complex as well, originated within the time-span of the Tuhua Orogeny, but no trace of contemporary volcanic material has yet been reported from the Devonian rocks of Reefton and Baton River.

New Zealand Geosyncline Volcanic Rocks

In Chapter Four it was noted that voluminous, dominantly mafic and ultramafic volcanic material makes up a large part of the geosyncline filling. From the outset, as indicated by the petrography of the Pelorus and Caples Groups, almost certainly as old as Carboniferous at the base, volcanoes were already contributing much of the sediment supply, and by early Permian times massive sequences of flows, pillow-lavas and agglomerate were going into the marginal zone of the geosyncline.

Fig. 8.8 (left)
A system of regularly-spaced, vertical joints breaks up this sill of basalt, near Duntroon, North Otago, into columns which commonly are five- or six-sided.

Fig. 8.9 (right)
An andesite dyke on the shore of Whangarei Harbour stands up as a wall because of its greater resistance to erosion compared with the rock through which it was intruded. It now provides a convenient natural jetty. (Photo: N.Z. Geological Survey)

Most descriptions give the impression of volcanoes on the foreland, though near the marginal trough shore. Lava certainly poured into the sea in large volumes, but not necessarily all from volcanoes on land. Pillow lavas occur also in the axial belt, many of these as isolated masses of pillows as though due to injection into the sea bottom sediments rather than to lavas flowing over the sea floor. More than likely, much of the lava was erupted from vents beneath the sea. Volcanic breccias in the marginal belt also suggest a local source, but the tuff which makes up a considerable proportion of the Hokonui Facies could have been either airborne from volcanoes on land or brought in by rivers.

Erosion has exposed bodies of more coarsely textured dioritic and gabbroic rocks, some of which may have been part of the feeding system of the geosynclinal volcanoes, along with many basaltic dykes of uncertain age, but the bulk of the coarser intrusive rocks probably date from the Rangitata Orogeny.

In contrast with the story of the Buller Geosyncline, the volcanoes in the New Zealand Geosyncline produced more mafic and ultramafic material than andesite and keratophyre (sodium-rich trachyte). This is consistent with the suggestion that the Buller Geosyncline should be classified with the less-vigorous 'miogeosyncline' type, whereas the New Zealand Geosyncline was a typically energetic, deeper 'eugeosynclinal' trough.

While the greater abundance of volcanic material is the outstanding feature of the Hokonui Facies, the Alpine Facies is by no means lacking in volcanic constituents. These are present in the low-grade metamorphic rocks at Kakahu, South Canterbury, from which came the first compelling fossil evidence for rocks as old as Carboniferous in the Geosyncline. In the same district basalt lavas and tuffs are interbedded with greywackes containing Permian fossils. Similar volcanic members appear at many places within the Torlesse Supergroup sequences of dominantly quartz-feldspar sandstone (greywacke). The pillow lavas referred to above are generally associated with variously coloured chert beds (jaspillite) and thin beds of pure limestone of limited extent. This association is in marked contrast with the turbid-water greywackes and argillites making up most of the Alpine Facies. It has been suggested that the piles of pillow lava built sea-mounts, on top of which chert and limestone could accumulate, out of reach of the streaming sediments on the main ocean floor. In the South Island they are more common in the eastern ranges of Canterbury and Marlborough than farther west; in the North Island, pillow lavas and jaspillite with grey tuffaceous greywacke are well displayed on the coast at Red Rocks Point about 5 km west of Ohiro Bay, Wellington, while coloured cherts are not uncommon in the axial ranges, on some islands in the Hauraki Gulf and in Northland.

It is difficult to appreciate the true magnitude and long duration of the volcanism along the south and west margin of the Geosyncline. Before the close of the Permian Period some tens of kilometres thickness of mostly mafic lavas and tuff had been erupted, to be followed during the Triassic and Jurassic Periods by repeated outbursts of andesitic and rhyolitic tephra making up a substantial part of the

next several kilometres above. The rest came largely from erosion of
the Permian volcanic accumulations on the foreland at first, but by the
beginning of the Jurassic Period these had been worn down to low
relief, and as there were fewer eruptions after that, and small ones, the
sediment in the marginal trough was to an increasing extent derived
from the granitic and metamorphic terrain beyond.

Igneous Activity during the Rangitata Orogeny

The history of geosynclinal volcanism just described has been inferred
from the deposits, for neither eruptive centres nor any kind of volcanic
edifices are now recognisable. Uplift and erosion while the Rangitata
Orogeny was in progress also minimised the chances of the structures
of any contemporary volcanoes being preserved. Sites where volcanic
products could accumulate and survive were restricted to eastern and
northern areas where oceanic troughs continued to subside. Evidence
of the kind of activity which took place during the Orogeny is therefore
indirect and incomplete.

At least some of the intrusive dykes and sills cutting through New
Zealand Geosyncline rocks should have reached the surface during the
Rangitata Orogeny, and some perhaps later. These intrusions range in
composition from dolerite (intrusive equivalent of basalt) through
porphyrite (intrusive equivalent of andesite) to quartz-porphyry
(intrusive equivalent of rhyolite) as well as more special varieties
including lamprophyre (like basalt or dolerite with a higher proportion
of darker minerals) and tinguaite (roughly, the intrusive equivalents of
keratophyre and phonolite, which are sodium-rich lava rocks related to
trachyte in many respects).

Porphyrite dykes intruding the Murihiku Supergroup rocks of
Southland have been assigned a Jurassic age, but as ages are revised, an
increasing proportion of intrusives are coming to be regarded as of
early or middle Cretaceous age. At or about the end of the Jurassic
Period, or early in the Cretaceous, an unusual kind of volcanic event
must have occurred to produce the peculiar Berlins Quartz Porphyry
of the Buller Gorge area. There has been much speculation as to how
this rock was formed. It contains abundant quartz phenocrysts in a
dense, often glassy matrix of rhyolite or dacite composition. Its
distribution is not like that of a sill or a dyke. Appearing from beneath
Brunner Coal Measures in the lower Buller Gorge, it extends
northwards in two main belts as far as Mount William. Within each belt
is a series of rounded masses up to 5 km across, and according to S.
Nathan, who has studied them extensively, the walls of each mass are
roughly vertical. Alongside the main highway about 4 km west of
Berlins Hotel the rock shows another of its highly distinctive features,
being literally crammed with angular fragments of the local Paleozoic
basement rock. This happens to be thermally-metamorphosed hornfels
derived from Greenland Group slate, but it is far from certain that the
alteration of slate to hornfels was due to heat when the
quartz-porphyry was intruded. It could have happened earlier. In some
places hornfels fragments are scattered through a fairly normal
quartz-porphyry; elsewhere the rhyolitic matrix is little more than a

film between hornfels fragments. Good examples of normal porphyry are not easily accessible, being on the north side of the Buller River, but plenty of boulders can be found on the river bed.

Among the various opinions that have been offered about this curious rock, perhaps the most attractive is that the quartz-porphyry masses were offshoots at a late stage in the history of the Paparoa Granite pluton, which encountered 'cold' rock not far below the surface at the time, and 'forcibly intruded it', to quote Nathan's words. It has even been suggested that the pluton itself broke through ('unroofed' is the usual expression) to the accompaniment of a massive Plinian-type explosion. This view gains some support from the presence of rhyolitic, glassy tuff in the immediately overlying Ohika Formation sediments. There are difficulties, however. For one, the amount of tuff so far identified does not seem to be enough to mark what would have been a colossal event. For another, the Ohika Formation has come to be regarded as younger than formerly, indeed younger than the quartz porphyry, but the question of the age of either is by no means settled. The unroofing hypothesis would fit in well enough with the inclusion of masses of hornfels fragments near the margin of the intrusion.

Mid-Cretaceous Volcanoes

From early in the Cretaceous through to the middle of that period was a time of widespread eruptions around New Zealand. While the upheavals of the Rangitata Orogeny were having their effects in the south and west, the geosynclinal kind of marine trough conditions were still firmly established, or had been renewed, in the far north, down the east coast of the North Island and across to Marlborough, accompanied by all the appropriate kinds of volcanic activity. Mainly mafic and ultramafic lava, pillow lava and agglomerate breccia were produced in great volumes, and later invaded by dykes and plugs of dolerite and gabbro and some keratophyre. Whether because the great piles of erupted material left little room, or because the supply was small, the associated sediments are said to make up no more than discontinuous lenses in among the volcanics.

It is not easy to picture what conditions were like in these troughs, but they must have been highly unstable and variable. It was not a favourable situation for accumulating evenly-bedded strata of wide extent, but rather one for irregular masses of rock and obscure stratigraphy. Over much of Northland the basement for younger Cretaceous and Tertiary sedimentary rocks is the Tangihua Volcanic Formation, erupted probably over quite a span of early or middle Cretaceous time. The same eruptive episode is represented in the East Cape district from Matakaoa Point to Cape Runaway and again south of Hicks Bay by a massive complex of basaltic rocks shown on the maps as 'Matakaoa Volcanics'. These have developed a landscape, soil colours and a vegetation rather distinct from the rest of that region. As far as can be judged from the rare fossils in the occasional sediment layer between the lavas, basalts began to be erupted here in the first part of the Cretaceous Period, but it is not easy to distinguish these from similar rocks south of Hicks Bay which are interlayered with late

Cretaceous and Oligocene sedimentary beds. So long a record of submarine volcanic eruptions must be connected with the long persistence of subsiding troughs and almost geosynclinal sedimentation in this region.

Mapped in the Awatere and Clarence valleys of Marlborough as the 'Gridiron Formation', mid-Cretaceous basalt lava piles from 200 to 650 m thick attracted interest because in at least one place it was certain that the eruptions had occurred on or very close to land, though the lavas soon afterwards had been covered by marine sediments similar to the Alpine Facies greywackes below them. It was a clear indication that a break in marine deposition did occur even in eastern districts during the Rangitata Orogeny. Indeed, thin coal beds and quartz sandstone under the Gridiron basalts were claimed by R. P. Suggate in 1958 to mark the very beginnnings of the Cretaceous-early Tertiary marine transgression over lands raised in the Orogeny.

Parts of the foothills ranges in mid-Canterbury are made up of andesite lava rocks and an unusual kind of rhyolite containing crystals of garnet. Being 'heavy' minerals, garnets can easily be separated from weathered rhyolite in a gold pan. Many cavities in the andesites have been filled with banded agate and crystalline quartz and amethyst to make geodes, much sought after by mineralogists and lapidarists. This belt of volcanic rocks forms prominent ridges and peaks from 1000 to 1700 metres high from Mount Misery near Whitecliffs across the Rakaia and Rangitata valleys to Ben McLeod. Rhyolite exposed under the Highway 72 bridge at Rakaia Gorge shows a fine, wavy layering attributed to the flowage of a stiff, viscous lava. Once regarded as Jurassic, the andesites and rhyolites are now mapped as Cretaceous. Though less strongly folded than the underlying Torlesse greywackes and the late Jurassic plant-beds, they must have been erupted, deeply eroded to produce coarse bouldery conglomerates, and then worn down to the level of the Late Cretaceous Peneplain, and all this *before*

Fig. 8.10
Rhyolite of probable Cretaceous age outcrops in the gorge of the Rakaia River where it is crossed by Highway 72. Close inspection of the rock shows a fine, wavy layering which has been ascribed to flow movements in a viscous lava. The rock contains tiny garnet crystals —an uncommon constituent in volcanic rocks.

Fig. 8.11
Dykes of basalt, some of them joining with sills, cut through Eocene and older rocks in the Broken River area of inland Canterbury. They were part of the feeder system to the submarine volcanoes which during the Oligocene Period produced pillow lavas and large amounts of stratified basaltic tuff containing marine fossils. This dyke crosses Broken River downstream from the Highway 73 bridge, cutting through greensand rock of Eocene or late Cretaceous age. It thins out a short distance behind the camera, and is cut off by a stream-eroded surface beneath gravels capping the terrace opposite. (Photo: R. Speight)

Fig. 8.12 (right)
The basaltic magma feeding mainly submarine volcanoes in the Oamaru district in early Tertiary times squeezed through still-wet sediments below the sea floor and cooled quickly to form dense, glassy dykes and small sills. This example is alongside the Kakanui River near Gemmells Crossing.

the late Cretaceous coal measures of Malvern Hills were deposited on top of them. Again, the timing is tight. The situation has not been helped by the radiometric dating of a sample of andesite from a drill core from beneath the Canterbury Plains—at 81 million years, that is, late Cretaceous. It would seem that we still have much to learn about the true tempo of geological events.

One more example of 'pre-peneplain' Cretaceous volcanism will be mentioned. While non-marine coal measures were accumulating in a trough on the site of the Paparoa Range, North Westland, basaltic volcanoes broke out near the eastern margin and contributed hundreds of metres of mudflow breccia and conglomerate. Only a few thin sheets of lava actually reached the trough and were interbedded with delta sands and gravels and coal beds of the Morgan Coal Measures. Only the eroded stump of the volcano remained when the Brunner Coal Measures were being laid down in advance of invading Eocene seas.

More Submarine Volcanoes in the South Island

Late Cretaceous and early Tertiary marine strata in Otago, Canterbury, Marlborough and Westland are interrupted at many localities by sills and dykes of basalt or dolerite. Usually they occur not far above the basal Quartzose Coal Measures. Proof that magma reached the sea floor is not always present, though pillowed forms with glassy margins suggest injection into sediment that was still wet and soft. Sediment was accumulating only slowly in the early stages of the marine invasion, and the sediments probably remained uncompacted and water-saturated for whole geological periods after deposition, so the basalt sheets and pillows could be significantly younger than the containing rocks.

While it cannot be proved that dolerite sills in Paleocene marine clays

Fig. 8.13
Undersea basaltic
eruptions at Oamaru
during the Eocene Period
produced these classic
examples of lava pillows,
exposed here on the shore a
short distance south of the
harbour breakwater. The
outer skins of dark,
basaltic glass are clearly
visible. Piling up amidst
sea-bottom ooze, the
pillows entrapped patches
of limey sediment
containing brachiopods
and other fossils.

at White Creek and View Hill near Oxford, Canterbury, and others
near Wharanui, Marlborough, actually supplied undersea volcanoes, a
hint that some did so comes from the presence of tuffaceous material
and bentonitic clay in the sediments. Clays of the bentonite group are
believed to result from alteration of finely divided volcanic tuff in a sea
water environment. More direct evidence for submarine eruptions is
found in South Westland between Paringa River mouth and Arnott
Point, where basalt lava, tuff and agglomerate beds are interbedded
with limestone and mudstone, also of Paleocene age. They have been
mapped as 'Arnott Volcanics'.

The case for contemporary eruptions is quite conclusive in eastern
Otago where at many places the Eocene and Oligocene marine strata
not only contain pillow lavas and water-quenched basaltic glass
('tachylite') breccias, but also well-bedded tuff layers with abundant
marine fossil shells. One of the finest and most accessible areas for
studying these phenomena is the coastal belt from Oamaru southwards
to Moeraki Peninsula, in beds mapped as 'Waiareka Volcanic
Formation' (Eocene) and 'Deborah Volcanic Formation' (Oligocene).
The most celebrated spot is Boatmans Harbour, a tiny bay just five
minutes walking time along the cliff-side path from Oamaru Harbour.
In this district there are several distinct volcanic piles that were built up
to contemporary wave-base and at times above, then to be cut down by
the waves to form flat-topped sea-mounts which afforded ideal sites for
mollusca, brachiopods, sea-urchins and bryozoa to live in abundance
and accumulate after death in the absence of land-derived sediment. In
this way, the purity of the Totara Limestone (the once-famous Oamaru
building-stone) and the McDonald Limestone (an excellent source of
agricultural and industrial lime) was assured. Basaltic breccia on the

Fig. 8.14
The cliffs around Cape Wanbrow at Oamaru expose a variety of kinds of stratified volcanic sediment, mostly the debris from undersea explosive eruptions during the Eocene and early Oligocene periods. In this outcrop they range in texture from very fine, laminated mudstone to coarse, blocky angular accumulations (agglomerate). The stratification in places shows evidence of contemporary scouring by currents and slumping of the deposits down the flanks of the volcanoes. Fossils are found in some of the sandy layers.

south side of Kakanui River mouth is attributed to undersea eruptions which brought up nodules of olivine and perfectly-shaped crystals of other minerals from deep levels in the lava feeding system. At Moeraki, a few kilometres farther south, basalt lava flows baked the underlying muds to a natural brick (porcellanite) and contain fragments of quartz broken from the underlying schist basement.

Basaltic undersea eruptions in Oligocene times left a striking record also in inland Canterbury—for example, the fossil-bearing basaltic tuffs alternating with limestone beds of the Thomas Formation at Castle Hill—and again farther north into Marlborough as the basalt lavas and breccias of the Cookson Formation. The upper part of the Matakaoa Volcanics of East Cape may represent the same episode, although, as we have seen, these are in a different setting.

Later Tertiary Land-based Volcanoes in the South Island

Professor W. N. Benson devoted more than half his lifetime to painstaking investigations of a very intricate volcanic history in the Dunedin district. The results were not published as a whole while he was alive, but his map, a monumental work in itself, has since been reproduced by the Geological Survey on a 1:50,000 scale as a memorial to him.

In Benson's view the Dunedin Volcano grew upon a land surface of but little relief which he called the 'Late Tertiary Peneplain'. This term was introduced in 1935 in the same paper as that in which he identified and named the Late Cretaceous Peneplain. It was supposed to be cut across earlier Tertiary and Cretaceous sediments, the Quartzose Coal Measures, the Late Cretaceous Peneplain and the schist underneath, all of which had previously been tilted, folded and faulted in mid-Tertiary times in the early stages of the Kaikoura Orogeny. The reality of a late Tertiary peneplain is more doubtful than that of the late Cretaceous one. It has been held by some that an erosion surface was carved across the rising alpine mass in late Tertiary times before the final Kaikoura upheaval, and it is true that late Tertiary volcanic rocks do rest on an erosion surface in the interior of Otago. However, a mid-Miocene trachyte tuff at Waipuna Bay containing marine fossils shows that the birth of the Dunedin Volcano took place under the sea.

The subsequent history of the volcano is usually presented in Benson's original terminology, but with modification of the timing he suggested in the light of more recent work which shows that the Dunedin Volcano was built up entirely within the Miocene Period. The following events are recognised:

(1) The First Eruptive Phase produced dominantly basalt lava flows and breccias, except to the west of Dunedin where the event is represented by greenish-grey phonolite lavas, containing the silica-poor sodic mineral nepheline as well as feldspar.

(2) Rocks of the First Phase were then eroded, and stream alluvium, lake deposits and peat of Miocene age accumulated in the valleys.

(3) The Second Main Eruptive Phase again produced chiefly basaltic lava flows and breccia beds, and phonolite in the north and west, including the extensive Logan Point and Waitati Phonolite flows.

(4) Basaltic lava eruptions of the Third Eruptive Phase brought to an end the main volcano-building story. Despite a minor mid-Miocene interruption, a large volcanic mountain had grown up which subsequent erosion has reduced almost, but not quite, to a volcanic skeleton.

The last phase, and perhaps earlier ones too, were echoed in inland Otago, though on a reduced scale. Eroded volcanic necks, dykes, agglomerates and tuffs with scraps of formerly extensive Miocene lava sheets survive on the uplands to the northwest and west of Dunedin as far as the northern slopes of the Kakanui Range. Mostly, these are mapped as 'Waipiata Basalt'. Variants from normal olivine basalt mainly involve sodium-enrichment and substitution of silica-impoverished 'feldspathoids' for feldspar; in short, what are termed 'alkaline' types.

Miocene volcanic rocks occur in inland Canterbury from the Rakaia River north to near Oxford. Harper Hills ridge near Hororata exposes the edges of basalt lava with breccia and tuff on its northern escarpment, and commercially workable bentonite clays are included in the sedimentary beds above the basalt on the gentler southern slopes. On the maps the lava appears as 'Harper Basalt' or 'Harper Hills Volcanics'. North of the Waimakariri River it reappears as isolated eroded remnants protruding through the Canterbury Plains at Burnt Hill and Starvation Hill near Oxford.

Banks Peninsula makes by far the most conspicuous, obviously volcanic feature in the South Island. Its broad dome-like aspect from the distance reminded early scientific explorers of the Hawaiian and Canary Islands volcanoes. The deep, central depressions ('calderas') now forming Lyttelton and Akaroa harbours have often been spoken of as 'craters', but essentially they are erosional features and not directly due to volcanic action. Canterbury's geological pioneer, Julius von Haast, established the structure of the Lyttelton volcano in 1861 when reporting on a scheme to drive a railway tunnel to connect Christchurch with Lyttelton—his careful and accurate cross-section through the Port Hills can still be seen in Canterbury Museum. Benson's counterpart in Canterbury, Robert Speight, published many papers on the Banks Peninsula double volcano in the first half of this century.

The Lyttelton volcano is based on rocks of a wide range of age from probably Triassic greywacke and chert and Cretaceous (or possibly Oligocene) andesite to mid-Miocene sandstone and rhyolite, which can be seen because the whole structure is now deeply dissected and eroded. Both the main lava domes were thought to have grown in the Pliocene Period, or possibly the early Pleistocene, but radiometric datings have confirmed an earlier beginning. Lyttelton dome was built up from late in the Miocene Period by repeated outpourings of andesite, trachyte and an odd kind of porphyritic basalt in which the phenocrysts are of plagioclase feldspar as well as augite, with minor amounts of tephra and thick rubbly interlayers between successive flows. Akaroa Volcano, built within the same period, eventually overwhelmed the eastern flank of its neighbour. Both domes were already deeply eroded before early Pliocene olivine basalt flows of the Diamond Harbour Lava poured down from somewhere on the upper flanks of Akaroa Volcano westwards into the central caldera of Lyttelton. The Summit Road around part of the perimeter of Lyttelton caldera exposes many dykes of trachyte, andesite and basalt which tend to be lined up radially with respect to a point near the head of the harbour, where a mesh of intersecting dykes may represent the main feeding system of the Lyttelton lavas. Likewise, gabbro and syenite on Onawe Peninsula at the head of Akaroa Harbour (scene of the last, hopeless resistance by local tribesmen against Te Rauparaha's marauders) may have crystallised at the end of the final eruption low down in the main feeding conduit. The two calderas were invaded by the sea in post-Glacial times.

North Island Mid-Tertiary Eruptions

The Miocene Period saw the beginning of a long chapter of volcanic events in the north distributed according to a pattern suggesting a relationship with the north-west structural trend of the Auckland Peninsula which lasted until the end of the Tertiary Era. The northern region was already substantially dry land by the time volcanism had resumed. Meanwhile, there had been a significant pulse of orogenic uplift, folding and faulting in the Oligocene Period, mild compared with the disturbances farther south at the height of the Kaikoura Orogeny but enough to strip late Cretaceous and early Tertiary strata from parts of Northland and to introduce unconformities at the base of the younger Tertiary sequences.

Miocene volcanic rocks in the north occur in two distinct belts. The

western one is traceable from the Marakopa Valley in south-west Auckland at intervals northwards to the Manukau Harbour, thence continuously as a vast pile of andesite lava flows, pillow lavas, agglomerate breccias and tuffs as far as Dargaville. A well known feature of the skyline in the northern Wairoa district is a conical hill called Tokatoka. Marking the neck of a long-extinct volcano, it is close to the northern end of the andesite belt.

Waipoua Kauri Forest is on part of a high, deeply dissected tableland between Kaipara and Hokianga harbours, reaching up to 600 m in altitude. This plateau is underlain by thick basaltic lavas which must have accumulated as a great lava lake, until recently believed to be of Pliocene age. By radiometric dating they are now known to be mid-Miocene, and most probably a later phase of the same eruptive cycle as that which produced andesites to the south. As they include pillow lavas and embody marine sediments, the western andesites were erupted under the sea, but by the time of the change to basaltic eruptions the North Auckland region was dry land. Apart from marginal flooding when sea level was high during the Pleistocene, most of it has been that way ever since.

Remnants of andesite volcanoes are small and scattered towards the southern end of the western belt, and probably reduced in area by erosion. They can be seen in the coastal cliffs about half way between Te Maika, south of Kawhia Heads, and Albatross Point. Marine strata

Fig. 8.15
Tokatoka, alongside the Wairoa River south of Dargaville, is a good example of a volcanic neck. The central plug of andesite lava which congealed in the throat of this Miocene volcano is now virtually all that remains. (Photo: N.Z. Geological Survey)

of similar age in North Taranaki contain andesitic tuff, so perhaps there were further eruptive centres south and seawards of Mokau River.

The main bulk of Miocene andesite volcanics lies to the north of Manukau Harbour. Though familiar to Auckland geologists for almost a century as 'Manukau Breccias', this term was not formally defined until about twenty years ago. Superb examples of pillow lava, including some very large specimens and one that is connected to a feeding dyke, can be seen along with breccias, flow-lavas and tuffs in the coastal cliffs between Whatipu and Muriwai. Detailed stratigraphic work by B. W. Hayward on these volcanic rocks and the related Waitemata Group of sediments has recently been published. A revised and rather complicated classification and naming of both volcanic and non-volcanic rocks is being developed, but the old name Manukau Breccias is likely to remain in use informally for some time.

The eastern belt of Miocene volcanic rocks appears from beneath younger igneous rocks at Otuhepe near Whakatane, and again near Te Puke and Tauranga. The main spread begins on the western side of the Kaimai Range near Matamata, broadens northwards to take up much of Coromandel Peninsula as far as Coromandel itself, and reappears on the southern end of Great Barrier Island. Isolated patches of similar andesites are recognised farther north as far as Whangaroa Harbour. Different names have been used in different areas. From Bay of Plenty to Great Barrier Island they are mapped as 'Beesons Island Volcanics', a name taken from a locality near Coromandel, in which region they were first described before the end of last century. Farther north, the name 'Wairakau Andesites' is used, for example, for rocks at Bream Head, the Hen Island and near Kaeo.

Both groups of the eastern belt rest unconformably upon the Mesozoic Waipapa Group and other sedimentary strata, or upon the Tangihua Volcanic Formation, and both differ from the western belt of rocks of similar age in various ways. Those of the eastern belt are more 'acidic', the andesites being accompanied by quartz-bearing types rather than by basalts as variants. They are intercalated with alluvial sands and coal layers in places and seem to have been erupted entirely on land, whereas at least the earlier western eruptions were under the sea. Again, the lower members of the eastern andesites have been greatly altered by metasomatic reactions with rising heated fluids. The date and cause of the alteration, referred to as 'propylitisation', has been in dispute for a long time. Chiefly at issue has been whether the propylitising fluids emanated from the same magma body as that which supplied the andesite lavas but at a late stage in its solidification, or whether from a later intrusion which fed the younger rhyolitic eruptions. The second alternative now seems to be favoured. Finally, the eastern andesites in the course of propylitisation are impregnated with minerals. The eastern Miocene andesites are the host rocks for the gold, silver, lead and other ores of the former Hauraki mining field.

Naturally there has been a good deal of speculation about the reason for these differences between two roughly parallel and contemporary volcanic belts. An ingenious suggestion by J. C. Schofield envisages a

Fig. 8.16
Hikurangi volcano, north of Whangarei, is a lava cone built of dacite. (Photo: N.Z. Geological Survey)

westerly-sloping slab-like body of magma at depth as having come into existence early in the Tertiary Period. Within it, because of the slope, the less mafic products of a long drawn out process of crystal fractionation tended to rise and be erupted from the eastern side, whereas the later eruptions from the western side were all from the basaltic fraction.

A younger series of late Miocene and early Pliocene rhyolite and dacite lava flows and plugs is recognised in Northland as the 'Parahaki Volcanics'. These seem to have erupted from a number of distinct centres near Maungaturoto, Whangarei, Kamo and Hikurangi. In the Hauraki region, too, younger rhyolite lavas above the Beesons Island Group are mapped under various local names, e.g. the 'Whitianga Group' in the Mercury Bay area. The rhyolite is usually glassy and 'spherulitic', that is, containing small spherical bodies resembling shot which, with the aid of a hand-lens, can usually be seen to consist of radiating glass or mineral fibres. As noted above, the mineralisation of the Miocene andesites may have coincided with the eruption of these later Miocene and Pliocene rhyolites.

Pleistocene Volcanic Trends

After eruptions had ceased at Banks Peninsula early in the Pliocene Period, no further outbreaks occurred in the South Island except in South Canterbury, where floods of basalt erupted near Geraldine and Timaru in early Pleistocene times. Subsequent up-tilting and erosion has removed all trace of their sources. Urban Timaru rises westwards on the surface of the flows, which are now covered only by a wind-borne silt mantle of loess. Solander Island to the south of Fiordland, is a remnant of an early Pleistocene andesitic volcano.

Although the character of volcanic activity in the far north did not change abruptly after the Tertiary Era, the influence of the north-westerly grain of Auckland Peninsula seems to disappear. Instead, a north-easterly orientation is suggested by the direction of the active Rotorua-Taupo Volcanic Zone. Egmont, which erupted repeatedly until within the last few hundred years, might perhaps seem to be on an extension of the western andesite belt were it not for the

apparently total quiescence of that belt for some twenty million years. It is more logical to associate Egmont with the central group of andesite-basalt volcanoes of Tongariro National Park, the history of which also really dates from the beginning of the Pleistocene Period.

The Auckland Basalt Eruptions

To return briefly to the far north, the tract of lower plateau country inland from the Bay of Islands is more obviously the remains of a basalt lava plain than the deeply eroded Waipoua uplands farther west. It was built up by voluminous outpourings of the Kerikeri Basalt early in the Pleistocene Period, and obliterated a pre-existing hilly topography. This is at the northern end of a string of basalt volcanoes located between the Miocene andesite belts. They erupted from time to time through the Pleistocene Period and almost up to the present day. Among the oldest are the larger volcanoes of the South Auckland district, including Pirongia and Karioi, dating from about the beginning of the Pleistocene, and smaller, somewhat younger basaltic cones of the Franklin district. The numerous small volcanoes of the Auckland Isthmus region represent a renewal of activity in a district that had been free of it since the time of the Manukau Breccias eruptions. This latest episode began less than 100,000 years ago, and culminated most recently in the building of the Rangitoto Island basalt lava cone within the last few hundred years. The eruptions produced lava flows of olivine basalt which flooded valleys previously eroded in Tertiary sandstones, blocked the drainage of others, and spilled over as minor flood-lava plains. Most are adorned by elegant cones of scoria which give the Auckland urban area its most distinctive geological character. Initial explosions produced shallow crater rings containing fragments of the underlying rocks as well as fine basaltic tuff, but in many cases these were buried beneath later products of the same or nearby eruptions. Examples still survive, however, as Panmure Basin, Orakei Basin and Shoal Bay. Readers are recommended to read E. J. Searle's excellent description of this volcanic terrain, so recently active, upon which Aucklanders have built a metropolis (*Auckland: City of Volcanoes*; Paul, Hamilton, 1968).

Ignimbrite—Product of Hot Sand Flows

Only after the city of St Pierre on the Antillean island of Martinique had been destroyed by a sudden avalanche of hot, dry volcanic ejecta blasted down the flanks of Mont Pelée volcano was it properly appreciated that the destructive force of an eruption is not necessarily directed up into the air. That was in 1902. Ten years later, though no one actually witnessed it (or lived to report it) the volcano Katmai on Kodiak Peninsula, Alaska, erupted violently, and at the same time a neighbouring valley was partially filled for several kilometres with sandy material up to 100 metres deep. The sand had arrived on the scene still hot enough to incinerate vegetation, and there were compelling arguments against its having fallen from the air. After cooling it compacted or agglutinated into a coherent, gritty rock of the composition of rhyolite. Water and other vapours distilled from the

Fig. 8.18
The small volcanoes of the Auckland Isthmus all emitted some kind of basalt. They vary in form because in some cases the main eruptive behaviour was lava outpouring while in others it was the building of a scoria cone above the vent. Some volcanoes are marked by broad, shallow craters, encircled by a low ridge of debris thrown up in an initial explosion as the lava broke through moisture-saturated rocks near the surface. (From E. J. Searle, 1964)

Fig. 8.19
An aerial shot of Rangitoto Island volcano, Auckland, showing the cluster of scoria cones on the summit of a broad basaltic lava cone which altogether give this volcano its distinctive profile. (Photo: N.Z. Geological Survey)

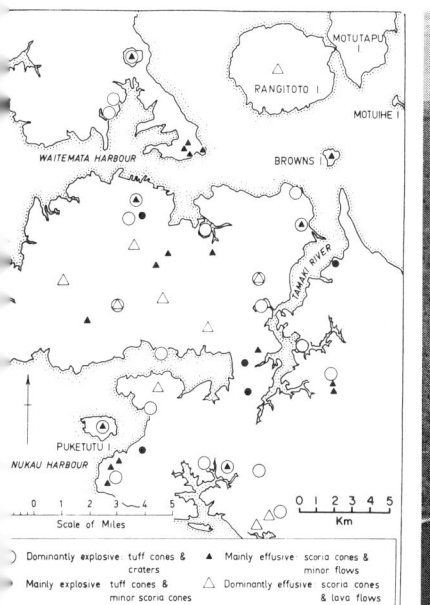

Dominantly explosive: tuff cones & craters   ▲ Mainly effusive: scoria cones & minor flows

Mainly explosive: tuff cones & minor scoria cones   △ Dominantly effusive: scoria cones & lava flows

Fig. 8.20
Horohoro Bluffs, in the
Rotorua district, the
eroded edge of a
plateau-forming sheet of
ignimbrite showing a col-
umnar effect due to the
presence of a system of
vertical joints. The colum-
nar structure, though also
ascribed to contraction of
the mass while cooling, is
notably less regular than
in the intrusive sheet of
basalt shown in Fig. 8.8.
(Photo: N.Z. Geological
Survey)

Fig. 8.21
Mayor Island in the Bay
of Plenty is a good exam-
ple of the humped form
typically shown by rhyolitic
volcanic mountains. It
reflects the relative
immobility of rhyolite lava,
which is usually emitted in
so viscous a state that it
cannot flow far from the
vent before cooling and
congealing. (C. A.
Cotton, 1922)

ground underneath and from its own cooling were emitted for many
months afterwards from countless fumarole vents on the sandy surface,
so the place came to be known as 'Valley of Ten Thousand Smokes'.
Having considered other possibilities, the American geologist C. N.
Fenner and his associates concluded that the material had been
discharged from a vent in the valley floor as a 'flow' of discrete
particles, solid or almost solid, and lubricated by films of hot gas. More
mobile than a heap of glass beads, the material flowed down gentle
slopes, and became more or less coherent after it came to rest. The
French geologist La Croix had already coined the term *nuée ardente*
('burning cloud') for the Mont Pelée hot-avalanche type of eruption,
and after Katmai the sand-flow was seen as a variant of the same
mechanism, both arising from explosive disintegration of acidic lava at
the point of extrusion.

The *nuée ardente* mechanism provided the answer to a puzzle in New
Zealand. We noted earlier that rhyolite, trachyte and sometimes

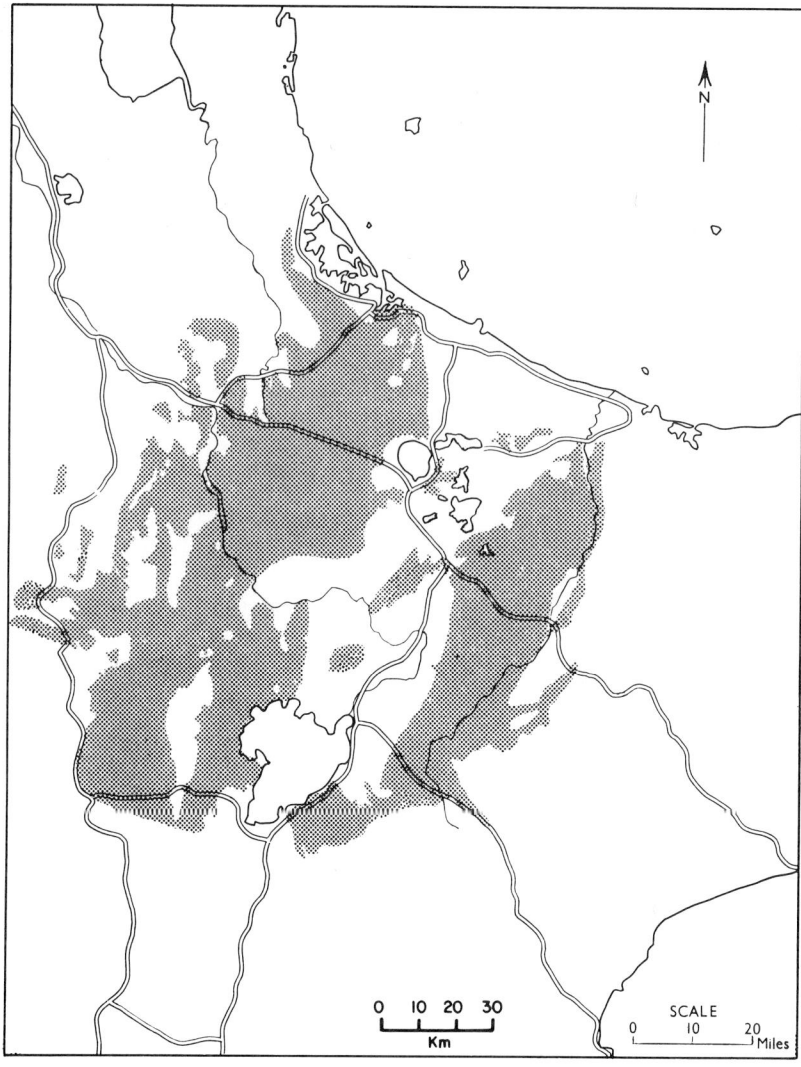

Fig. 8.22
The enormous extent of the ignimbrite sheets of rhyolite in the north-east of the North Island is evident from this map compiled by B. N. Thompson (1964).

SCALE

0 10 20 30
Km

0   10   20
Miles

andesite lavas are emitted in a highly viscous state, tending to heap up into steep-sided dome-like masses around the vent, or to reach a short distance away from it as a strongly convex flow, or perhaps merely to protrude a few tens of metres upwards as a jaggedly-rough plug called a 'tholoid'. Pumice, made of viscous lava frothed up by expanding gases, is another common product of such eruptions. The puzzle was this. If salic ('acidic') lavas are so viscous, how then to explain the occurrence of massive, flat-lying rhyolitic rock in sheets 100 metres thick making up plateaus over some 25,000 square kilometres of the central North Island?

P. Marshall recognised that the *nuée ardente* kind of eruption offered an explanation, and in 1935 he gave the name 'ignimbrite' to rock believed to have been formed in this way. Detailed mapping has since identified a considerable number of distinct ignimbrite units of different ages and distributions. It became essential to understand them thoroughly when a string of electricity-generating stations came

to be built along the Waikato River, their dams partly or wholly sited on ignimbrite.

Ignimbrite landscapes are distinctive. The broad arch of the Mamaku Plateau north-west of Rotorua and the even profile of the Kaingaroa Plain are essentially ignimbrite constructional surfaces, modified a little by later pumice showers and marginal gullying. The ignimbrites were emitted from many separate sources, now buried. They began to be erupted in the Rotorua-Taupo region before the end of the Pliocene Period, and the last of them could be less than 100,000 years old.

Volcanic Mudflows—'Lahars'

The coarse, rubbly material made up of andesite lumps in a sandy-muddy matrix which can be seen in roadside and railway cuttings around the south-western perimeter of Ruapehu was once mistaken for glacial moraines. It was in fact an important factor in James Park's case earlier in this century when he claimed that an ice cap had covered the centre of the North Island. P. Marshall, L. I. Grange and others showed that they were the same as deposits described in Indonesia by Dutch geologists, under the name 'lahar'. Their origin is as mudflows of volcanic debris generated by heavy rainfall upon recently deposited tephra, by outbreaks of water from crater lakes, or by contact of hot volcanic material with glacier ice and snow. The Tongariro group of volcanoes and Egmont all have extensive outer fringes of mudflow sheets with a characteristic hummocky surface and fields of curious, conical hillocks, seen for example at Opunake. Most readers will recall well enough that in December 1954 the worst train disaster in New Zealand resulted from

the demolition of a bridge by a lahar in the Whangaehu valley caused by the sudden escape of water from Ruapehu's crater lake.

Pumice and Tephra Eruptions

Soils over a large part of the North Island have been greatly affected by repeated showers of mostly fine-grained fragmental volcanic material. Most of it was thrown out by various volcanoes in the Rotorua, Taupo and Bay of Plenty areas and from Mount Egmont at different times through the Pleistocene Period and right up to the historic period. Within the past few decades there have been minor showers from Ruapehu and Ngauruhoe. Naturally they are most important in the recently active regions, but one particular shower (Aukaotere Ash) can be picked out easily in soils around Wellington. Minute pieces of volcanic glass are detectable microscopically in Marlborough soils and as far away as the Chatham Islands.

Approaching the main region of later volcanism, the effects change from minor traces of extraneous volcanic matter to the development of soils being entirely, and in places repeatedly, from surface mantlings of tephra. W. A. Pullar and other soil scientists have worked out the sequences of tephra layers and soils developed in them in different areas. The most notable volcanic event thus recorded was the sequence of Taupo Pumice Eruptions which occurred about A.D. 150. These probably originated from somewhere near or within the perimeter of Lake Taupo, and were responsible for the tens of metres of blocky, rubbly and granular rhyolite pumice exposed in cuttings along the Taupo-Napier highway. As far away as Gisborne and Napier, the Taupo Pumice is easily recognisable as a whitish-grey mantling of the

Fig. 8.25
The Tongariro volcano group viewed from the north. Nearest to the camera is Tongariro itself, then the active cone of Ngauruhoe and the massive bulk of Ruapehu in the rear. An explosion crater is conspicuous on the filled-in caldera of North Crater and a fault scarp breaks the western slope of Tongariro on the right. The steaming springs of Ketetahi can be seen in a gully on the near slope. (Photo: N.Z. Geological Survey)

soils up to 20 cm thick. Obviously thicknesses were greatly affected by the direction and strength of the wind during the eruptions. Southerlies must have prevailed during a rhyolitic blast from the long-active Okataina eruptive centre about 40,000 years ago, which is recorded by ash in soils around Auckland City.

Tephras and tephra-derived soils have great agricultural importance. Not only do they affect the water-retaining properties and erosion-resistance of soils, but also their chemical character. An early success of soil science in this country was the discovery that a form of cattle sickness common in the central North Island was in fact a deficiency disease due to lack of particular elements, especially cobalt, even as minute traces in some volcanic soils.

The Tongariro Volcanoes

Travellers crossing the middle of the North Island, given reasonable weather, never fail to be impressed by its crowning glory, that group of volcanic mountains centred in Tongariro National Park. Rising from a broad plateau between the Kaimanawa Range to the east and the Hauhungaroa Range west of Lake Taupo, the group is dominated by one or more of the three highest, Ruapehu, Ngauruhoe and Tongariro, but there are also six more andesite and basalt mountains in excess of, or approaching, 1000 m in height which belong to the same group.*

An obvious alignment of the main centres of recent activity shows that the group as a whole is related in origin to the same system of faults as that defining the Rotorua-Taupo Volcanic Zone north-eastwards to the Bay of Plenty shore and beyond. Next in importance after the three highest come Pihanga and Kakaramea at the south end of Lake Taupo, and Hauhungatahi to the west of Ruapehu, although the last is merely a basaltic volcano capping a mountain of late Tertiary strata. The central summit cones attract the eye, but in fact the vast peripheral pediments of lahar deposits and alluvium make up a substantial part of the total bulk.

The youngest sedimentary rocks of surrounding regions contain evidence, in the form of andesitic tuff, that andesite and rhyolite eruptions began to build the Tongariro volcanoes at about the beginning of the Pleistocene Period. The final upbuilding of the present cones has taken place since the late Pleistocene. It is still doubtful whether Ruapehu had, by the beginning of the last glacial phase of the Pleistocene Period, grown high enough to be above the level of the snow line, lowered though this would then have been. Tongariro almost certainly had not. Both achieved their full growth within the last few ten-thousands of years, and most recently Tongariro has added Ngauruhoe, a composite cone of lava-flows and fragmental

---

* Confusion over the naming of the three highest volcanoes and their individual peaks arose initially from inconsistent Maori appellations recorded by the earliest white explorers. Though not originally applied this way, 'Ruapehu' now attaches to the whole mountain rather than to any one of its peaks; 'Tongariro' to the whole mountain and also to its northern summit. 'Ngauruhoe' is the younger cone built up from the southern end of Tongariro and now exceeding it in height.

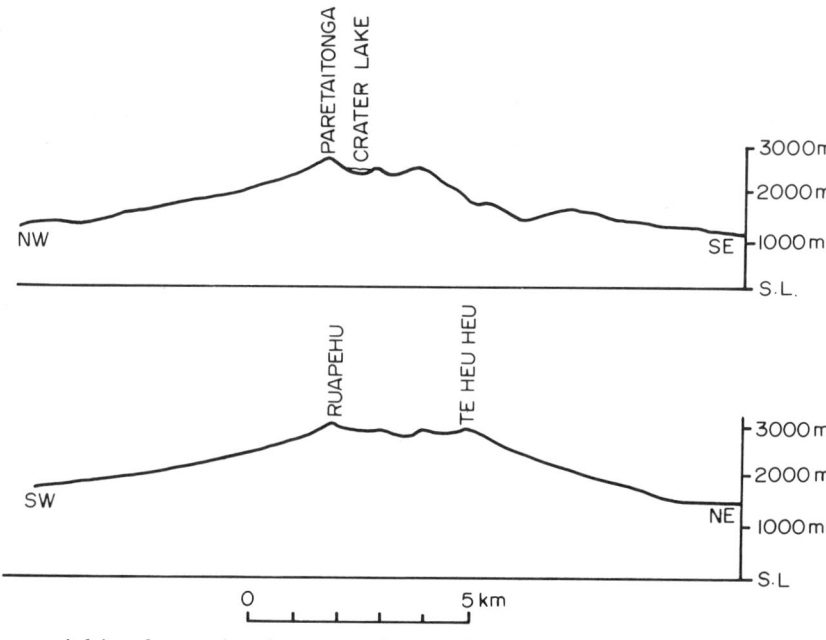

Fig. 8.26
Two characteristic profiles
of Ruapehu from the
Ohakune side (top) and
the Waiouru side. The
upper slopes reflect vol-
canic processes of lava
flow, ash fall and dry, hot
debris sliding; the lower
slopes of mudflows and
alluvial fan-building.

material in alternating layers, at its southern end. The oldest volcanic
rocks of the group are of basalt with some dome-forming rhyolite,
dating from early in the Pleistocene Period. The main bulk of the
mountains is of andesite, but some olivine-bearing varieties have been
called basalt in the past.

The Ruapehu profile, its concave slopes steepening upwards
majestically to an irregular ring of summits up to 2800 m in altitude,
enclosing an elliptical area roughly 2.5 by 3.2 km across, invites
speculation that the mountain once rose perhaps 1000 m higher. The
breadth of the summit is partly due to the building of the mountain by
outpourings from more than one central vent, yet for ninety years it
has been suggested repeatedly that the top of the mountain has been
cut off. Whether this truncation (if true) was the result of a great
destructive explosion, or of an internal collapse at the end of the
building-up period, or a combination of both, is far from clear. Not
enough older rock debris is lying around the mountain to give full
support to the 'mighty blast' theory. The more attractive hypothesis is
that of a summit collapse-caldera filled up subsequently by recent
eruptions from several smaller vent craters within it. The Tongariro
story is probably similar.

All three of the largest volcanoes have been active continually
through the historic period of Maori and European occupation of the
region. Ruapehu has erupted steam and tephra more than thirty times
since 1861 and lava has appeared on at least two occasions. Ngauruhoe
has had more than sixty tephra eruptions, with lava at least twice. The
activity of the several Tongariro craters has been mainly steam, but
tephra has been reported from Te Maari and Red Crater, and lava may
have been present in Red Crater within the last century or so.

Most of the Ruapehu eruptions of the past one thousand years must
have come from one or other of the summit craters, though a few are

Fig. 8.27
Although Ngauruhoe
probably emitted lava in
1870 (reports are some-
what contradictory) there
is no doubt that the series
of eruptions from May
1954 to March 1955 pro-
duced the most voluminous
lava flows yet witnessed by
pakeha. Gregg estimated
that about 6,000,000
cubic metres of lava were
emitted. (D. R. Gregg,
1940)

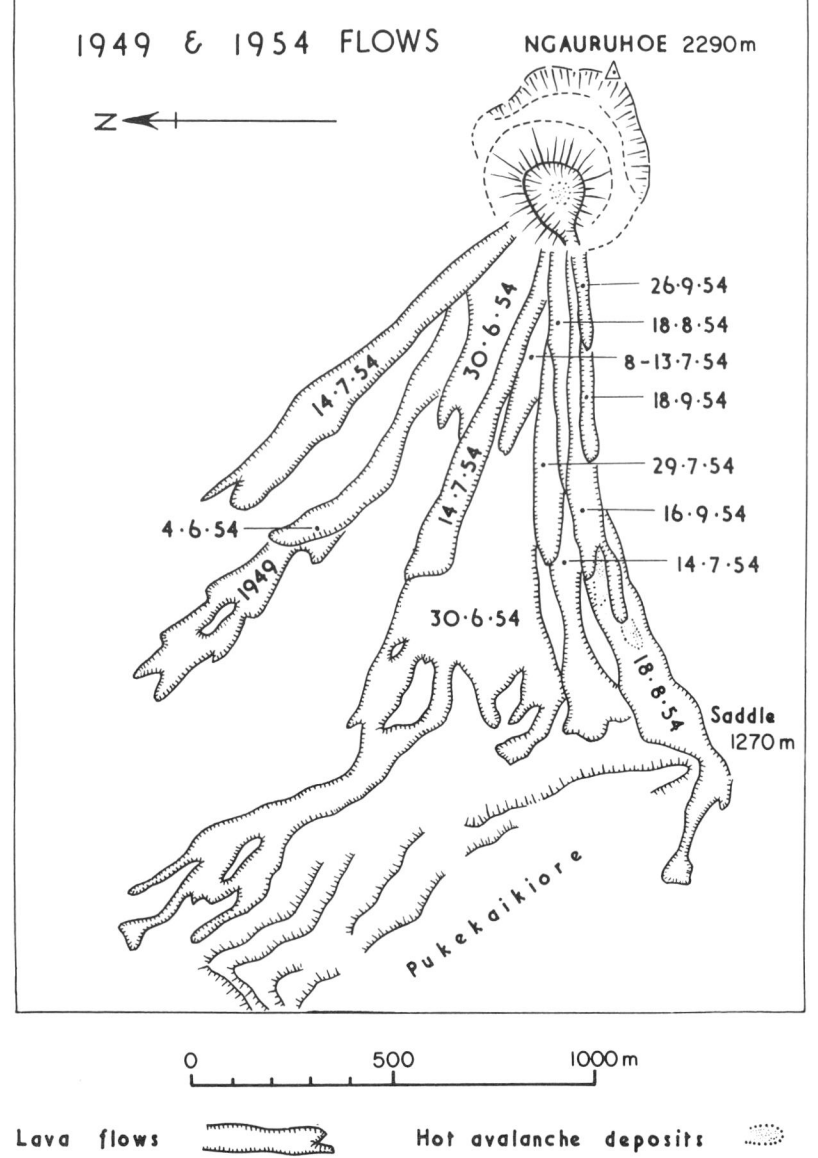

known to have broken out on the outer slopes. Since the 1861 outburst the summit crater lake has spent part of the time frozen, part of it in a steaming or boiling state, and part of it as open water but cool. Several times it has been partially or entirely emptied of water, either as an accompaniment to steam and mud eruptions, or by leakage through marginal ice-caves leading to floods in the Whangaehu River. Active lava in the form of a tholoid or 'plug-dome' was seen in the crater for the first time in March 1945. Three months later the growing tholoid had completely replaced the crater lake. Eruptions then increased in vigour, emitting solid debris and vapour until a large ash explosion in August initiated a decline. Tephra eruptions continued until the end of the year, and the lake began to fill again. Since then there have been steam and tephra eruptions every few years.

Ngauruhoe is the most continuously active New Zealand volcano, emitting steam plumes and darker clouds at frequent intervals. Lava has been present in the crater several times in the past century. Tholoid domes built up in 1949 and again in 1954, when viscous andesite lava overtopped the north-western crater rim and flowed sluggishly, to the accompaniment of much avalanching, down the outer slopes. In August 1954, four months after the start of this eruption, the main lobe of the flow had reached the base of the main cone, and then after more explosive activity the tholoid collapsed below the crater floor. These eruptions, the first known to have produced a lava flow in New Zealand in historic times, were over by 1955. The most notable recent eruption occurred in late 1974 and early 1975, when incandescent lava blocks and finer debris were ejected. Overtopping the crater rim, the material descended the outer slopes in fiery avalanches which possessed much greater mobility than that of the viscous andesitic lava itself. But perhaps Ngauruhoe's finest achievement to date was in 1949, when it timed its first major eruption involving lava in eighty years to begin just as a party of visiting geologists, in New Zealand for the Seventh Pacific Science Congress, happened to be passing on their way to the Chateau Tongariro. One of the visitors later published an account of the eruption in the magazine *Scientific Monthly*.*

Tarawera, 1886

Minor tephra eruptions have been witnessed in New Zealand by Europeans since 1839, and no doubt the Maori people previously had seen many more emanating from the Tongariro volcanoes, but neither before that date nor since has man observed in this country a volcanic event approaching the scale of the Tarawera eruption of 10 June 1886.

Strictly, the name Tarawera applied to only one part of the summit of this 1100-m mountain; there were other names (Wahanga, Ruawahia) for other prominences but no name for the mountain as a whole. It was built up mainly as two adjacent rhyolite domes and associated pumice beds, all resting on a foundation of Pleistocene ignimbrite.

Although many stories have been told about premonitory symptoms, many of them are contradictory, according to L. I. Grange, who in 1937 compared various contemporary and later accounts. Variations in the vigour of hot springs and geysers in the area were quite usual, but naturally enough any increases that had been noted prior to the eruption were afterwards seen as significant. Only one event is accepted by Grange as probably connected with the impending catastrophe. This was the occurrence of 'seiches', i.e. waves or surges, on lakes Tarawera and Rerewhakaitu a few days before.

Preceded for about an hour by strong, local earthquakes, the main

* The foregoing account of the Tongariro group of volcanoes has been summarised mainly from 'The Geology of Tongariro Subdivision' by D. R. Gregg, *N. Z. Geological Survey Bulletin 40* (1960).

eruption began with a great explosion and a flash of brightness at 1.30 a.m. An hour later, a fissure had opened up in a NE-SW direction across the mountain, and was active along its full length, throwing out white incandescent rocks under an enormous cloud of tephra and steam. By 3.30 a.m. the fissure extended southwards for fifteen kilometres and was erupting simultaneously from many points. The former Lake Rotomahana blew out at about this time, taking with it the famous Pink and White Terraces and covering the surrounding region with mud and rocks. By 10 a.m. of the same morning the eruption was nearly over except at the western end of the rift. By midnight all was calm apart from the innumerable new steam vents around the much-modified and lowered Rotomahana.

The main eruption thus had lasted less than twenty-four hours. It had left a line of craters along the full 15 km of the 'Tarawera Rift', and had buried the neighbouring region beneath mud and tephra up to 45 m thick. Vegetation was destroyed, the landscape considerably modified, and an estimated 102 people had died. Geologically, the eruption was interesting, firstly because it is unusual for violently explosive activity to occur simultaneously along such a long, straight

Fig. 8.28 (opposite) Hot debris-avalanche flows reached the foot of the Ngauruhoe cone during the 1974–5 eruption. (Photo: N.Z. Geological Survey)

Fig. 8.29 (above) The toe of the main hot-avalanche flow from the 1974–5 Ngauruhoe eruption. (Photo: N.Z. Geological Survey)

Fig. 8.30
An aerial view of a linear crater, the so-called 'chasm' across the broad crest of Tarawera rhyolitic dome, from which incandescent lava blocks and ash were discharged with great force for several hours on the night of 10 June 1886. The cliffs on the near side mark the limits of an older rhyolitic lava flow. Lake Tarawera lies beyond. (Photo: N.Z. Geological Survey)

line; secondly, because the fresh magma produced was basaltic, emitted from a fissure cutting across an older rhyolite volcano; and thirdly, because although basaltic it yielded explosive products only, and no lava.

Mud eruptions occurred from time to time afterwards at the southern end of the rift. The best known was Waimangu 'Geyser', strictly a recurring mud eruption rather than a true geyser. It threw out mud and stones and a lofty vapour column at intervals of about thirty-six hours between 1900 and 1904, then became erratic and expired with a locally disastrous eruption in 1917.

In 1886 the population around Tarawera was meagre. That is not the case today, and vital industries have developed in the region. While there is no particular reason why a similar blast should happen again at the same place, the whole district has an unbroken history of repeated volcanic eruptions since the remote past. Human occupation of it must therefore always be at some risk, hopefully small but duly reckoned with.

White Island Volcano

Apparently on the north-easterly continuation of the Taupo-Rotorua
Volcanic Zone into Bay of Plenty, White Island is merely the tip of a
sizeable andesite volcano rising from the ocean floor. Although Captain
Cook reported no activity when he discovered the island in 1776, it is
seldom without a plume of vapour, emitted from a group of
recurrently very active fumaroles in the main crater. The gases
sometimes issue at temperatures as high as 900°C. Tephra eruptions
occur from time to time, mainly fine ash being erupted with steam
clouds rising to more than 1000 metres. Experiments by R. H. Clark
and others suggest that the crater floor levels tilt and bulge during the
build-up period before an eruption. The fumaroles deposit alum,
gypsum and sulphur crystals around their vents. Attempts to exploit
the sulphur deposits in the early part of this century came to an end in
1914 when a landslide from the crater wall and the resulting mudflow
overwhelmed the operations and killed a number of men.

The latest important eruption began in December 1976. It is
interesting not only because a new outbreak was expected on the basis
of evidence from ground deformation, changes in the local magnetic
field and in the temperature and composition of the fumarole gases,
but also because large plastic or semi-molten lava blocks have been
emitted as well as smaller pumiceous fragments and finer ash. The
eruption continued through 1977 and activity was still at a high level
early in 1978.

Chapter Nine

GLACIATION

The Ice Ages in New Zealand

Land of Ice and Fire

In Iceland there lives a remarkable man, geologist by profession, who is deeply interested in the ethnic and cultural history of the Norse people—a poet, too, and a fine singer of the traditional songs of his homeland. His name is Sigidur Thorarinsson, and some readers no doubt will have seen his dramatic pictorial books and documentary films about Icelandic volcanoes in a glacier setting. It was he who coined the appropriate phrase 'Land of Ice and Fire' as the title of one of his books, though visitors to that fascinating region may be surprised to find how little of Iceland is under permanent ice and how much is at least as green as northern Scotland. It is, in fact, a far greener land than is barren, ice-bound Greenland.

Thorarinsson's term almost sums up the Pleistocene and later history of New Zealand as well. We saw in Chapter Eight that volcanic happenings dominated the Pleistocene history of the North Island, and this chapter will describe expansions of glacier ice over much of the South Island and small North Island areas as well. The one important aspect of New Zealand Pleistocene history not embodied in the phrase 'land of ice and fire' is the Pleistocene upheaval of our mountain areas.

What is an 'Ice Age'?

In some minds these words may convey pictures of hairy elephants ('woolly mammoths', if you prefer), overwhelmed by the menace of advancing ice sheets in northern Asia, freezing to death as they stood, petrified apparently by fear, their bodily flesh then to be preserved in a kind of natural deep-freezer until modern times. The notion is absurd, of course, and for several reasons. For one thing, warm-blooded animals are today enduring, in sub-polar regions, frigid conditions that are probably no less severe than those in the Pleistocene Period, in which mammoths and other large creatures were nevertheless able to sustain their massive forms on vegetable diets and reproduce their kind. The idea of sudden extinction has timing problems as well. Exactly how long an ice age would take to set in is still a good question, but the extreme minimum seems to be centuries, so presumably the entire populations of large animals whose frozen remains in Siberian soils seem to pose such a puzzle had the time and opportunity to migrate—and many did. Not a great deal of the Siberian land mass went under glacier ice at any stage, and it would have been increasing inaccessibility of usual food supplies that mainly extinguished those populations.

The most important factor was that the ground was permanently frozen, except where it thawed to a shallow depth in the summer. The preservation of frozen corpses of species of animals now extinct was a fortuitous thing, yet easily enough visualised in the widespread, permanently frozen soils and subsoils of the higher latitudes today, where the possibility still exists for accidental preservation of animal corpses by freezing.

The concept of an 'ice age' is of course really much broader than is popularly understood. With regard to certain regions it meant a period of extended glaciers and ice caps; in others it meant a time of prolonged sub-freezing temperatures leading to permanently frozen subsoil, accumulation of solid ice masses in the subsoil, and important changes in the distribution of plant and animal species. The higher latitudes are experiencing an ice age at present. From the geological standpoint, the definition of an 'ice-age' means that in previously temperate regions an appreciable lowering of the mean annual temperature continued long enough to leave its environmental mark in the geological record.

Most of this chapter will be dealing with the comings and goings of our glaciers during the Pleistocene ice age. It will look also at the evidence for frigid conditions in some regions never reached by glaciers, and at other events associated with climatic fluctuations, including rise and fall of sea level.

Does 'glaciation' adequately cover matters to do with expanding and contracting glaciers and ice caps? Purists have tried more than once to sharpen the meaning of the word by limiting it to erosional and other effects of glacier ice upon the country over which it flows, as distinct from the act of invasion of territory by glacier ice (which is 'glacierisation'). The distinction found few supporters, and I shall be using 'glaciation' in the more common way which embraces both glacier advances (looked upon as geological events) and the distinctive deposits and landscape features from which past glacial events have been inferred.

From the discussion of the terms 'Pleistocene' and 'Quaternary' in Chapter Seven it will be apparent that these have been mixed up in varying ways with the notions of 'Ice Age' and 'Glacial Period', but that there is no longer a close linkage between geological time terms and the occurrence of glaciation.

The Great Glacial Controversy

Few people nowadays have not heard something about former spreads of ice over lands now clothed in vegetation and inhabited by man. Even fewer seem to find such an idea incredible, yet it originally encountered much opposition. Serious suggestions about prehistoric episodes of severe cold date from well back in the eighteenth century, and the concept of a vast expansion of glaciers over northern and central Europe and North America was advocated vigorously from about 1830 onwards. Resistance arose largely from the fact that the new theory would supplant one which depicted the submergence of lands beneath iceberg-bearing seas, and which therefore appealed to religious fundamentalists as a manifestation of the Flood. It seems strange today that thirty years were to pass before the 'Glacial Theory' was generally accepted and made respectable. There has been no rational opposition to it since early in the present century.

It has been estimated that the greatest extent of ice cover on land during the Pleistocene Period amounted to 44 million square kilometres (some would put it a little higher), while the present extent of ice, not counting the sea-borne ice shelves around Antarctica, totals nearly 15 million, that is, about one-tenth of the land area of the earth. According to R. F. Flint, ice at the maximum covered 29 per cent of the present land area, but it must be remembered that the total land area would then have been greater because sea level was perhaps as much as 130 m lower than now owing to the abstraction of an extra 47 million cubic kilometres of water to make the extended glaciers and ice sheets.

Estimates of the temperature drop range from a few degrees up to ten degrees Celsius, and undoubtedly there was a wide variation across the latitude belts and in different climatic situations in each land mass. The maximum depression suggested for New Zealand is six or seven degrees, but some consider this an excessive estimate.

Ice Ages in the Remote Past

The classical Glacial Theory was concerned with events of the latest geological periods. Signs of more ancient glaciations are unmistakable in sedimentary rocks of late Paleozoic and Precambrian ages in every continent, and have been claimed for other periods as well.

No single type of sedimentary deposit on its own can be identified with complete confidence as a product of ancient glaciation. Most commonly cited as evidence is 'till' (also formerly called 'boulder-clay'), composed of stones of all sizes and shapes, angular and rounded, some showing surface scratches ('striae'), all set in a matrix of silt, sand or clay and generally compacted. Different sizes and shapes of stones are irregularly distributed, and stratification is not usually obvious. The decay of glaciers nearly everywhere in recent decades has given opportunities for studying how deposits of till formed from debris previously carried by the ice. Mudflows and other mass-movements of

Fig. 9.1 (left)
Origin of glacial till (1). The advancing snout of Fox Glacier in 1965, as it was thrusting over recently formed till made up of debris recently released from the base of the glacier by melting.

Fig. 9.2 (right)
Origin of glacial till (2). A close-up, showing the glacier snout overriding still-wet, uncompacted, muddy till which is being squeezed out from under the sole of the glacier.

Fig. 9.3
Transporting power of a glacier (1). When this picture was taken about 1941 the Fox Glacier had already transported an enormous mass of rock debris a distance of several hundred metres from where it had fallen on to the ice from the steep rock face on the right. In the succeeding years the glacier snout receded and the debris was dumped, most of it being carried away by the Fox River.

Fig. 9.4
Transporting power of a glacier (2). This enormous block of schist, measuring 15 m by 12 m by 9 m, appeared while glacial gravels were being sluiced for gold near Kumara about 1880. It is known as the Londonderry Stone, from the name of the sluicing claim. Parallel scratch-marks (glacial striae) can be seen on the surface. (Photo given to the author by the late Mr W. F. Heinz; photographer unknown)

waste may produce similar-looking deposits, so it is necessary to seek other distinctive deposits of the near-glacier environment, including the laminated silts which form in ice-dammed lakes and usually contain scattered pebbles dropped from floating ice blocks, and outwash gravels deposited by glacier-fed rivers. The latter contain a wide range of pebble and boulder sizes with much fine silt in the interstices.

Till compacted into solid rock is 'tillite'. When found together with laminated, pebbly silt-rock and thick conglomerate beds, a glacial origin is probable, and all the more certain if the strata rest unconformably on a worn, grooved and striated surface of more ancient rock. Only then is it really safe to infer that cold, glacial conditions existed when the beds were formed. Only exceptionally have moraine mounds and other surface features survived from pre-Pleistocene glaciations.

Late Paleozoic rocks in Australia contain undoubted evidence of an

Fig. 9.5
A close view of typical glacial gravel, showing the usual scattering of larger cobbles. The uppermost metre or two of gravel beneath a wavy surface is believed to be till under a capping of loess which may have accumulated during later ice advances. The lower portion is typical of outwash gravel. The locality is on a high terrace above the Midland Railway line near Avoca, and the deposit marks the first phase of the Blackwater Advance (Otira Glaciation) by the Waimakariri Glacier system (pp. 258–9).

Fig. 9.6
The oldest known glacial deposits in New Zealand are those due to the early Pleistocene Ross Glaciation. Tilted and folded during the Kaikoura Orogeny, they formed part of the 'bottom' for gold-sluicing in Jones Creek, a small stream behind the township of Ross. On the left is till made of mainly angular fragments of greywacke and hornfels; granite and schist fragments are more common at other outcrops. On the right, steeply-dipping laminated lake silts contain scattered pebbles. The silts are believed to have accumulated in a body of still, fresh water alongside (perhaps underneath) the glacier in its declining stages. The pebbles could have been brought into the lake on or in floating ice blocks—a common enough occurrence in ice-margin lakes today.

ice-sheet which rested at least partly on land, but in New Zealand the rocks of equivalent age are almost entirely of marine origin and contain nothing to indicate ice near at hand. The earliest glacial event recorded here was the Ross Glaciation, at about the boundary between the Pliocene and Pleistocene Periods.

How Many Ice Advances?

At first the Glacial Period was thought of as a single event, but by the end of the nineteenth century it was accepted in Europe and North America that sequences of glacial deposits separated by buried soils or zones of weathering indicated that there had been not merely one period of ice advance but several. Four distinct glaciations were recognised on both sides of the Atlantic.

When it had been established that older till and outwash deposits had been to some degree weathered and eroded before the next younger set were formed, and that similar sequences could be found in separate areas, there could be little doubt about intervening periods of substantially warmer climate. Confirmation of this came when the plant and animal remains in peats formed during the intervals were studied. For a long time it remained purely a tacit assumption that the episodes of glacier advance and retreat in different areas were synchronous, reflecting a succession of world-wide climatic oscillations. In mountain regions deposits from successive ice advances tend to form flights of terraces, and it was on this basis that the chronology of alternating glaciations and non-glacial ('interglacial') intervals was first established early in this century in the western European Alps by A. Penck and E. Brückner, extended to North America, and eventually applied in other lands as well.

On the lowlands of Northern Europe and the mid-western prairies of North America, where glaciations had been of the nature of spreading ice sheets rather than separate, valley-confined ice tongues, the chronology of events was worked out from sequences of till and associated sediments with intervening fossil soils or weathering zones signifying warm interludes. Often these were in clear stratigraphic succession, but there were difficulties in linking lowland with alpine chronologies, let alone those of one country with another. Even where the same number of glaciations seemed to have occurred in different areas, this did not mean that sequences could be matched event for event from one region to another, and still less was it justifiable to correlate sequences across the oceans. Yet the original four Alpine glaciations as named by Penck and Brückner were enthusiastically recognised around the world. It seemed all too easy to find four glaciations.

Pleistocene deposits on land sometimes yield fossil remains of vertebrate animals but morainic deposits, till and outwash gravels are not particularly favourable situations for their survival. Fossils are more abundant in the lake and stream deposits formed in warm intervals. They include mammalian bones, insects, freshwater molluscan shells and plants, but the normal uses of fossils for correlation of deposits and age-determination are restricted. This is partly because of the shortness of time-intervals between events, too short to allow for evolutionary changes that could be detected.

In recent years more information has become available about the climatic history of the Pleistocene from the study of microfossils extracted from sea-bottom drill cores, and it is now possible to attach dates to some of the temperature fluctuations thus indicated. A picture of world-wide climate oscillations through the Pleistocene is emerging, but, alas, it is far from being a straightforward picture against which the climatic evidence on land can easily be matched, and there is room for differences of opinion. Nevertheless it remains highly probable that the major climate oscillations which are the most distinctive characteristic of the Pleistocene Period were synchronous around the world, even though the shorter and weaker fluctuations seem to be harder to match from region to region as we learn more about them.

There are problems in applying the usual methods of

age-determination to the Pleistocene. As noted above, for most groups
of organisms found fossilised the period concerned is rather short for
methods based on evolutionary successions, and the ranges of the
so-called 'radiometric methods' are not very appropriate either. The
one which depends upon assaying relative amounts of the radioactive
isotopes of carbon (referred to as 'the radiocarbon method' or just '14
C') is available for samples of suitable carbonaceous materials less than
about 45,000 years old, and with special techniques but decreasing
precision back to about 70,000 years. It therefore does not take us very
far back into Pleistocene time. Sometimes there are errors, due to
contamination of the original carbon of the once-living tissue in the
sample, which are detectable usually only by inconsistent results, and
which are sometimes impossible to overcome. Where volcanic lava and
tephra are present, radiometric methods which range from the past
into early Pleistocene times can be used, and completely new methods
now being evolved hold promise of spanning the rest of the period. But
of course the deposits cannot be dated by any method if they do not
contain the necessary kinds of materials.

Swinging Sea Levels

The above heading was suggested by the title of a book by Reginald A.
Daly, an influential American writer and teacher of geology in the
inter-war period. The phrase has a ring about it to remind us that sea
level fluctuated throughout the Pleistocene Period in sympathy with
alternate glaciations and deglaciations of the land masses, by amounts
of the order of 100 metres. When ice covered the land thickly and
extensively, sea level was down; when the ice melted again, sea level
rose. It came up finally, by about 130 m, between 10,000 and 4700
years ago while the last of the great Pleistocene ice sheets were
disappearing.

Changes in the relative levels of land and sea ascribed to rising and
falling water level rather than to vertical movements of the crust affect
the whole world almost alike and synchronously; they are called
'eustatic' changes. Sea level shifts due essentially to expanding and
contracting glaciers and ice sheets are called 'glacioeustatic' fluctuations.

Signs of both higher and lower stands of sea level (relative to the
land) during the Pleistocene Period are to be found on the coasts of
regions where significant crustal movements are unlikely to have
occurred. In these cases the higher and lower strand-lines are accepted
as marking former eustatic ocean levels, and the higher ones related to
intervals of warm climate. Beginning in the Mediterranean region
more than forty years ago, a scheme for classifying raised beaches,
wave-eroded rock platforms and sea cliffs purely on the basis of their
vertical distances from present sea level and tying them into the Alpine
succession of glaciations and interglacials was soon being applied
around the world. Many anomalies were later found as the fossils in the
Mediterranean terrace deposits were studied more closely and as more
detailed knowledge accumulated about the glacial events inland. By
1950 confidence was waning in correlations of interglacial events and
deposits from one land to another based merely on heights of raised

shoreline features. It was no longer possible to ignore the fact that the Mediterranean basin itself has by no means been free from crustal disturbances. Yet although intercontinental correlations based on this kind of evidence alone are unacceptable, the whole principle of glacio-eustatism remains valid and, in places where vertical crustal movements have not interfered, it has been amply confirmed that deposits on coastal terraces were formed during times of warmer climate.

Glaciations, Interglacials, Stadials, etc.

When the Glacial Theory was young, it was just a matter of the 'Ice Age' or the 'Glacial Period'. Now there is such a mass of information about Pleistocene climatic history and so many complex ideas that many terms and definitions have had to be invented to express it all without ambiguity. Some of the distinctions may be subtle, but they can certainly cause confusion if they are not borne in mind at all times in Pleistocene work. Here are just a few of the more important distinctions:

    (a) world-wide climate change v regional and local climate change;

    (b) major glacier advances, retreats v minor glacier fluctuations;

    (c) magnitude of climate oscillations v duration of oscillations;

    (d) glacier advances v other, non-glacial cold climate effects;

    (e) climate change v glacier fluctuation v the evidence (deposits etc.)

    Some of the concepts and terms introduced in Chapter Four (p. 107) can be applied without difficulty in Pleistocene matters but others cannot. For years, academies, national geological institutions, international congresses and specially appointed commissions have been grappling with the problems of Pleistocene stratigraphic terminology but they have not solved all of them. Meanwhile, ambiguities and inconsistent uses continue. The most persistent ambiguity lies in the practice of referring to all the major cool-climate episodes as 'Glaciations' even in relation to places where glaciers did not appear. The alternative, neutral term 'cool oscillation' (of climate) has never caught on. Important episodes of warmth between major coolings are, with equal inconsistency, called 'Interglacials', a term often used as a noun, as well as adjectivally. In the past the term 'Stage' was applied loosely to both cool and warm events, but now it is required to be defined more formally with reference to the geological evidence, as in the case of time-stratigraphic units. Minor cold or warm oscillations within the major events are called (cool) 'stadials' when they interrupt a (warm) interglacial period, and (warm) 'interstadials' when they interrupt a cold period. The responses by glaciers in different valleys or regions to a cold episode are 'advances', but there are no formal names for 'retreats'. By no means all possibilities are covered by these terms.

    Universal agreement has never been reached on how best to *define* 'interglacial' and 'interstadial'. How long must a warm interval have lasted, and how warm should the climate have become to qualify as a full interglacial? We will see later that the answers to some important questions in New Zealand hinge upon these distinctions.

The 'Glacial Period' in New Zealand

The early attention given by our pioneer geologists is a reflection of the interest still attaching to the 'Glacial Controversy'. Hochstetter, Hector and Haast when making their earliest explorations about 1860, and Hutton a little later, were all anxious to find how important the Glacial Period had been in these southern temperate latitudes. Their writings show how much they were impressed by glacial evidence in Nelson, Canterbury and Otago. There was debate as to whether the period of glaciation was Pliocene or Pleistocene, but this question could not possibly have been solved on the information available at the time. Enthusiastic descriptions were given of the massive mounds of moraine and thick sheets of outwash spread out in many South Island valley systems and forming impressive flights of terraces. Though these features are most conspicuous in the more open country east of the main divide, they occur on such a grand scale in Westland that the bush could not hide them from mountain viewpoints. The activities of gold miners and prospectors soon opened up countless exposures of glacial deposits which were examined and described at different times by Haast, S. H. Cox, Alexander McKay and others. More recent observers have been less fortunate since the decline of gold-mining. Even new road works provide only a brief opportunity to study the deposits before they are covered by a fresh growth of vegetation.

The Ross Glaciation—Our Earliest

Fig. 9.7
Locality sketch of the Jones Creek section through Ross Glaciation deposits.

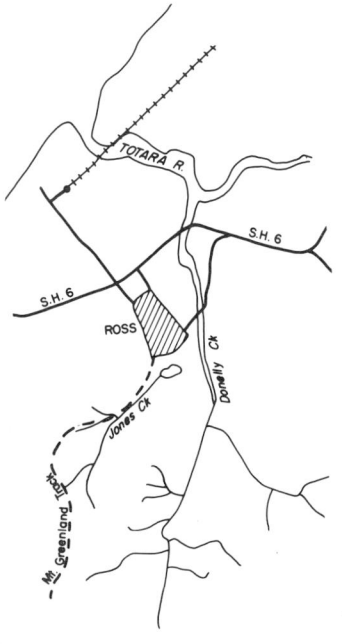

A little must be added to what was given in Chapter Seven on this subject. In Jones Creek just south-east of Ross township, where the original discovery was made in 1945, tillite composed of large and small blocks and cobbles of greywacke, hornfels, schist and granite occurs with laminated lake silts within the Old Man Group. The beds are not always well exposed, but enough can usually be seen to justify the effort of a scramble through bush down to the stream from a point on the Mount Greenland Track about 100 m from the start. The glacial sequence and the underlying Tertiary marine sandstone and conglomerate are involved together in folding which occurred during a climax of the Kaikoura Orogeny. The younger, 'post-Kaikoura' glacial deposits of Westland rest unconformably on upturned and eroded Old Man Group beds. Deposits of the Ross Glaciation have been found also in the Arahura Valley near Humphreys Gully (where recently re-exposed on forestry roads) and in Findlay Creek, a branch of Nelson Creek in North Westland.

The evidence for the Ross Glaciation is purely geological, from strata interbedded in a folded sequence, so that there is no direct evidence as to the form of the land surface at the time. The coarse texture of the tillite, however, implies that the glaciers came from a mountainous region close at hand, and the presence of schist fragments in the tillite also confirms that the Southern Alps had already risen appreciably. In the New Zealand geological time scale, the Ross Glaciation belongs in the Lower Nukumaruan (Hautawan) Substage, approximately at the Pliocene-Pleistocene boundary if not a little older.

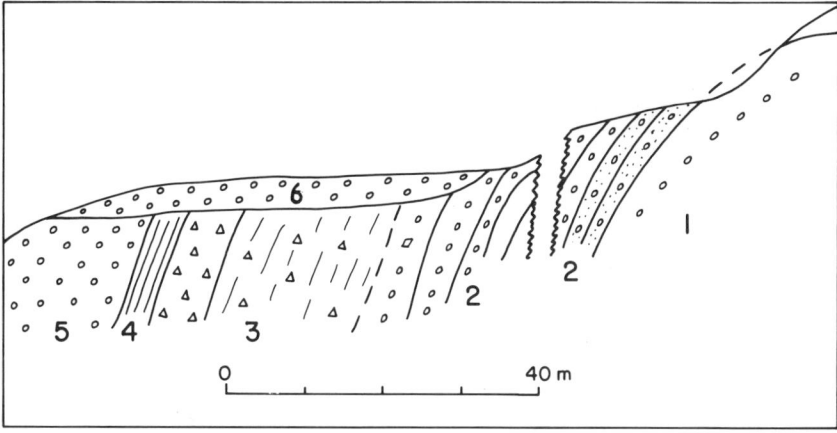

Fig. 9.8
Diagrammatic NE-SW section along Jones Creek from behind Ross township (left) and up a small western tributary crossing the Mount Greenland Track. 1. Pliocene marine sandstone with a layer of scattered boulders near the top grading upwards to finer conglomerate, coarse sand, and thin lignite beds. 2. Conglomerate of the early Pleistocene Old Man Group, including near the top some material that must have been transported from some 15 km to the east, most of the way by ice. 3. Angular and rounded boulders of schist, granite, hornfels, etc. in a fine, compact grey silt matrix—the tillite of the Ross Glaciation. 4. Laminated glacial-lake silts, formed probably in the waning phase of the Ross Glaciation. 5. Return to non-glacial Old Man Group conglomerate. 6. Unconformable capping of late Pleistocene gravels. (Adapted from a marginal inset diagram on the map Quaternary Geology of New Zealand—South Island, *1973, N.Z. Geological Survey)*

Ice Cap or Valley Glaciers?

The earliest accounts of New Zealand glaciation by Hochstetter and Haast spoke of glaciers expanded within the valleys, rivers of ice connected at the source by sharing the supply from common icefields but not of a continuous ice cap. In 1902 James Park, then Professor of Mining at Otago University, mistook some mudflow debris on the hills around Dunedin for glacial moraine and on that evidence postulated a capping of ice there. Others were not impressed, but with characteristic independence Park went on describing what he preferred to believe were morainic deposits in various unlikely places. By 1908 he had built up an extraordinary picture of an ice cap covering practically all of the South Island and Stewart Island and a good deal of the North Island as well. Great rivers of ice were supposed to have flowed through Cook and Foveaux straits, like those which must once have occupied the Irish Sea and the northern part of the English Channel. Although he identified many moraines correctly, the more remote ones were non-glacial breccias of various kinds, such as the Henley Breccia of southeastern Otago ('Henley Moraine' to Park), now known to be Cretaceous in age, and lahars (volcanic mudflows) on the perimeters of Tongariro and Mount Egmont volcanoes.

This time Marshall was on the side of the angels. He was prominent in a long campaign to persuade Park, by that time a very influential man in the mining and geological worlds and an author of successful textbooks, to give up his idea of a 'Great Ice Age of New Zealand', but with little effect. As late as 1926 in one of his last papers Park was still describing the lahar deposits near National Park railway station (then called 'Waimarino') as morainic mounds.

Before condemning Park too severely for his obstinacy, it would be well to recall that long and bitter arguments have raged also in other lands on similar questions of whether ancient breccia formations record ancient glacial action or something else.

The maximum spread of Pleistocene ice in New Zealand is not easy to estimate. To begin with, we still know so little about the extent of the Ross Glaciation deposits that no estimate is possible. As regards late Pleistocene glacial events, the outlines of the many separate ice-streams occupying mountain valleys and piedmont lowlands were intricate and would have varied so much over short spans of time that accurate measurements of their maxima are not worth attempting. To say that glacier ice covered 4000 square kilometres at most would be putting the matter in the right perspective. For comparison, the present ice extent, diminished as it has been by recent glacier recessions, can amount to barely 1000 square kilometres.

The Number of Glaciations

Early New Zealand geologists were divided into two camps on this question. From as far back as 1875 F. W. Hutton was advocating two glaciations, the earlier of them in the Pliocene Period. McKay in 1894, when describing the gold-bearing gravels of Westland, mentioned two glaciations and a similar belief was expressed by J. Henderson later as regards southeastern Nelson. In 1926 P. G. Morgan predicted correctly that more than two Pleistocene glaciations would be recognised eventually in New Zealand.

It is not easy to find evidence for the existence of intervening periods of warmth in deposits between glacial beds. Speight, who was looked upon as an authority in glacial matters, could see none and was inclined to regard the separate moraines spaced out along many South Island valleys as marking, not different ice advances, but merely a succession of pauses in the retreat from the maximum of a single advance. The clear signs of intervening erosion and weathering were overlooked, and it was not until 1938 that Speight conceded there had been a significant retreat between two advances of the Waimakariri Glacier—and that was on faulty evidence.

A younger generation of geologists from 1940 onwards was convinced of multiple glaciation, so that in 1957, when at a meeting in Dunedin R. P. Suggate and the author put forward the suggestion that the successive glaciations recognisable on both sides of the Southern Alps should be given formal names, no voice was raised in opposition. Since then, the chronology of Pleistocene glacial events in New Zealand has interested many geologists, geographers, botanists and soil scientists. New Zealand being so far from the countries where Pleistocene glacial chronologies were set up, and safe grounds for correlation being absent, it was quickly realised that our chronology would have to be established independently, and our own subdivision of Pleistocene time based upon climate fluctuations. Correlation of New Zealand events with those of other lands became the secondary objective of glaciation studies, not the prime one.

The Warmer Episodes

As noted above, interglacial evidence in the form of deposits between sets of glacial beds is rare in New Zealand, largely because continual orogenic uplift and erosion tended to put some appreciable vertical distance between them, and to reduce the chances for survival of deposits that have accumulated during the warmer phases. Evidence for interglacial periods was therefore mainly circumstantial in inland areas.

Fig. 9.9
The distinctive profile of
Point Elisabeth is familiar
to travellers from Westport
to Greymouth along the
coastal highway. Its gentle
seaward-sloping upper
surface, cut by wave action
across early Tertiary strata
during the last (Oturi)
interglacial episode, is
capped by sands and
gravel which have been
mined for gold. A former
seawards extension of the
surface is indicated by the
rocky pinnacle ('stack') to
the right of the picture.
(There are others out of
sight, farther to the right.)
There is good evidence
that this coastal area has
been raised by about 5 m
over the past 4500 years,
and the present height of
the Point Elisabeth terrace
therefore gives an
exaggerated idea of the
height of the ocean during
the Oturi Interglacial.

The terraces and raised beaches that are so prominent on New Zealand coasts were noticed in many papers and Survey *Bulletins* through the years, but little attention was paid to the deposits, and heights above sea level were reported in various inconsistent ways. While mapping the geology of the Wanganui district in the 1940s C. A. Fleming studied in detail the sediments and fossils of the deposits on the coastal terraces there, and was able to draw inferences as to the climate that prevailed when the sea stood at each of the higher levels. The outcome was an important paper in 1953, in which Fleming proposed to set up two stages within the late Pleistocene Hawera Series (Terangian Stage; Oturian Stage) with the suggestion that sea temperatures were warm at those times, and a hint that they could be regarded as interglacial stages, but only in the New Zealand context. Long-distance correlations based solely on altitude were already in disfavour, and in any case the Wanganui region had suffered warping during the Pleistocene. The fossils, though giving some indications of temperature, provided no grounds for overseas correlations.

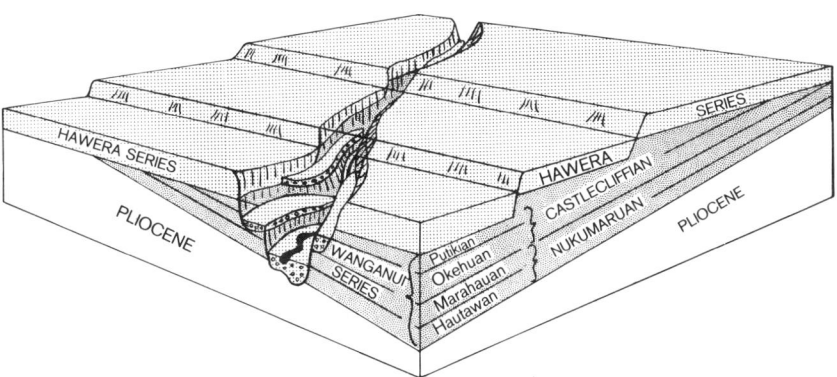

Fig. 9.10
A block diagram drawn by
C. A. Fleming to show the
relationship between late
Pleistocene upland terrace
deposits (Hawera Series),
slightly tilted Pliocene-
early Pleistocene marine
strata, and younger allu-
vial deposits in valleys
excavated in Holocene
times. The diagram repre-
sents no particular place
but depicts the general
situation in coastal areas
north of Wanganui. (C.
A. Fleming, 1973)

Ten years later, the first New Zealand glacial chronology proposed by Gage and Suggate in 1957 (published 1958) having in the meantime been improved and elaborated, Suggate took the bold step of dovetailing Fleming's interglacial stages of the Hawera Series, established at Wanganui, with the cool-climate stages of the South Island glacial succession. Thus arose the first scheme for classification of the late Pleistocene in New Zealand in which properly defined cold-climate and warm-climate events were integrated. Having proposed a further modification in 1965 which involved adding another glaciation, Suggate also added a third warm-climate stage, based this time upon high-level coastal terraces on the West Coast of the South Island between Westport and Greymouth, without fossils but assumed to have originated during a glacio-eustatic high stand of sea level earlier than that of the Terangian Stage. The revised definitions for the climatic stages embodied indirect evidence in the South Island for warm conditions during times supposedly equivalent to the Oturian and Terangian stages. This evidence came from plant fossils in peats that had been deposited in lagoons on some coastal terraces while sea level stood high.

More than twenty years ago M. T. Te Punga drew attention to the presence of another kind of evidence for warm climate interludes in the New Zealand Pleistocene. He noted in some Wellington localities that the greywacke bedrock was intensely and deeply weathered to a bright red soil colour instead of the usual yellowish browns. Subsequent work on the soils established that the weathering must have taken place beneath an old land surface at times when the climate was considerably warmer but with well-defined rainy seasons, as in subtropical regions today. The warm episodes have to be assigned to one or more of the interglacial stages in mid-Pleistocene times. Red weathering has been found in a number of places in Wellington Province at altitudes of up to about 200 m (with a few higher exceptions), and is seen affecting Haast Schist outcrops at a few places on the Grove Road between Picton and Havelock, Marlborough.

Dividing up the New Zealand Pleistocene—and Some Problems

The object of the 1957 glacial chronology and subsequent amendments has always been to provide a reliable, independent timetable against which the supposed timing of various local events could be shown and compared. Individual investigators of the glacial record have always been encouraged to give separate sets of names to the ice advances recognised in different areas, so as to avoid too much confusion if the correlations or the classification itself had to be changed, as indeed has happened.

The 1965 version, as put out in *N. Z. Geological Survey Bulletin 77* ('Late Pleistocene Geology of the Northern Part of the South Island, New Zealand', by R. P. Suggate) is outlined in Table 9-1 (p. 247). It is the one now used by practically all scientists whose work concerns the Pleistocene Period, including biologists and soil scientists. We should therefore now consider how well this scheme serves, not merely as a tabulation or calendar of climatic events, but rather as a sound and adequate basis for the classification of the Pleistocene Period.

Suggate has always argued that it was safe to assume that at least the major climatic oscillations were world-wide in their effects, and synchronous; and further, being unable to use fossils for time-markers in the same way as in earlier periods, that we had to adopt climatic oscillations as the only practicable means of classifying and subdividing Pleistocene time. It is indeed the most widely acceptable basis, but with the enormous recent advances in knowledge about the sediments beneath the ocean floors, where the chances of the record being complete are better than on land, the glacial successions on land are losing their position of prime importance.

One of the problems inherent in a glaciation-based classification is uncertainty as to whether every major climatic deterioration was recorded by glacial advances, and whether the record has survived. Are there hidden gaps and overlaps in the sequence as we know it? Can we be sure that Table 9-1 lists all the important glaciations, and how great is the time-gap, due to the Kaikoura Orogeny, between the Ross and Porika glaciations?

Table 9-1

CLASSIFICATION AND CHRONOLOGY OF THE NEW ZEALAND PLEISTOCENE AND HOLOCENE

Earlier Classification	STAGES as in 1965 Classification		N.Z. Series	International Geological Time Units	Estimated Age (years)
	COLD	WARM			
		ARANUIAN		HOLOCENE	— 10 000
OTIRA GLACIATION	OTIRAN				70 000 or 100 000
		OTURIAN			— 130 000?
	WAIMEAN		HAWERA SERIES		
		TERANGIAN			— 250 000?
WAIMAUNGA GLACIATION	WAIMAUNGAN			PLEISTOCENE	
		WAIWHERAN			
PORIKA GLACIATION	PORIKAN				— 500 000?
Unconformity	Time-gap ?		?		
	CASTLECLIFFIAN (2 substages: cool, warm)				—1 200 000
	NUKUMARUAN (3 substages: cool, warm, cool)		WANGANUI SERIES		
ROSS GLACIATION	ROSS GLACIAL				—1 800 000
				PLIOCENE	

It is not easy to connect up the histories suggested by different lines of climatic evidence. Take, for instance, the raised coastal terrace deposits that yield warm-climate evidence of the kind upon which the Oturian and Terangian Stages were defined. Such deposits do not generally occur in direct continuity or in any straightforward stratigraphic relationship with cold-climate evidence, mainly for reasons which follow. During the times of intense frost and glacier erosion without the moderating influence of a good cover of vegetation, rivers become so heavily charged with debris that they aggrade their beds with thick gravel deposits, whereas at the same time sea level is falling and the lower reaches may actually be cutting down. Then, when warm climate returns, hills become re-forested and erosion is reduced, sediment load drops away and the rivers cut down into the gravels that accumulated in the upper reaches in the cold phase while at the same time sea level is coming up again. Aggradation will then occur in the lower reaches if enough sediment is available. River profiles of glacial and interglacial times thus tend to have different gradients and the deposits are related to different base levels (p. 325).

In other lands that were inhabited by mammals throughout the Pleistocene, vertebrate paleontology helps in the classification of cold-climate deposits, but in New Zealand this was not the case. Fossils are by no means common on the coastal terraces. Inland deposits have yielded little evidence other than plant remains, and indeed the microscopic pollen grains and spores have proved particularly useful, but as climatic indicators rather than as time-markers. Apart from pollen in peats, fossil material is rare in our cold-climate deposits on land.

Much climatic information has been obtained from fossil pollen studies ('palynology'). Pollens are extracted, identified and counted in samples of Pleistocene peat that accumulated in lakes, stream backwaters, swamps and coastal lagoons. The procedures are time-consuming and call for skill and patience. Nevertheless, since the pioneering work in New Zealand in the 1930s and 1940s by Drs Lucy Cranwell (Mrs Watson Smith) and W. F. Harris, the method has been carried to the stage of becoming an indispensable tool in our Pleistocene studies, notably by N. T. Moar and D. C. Mildenhall.

Palynological studies frequently yield valuable evidence as to the climate at the time the deposits were laid down. Age is not indicated directly as a rule, but the climatic information can usually be fitted in with other data, the stratigraphic position of the peat deposit relative to other deposits, radiometric datings of the carbon content of fossil wood and peat (where within the time-range of that method), and sometimes evidence from associated volcanic ash deposits the ages of which are independently known. The palynological results, giving details of usually disjointed climatic sequences, are thus fitted on to an historical framework of successive geological events.

Volcanic ash (tephra) mantles have helped considerably in the task of correlating gravel deposits in different parts of the North Island. It is not always possible to identify tephra deposits, especially where they were originally thin or have since developed a soil, but very often the mineral composition is distinctive enough to tell that a tephra originated during one of a series of eruptions from a particular volcano. This may reduce the span of possible age for another kind of deposit of a surface either above or beneath the tephra layer. In north-western North America, Iceland, and in the Andes particularly, ash showers and glacial deposits conveniently occur together, but, alas, in New Zealand the Pleistocene volcanoes were in the North Island and the glaciers nearly all in the South.

The difficulty of linking South Island glacial with North Island non-glacial evidence is a serious weakness in the current scheme. Coastal terraces and deposits at similar levels do not necessarily relate to the same climatic phase, for vertical crustal displacements have affected the Wanganui region where interglacials are defined as well as the North Westland region furnishing the type glacial sequences, and the possibility of differential movements along the whole intervening distance of 400 km seems to be strong. It would be better if the interglacial stages could be redefined from deposits closer to the glaciated South Island areas, best of all within the western region from where the glacial stage names are derived. This may yet be achieved.

The kind of field evidence upon which an interglacial stage might better be defined is illustrated in Joyce Stream, a tributary of the Waimakariri River near Kowai Bush in the Springfield district. When clear of gorse and broom, this stream exposes the following series of deposits, in upwards succession: glacial outwash gravels; peat; lake silts; more outwash; coarse morainic gravels (Fig. 9.12). The lower outwash, from the Woodstock Advance, had been well weathered and deeply eroded under mild, interglacial conditions before the peat and lake silt were deposited. Pollen in the peat and silts shows a progressive change from forest to grassland and barren conditions in the surrounding area, and a piece of wood in the peat bed was dated radiometrically as more than 45,000 years old. The upper outwash heralds the Otarama

Fig. 9.11
This section in Joyce Stream, near Kowai Bush, Canterbury, covers the transition from glacial outwash of the Woodstock Advance of the Waimakariri Glacier upwards through interglacial lake-silts and peat to outwash and moraine of the Otarama Advance (early Otira Glaciation). Fossil wood fragments in the silts gave a radiocarbon age of more than 45,000 years.

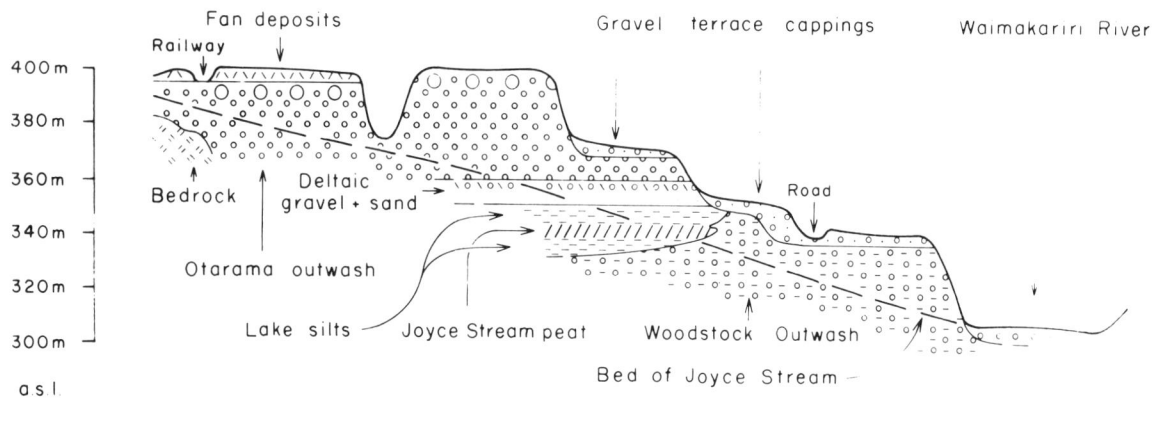

Fig. 9.12
Moraine deposit built up at the snout of the Waimakariri Glacier at its maximum during the Otarama Advance, early Otira Glaciation. This was the last occasion when the ice thrust down the Waimakariri Gorge and out to the mountain front. The exposure is in a railway cutting near Kowai Bush.

Advance, and marks the beginning of the Otira Glaciation. The build-up of gravels in the Waimakariri valley obliterated the lake, and eventually the ice reached within a kilometre or less of Joyce Stream. The lower part of the peat, together with the episode of weathering and erosion of the Woodstock deposits below could be embodied in the definition of a 'Joyce Stream Interglacial' preceding the Otira Glaciation.

Compared with the previous version introduced in 1961, the 1965 glacial chronology proposed to make some important changes with far-reaching consequences, as follows:

(a) The first stadial of the Otira Glaciation was upgraded to the status of a full glaciation (Waimea Glaciation).
(b) As a result the preceding warm interstadial had to become an interglacial stage, which—
(c) had to be identified with the Oturian Stage.
(d) The interglacial event preceding the Waimaunga Glaciation became identified with the Terangian Stage; and—
(e) there were consequential changes in previously accepted correlations between events in various parts of the South Island.

At Joyce Creek, for example, the interglacial episodé preceding the Otarama Advance became identified with the Oturian Stage and the Woodstock Advance with the Waimea Glaciation (Waimean Stage) rather than with the Waimaunga Glaciation as previously correlated. This in turn affects otherwise satisfactory correlations between events east and west of the Southern Alps. Such upsets form part of the basis for objections to Suggate's revision.

However, although not everyone agrees with the changes proposed in 1965, and followed in all subsequent Geological Survey publications,

the namings of the glacial advances in different areas, and of the deposits and features that provided the evidence, remain unaffected thus assuring the stability essential to prevent total confusion.

The 1965 changes involve a question upon which general agreement is still lacking. What are the proper criteria for distinguishing minor interstadial from major interglacial events? Should it depend upon how warm, how long it stayed warm, or a combination of both? The majority of opinion seems to agree that warmth comparable with that of the present day, lasting long enough to be reflected by a return to high sea level, indicates an inter-glacial event. On land, temperature indications come mainly from the evidence of fossil plants and pollens (in New Zealand, at any rate) but they are far from precise. The palynologists themselves never fail to stress the sources of possible error in their interpretations of what the climate was like, regionally and locally, while the buried peat deposits were being formed.

Fig. 9.13
The coarse, bouldery layer of moraine shown in Fig. 9.12 is seen again here in Joyce Stream at the top of about 70 m thickness of glacial outwash gravels that were deposited by meltwater torrents ahead of the advancing Waimakariri Glacier during the Otarama Advance. (see pp. 249–50)

Terminology

It has been the practice among New Zealand geologists for the last two decades to maintain distinction between the following kinds of information concerning glaciation:

(1) An advance of ice is described and named as a separate event in each valley system, e.g. the Otarama Advance in the Waimakariri River catchment, Canterbury.

(2) Major advances that appear to have affected the whole glaciated region at the same time, as far as we can tell, are a 'Glaciation' (e.g. Otira Glaciation), and the warmer intervals when all glaciers are supposed to have retreated and perhaps disappeared are

'Interglacials' (e.g. Oturi Interglacial). (Both (1) and (2) are regarded as climatically-induced geological events.)

(3) To identify such events with the geological evidence for them in some designated type area, we recognise a kind of time-stratigraphic unit in which climate-oscillation is the time-marker; thus climate-stratigraphic 'Stages' (e.g. Otiran Glacial Stage).

(4) The glacial and interglacial deposits themselves are named and described in the usual way as lithological rock units (e.g. 'Burnham Formation' for some parts of the glacial gravel of Otiran age in Canterbury).

This explanation should assist readers to appreciate how the Pleistocene deposits were classified and represented on the 1:250,000 scale geological maps. It may be helpful to refer back to the discussion of stratigraphic units in Chapter Four (p. 107). So far we have not consistently given formal names to stadial and interstadial events.

Dating the Glacial Events

Earlier glacial episodes in New Zealand beyond the range of radiocarbon dating methods are still of uncertain age. As shown in Chapter Seven, the date of the Ross Glaciation is known from the stratigraphy of the Westland region to be towards the close of the late-Tertiary/early-Pleistocene chapter of our history, that is, about two million years ago. In its type area, a major unconformity divides off the Ross Glacial evidence from that of all the subsequent glacial events.

Deposits of the succeeding Porika Glaciation, though warped, faulted and deeply eroded, are nowhere as intensely deformed and deeply eroded as are the Old Man Group of conglomerates, etc., which contain the Ross Glaciation deposits in Westland. This contrast has always been part of the grounds for assuming a substantial lapse of time between the two glaciations. Recent palynological work by D. C. Mildenhall and others has, however, raised some doubts about the magnitude of the break between them. Certain lake deposits now found as remnants on ridge summits near Lake Rotoroa in Nelson, believed to have formed during the Porika Glaciation (in fact, part of the original evidence for it), may prove to be little if at all younger than the Old Man Group. If this is true, there would be no significant time-gap between Wanganui Series and Hawera Series here—in fact there may be overlap. Because of the problems it brings up with regard to the field relationships of the two groups of deposits—the younger set within a recognisably post-Kaikoura landscape, the older in nearby regions involved in the Kaikoura movements along with older formations—more work is needed to confirm the suggestion.

Unfortunately, no means are available for obtaining directly the true ages and time-spans of the Porika and Waimaunga glaciations, nor of the Waimea Advance. These events are all beyond the range of radiocarbon dating, and the deposits contain no materials capable of being dated by other available methods. Taking into account North Island datings of Pleistocene events involving evidence from tephra deposits, and accepting the current interpolations of glacial and

Fig. 9.14
Lake Ohau, like the other large lakes of the upper Waitaki catchment, is impounded behind high ridges of moraine built up during the last major ice advances. (Photographed in 1947)

interglacial stages as correct, the Porika Glaciation appears to be about 500,000 years old, and the Waimaunga Glaciation about 250,000 years old. This is in accord with suggestions by C. A. Fleming in 1973. The Oturian Stage is thought to have begun about 130,000 years ago, and is generally regarded as the New Zealand equivalent of the 'Last Interglacial'.

The Otira Glaciation commenced prior to the earliest times within reach of radiocarbon dating. There is wide disagreement as to when the Last Interglacial Period, in the global sense, came to an end, opinions ranging between about 70,000 and 116,000 years ago. Current views in this country as to when the Otira Glaciation began depend of course upon whether or not the Waimea Advance is regarded as part of it. If so, Otiran time began more than 100,000 years ago; if not, perhaps no more than 70,000 years. The younger age comes close to the extreme limit of range for radiocarbon dating. There are some dates relating to the times of the Waimea Advance, but unfortunately they are under some suspicion of being in error because of contamination.

Later advances of the Otira Glaciation occurred between 23,000 and 18,000 years ago, and again between 16,000 and 13,000 years ago.

Some Valley Glaciation Sequences in the South Island

Effects from late Pleistocene ice advances are conspicuous in most mountain valleys from the Tasman and Spenser ranges southwards to Foveaux Strait, and on the summits of Stewart Island. With the help of the 1:250,000 scale geological maps one can locate and identify systems of moraines, outwash gravel plains and terraces on varying scales of magnitude and grandeur. A 1:1,000,000-scale map, 'Quaternary Geology—South Island' (*N.Z. Geological Survey Miscellaneous Series Map 6*, 1973) provides a highly-condensed and valuable guide to all kinds of glacial phenomena. In greater detail, descriptions of glacial features in the northern parts are to be found in R. P. Suggate's 'Late Pleistocene Geology . . .' (*N.Z. Geological Survey Bulletin 77*, 1965). A number of the earlier Survey *Bulletins*

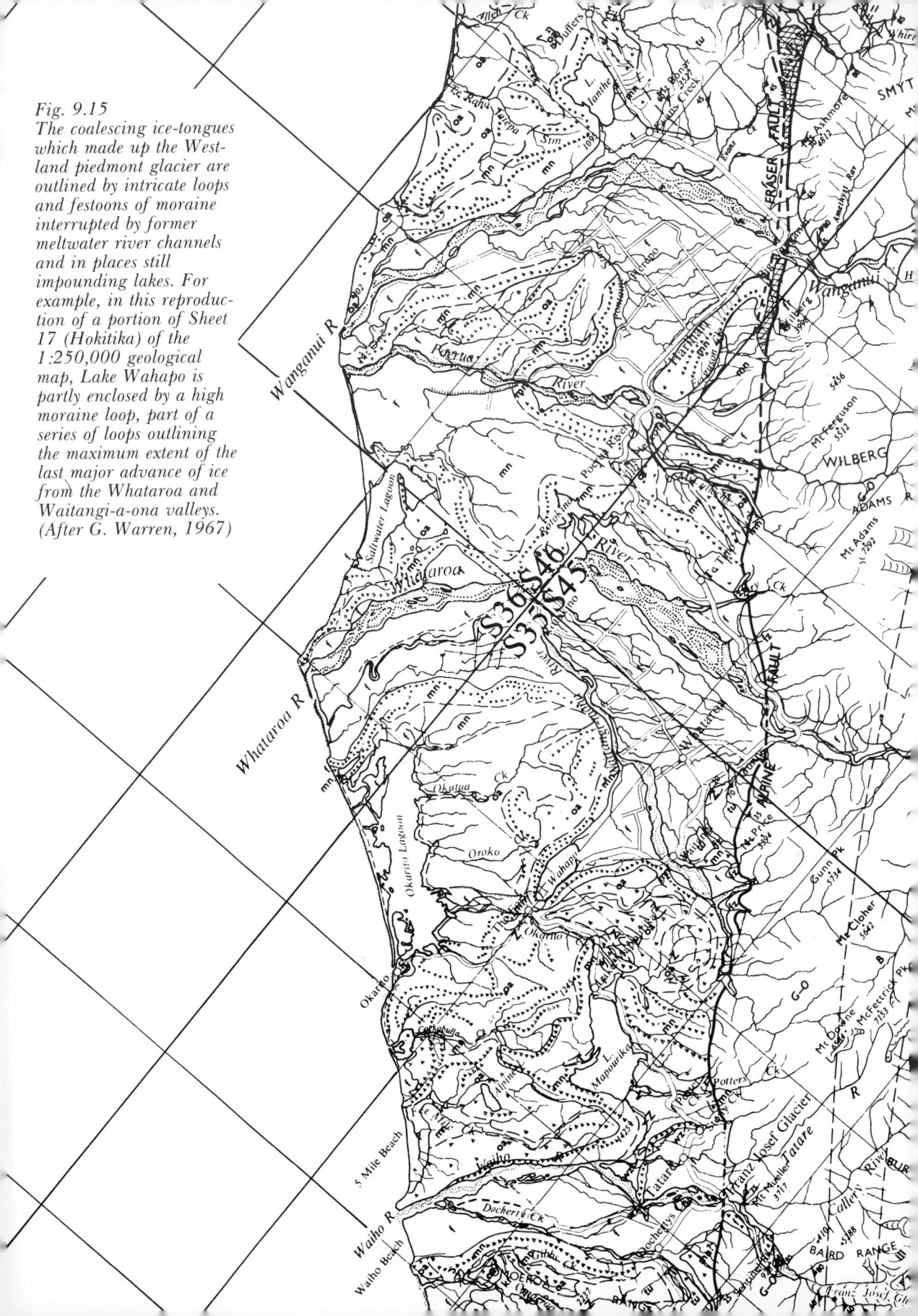

Fig. 9.15
The coalescing ice-tongues which made up the West-land piedmont glacier are outlined by intricate loops and festoons of moraine interrupted by former meltwater river channels and in places still impounding lakes. For example, in this reproduction of a portion of Sheet 17 (Hokitika) of the 1:250,000 geological map, Lake Wahapo is partly enclosed by a high moraine loop, part of a series of loops outlining the maximum extent of the last major advance of ice from the Whataroa and Waitangi-a-ona valleys. (After G. Warren, 1967)

Fig. 9.16
Bouldery till with a sand/ silt matrix near the Dilmanstown Dam, Kumara. It was deposited by the Tarakamau Valley glacier during the maximum of one of later advances in the Otira Glaciation.

accompanied by maps on the 1:63,360 scale covered areas of Pleistocene glaciation, but in nearly every case the mapping was done before the present glaciation chronology was evolved.

Table 9-2 presents a list of the named, major ice advances in six different South Island areas, showing how they are believed to correlate with one another, and with the glaciation chronology. Correlation charts of this kind are always subject to being outdated by later work, and must therefore be regarded as tentative.

The Mackenzie Basin of inland South Canterbury is outstanding for the variety and the grandeur of features due to the Pleistocene expansion of large glaciers arising on the highest parts of the Southern Alps. From the viewpoint of the top of Mount John, Tekapo (which can be reached by car), no great challenge to the imagination is required in order to visualise the entire landscape below as it would have looked when buried beneath coalescing rivers of ice pouring out from the Mount Cook region. The successive moraine systems and outwash surfaces connected with them are well defined, and a variety of kinds of glacial deposits are exposed in excavations and cuttings around the hydroelectric works. Glaciation on a grand scale is displayed also in the Te Anau-Manapouri area, and indeed throughout the Southern Lakes region.

In central and southern Westland separate ice streams disgorged across the Alpine Fault from all the larger alpine valleys at times joined up along the mountain front to form a broad 'piedmont glacier', which in places extended westwards beyond the present coastline. Lakes Ianthe, Wahapo and Mapourika today lie inside high loops and ridges of very coarse moraine following the intricately lobed and scalloped

SUCCESSION OF GLACIAL ADVANCES IN SOME SOUTH ISLAND VALLEYS			Table 9–2
GLACIATION	**UPPER BULLER ADVANCES**	**NORTH WESTLAND ADVANCES**	**WAIMAKARIRI ADVANCES**
		(South Westland: Waiho)	(present)
OTIRA	St. Arnaud	Kumara-3	Poulter 1, 2, 3
	Black Hill	Kumara-2(2) Kumara-2(1)	Blackwater 2 Blackwater 1
WAIMEA	Tophouse	Kumara-1	Otarama
WAIMAUNGA	Kikiwa	Hohonu	Woodstock
PORIKA	Porika	Porika	Avoca
ROSS	(present?)	Ross	

Fig. 9.17
These laminated silts in a branch of Nelson Creek, North Westland, accumulated during one of the later advances of the Otira Glaciation while ice was blocking the normal drainage.

RAKAIA ADVANCES	UPPER WAITAKI ADVANCES	UPPER CLUTHA ADVANCES
present?)	Birch Hill (and others)	
Acheron 1, 2, 3	Tekapo Mt. John	Hawea
Bayfield 2 Bayfield 1	Balmoral 1, 2	Albert Town
Tui Creek	Wolds (?)	Luggate
Woodlands	Wolds (?)	Lindis
		Clyde

Fig. 9.18
A close-up view of laminated glacial lake silts found in Winding Creek, a branch of Broken River, Canterbury, formed during a period of ice retreat in the Otira Glaciation. Such layered glacial lake silts have been called 'varve silts', but strictly this term should be reserved for very regularly banded glacial lake silts, each layer of which shows graded bedding (p. 89), marking seasonal cycles of sedimentation. No truly seasonal varves have been found yet in New Zealand.

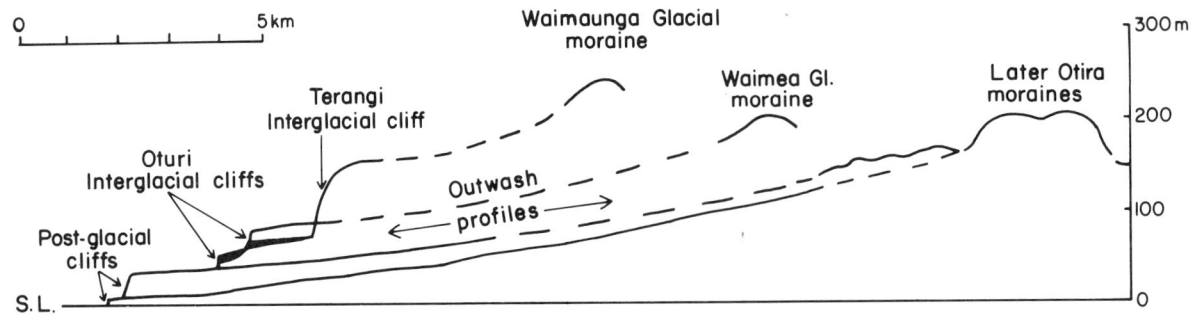

Fig. 9.19
Profiles of moraines and their connected outwash gravel surfaces in central Westland, built up in successive late Pleistocene ice advances, and their relationships with sea cliffs and coastal benches cut by the sea during warm, intervening periods of high sea level. Thus, after the Waimaunga Glaciation sea level rose and in the ensuing interglacial period (Terangi) a cliff was cut and beach sand and lagoon peat deposits accumulated on the wave-cut bench at its foot. Outwash gravel from the Waimea Advance overwhelmed these features but this in its turn was cut back by wave erosion at two different levels marking high stands of the ocean in the Oturi Interglacial. The evidence is well preserved in this region because the oscillations of sea level due to climate changes were superimposed upon slow, continual uplift of the land. (After R. P. Suggate, 1973)

line of the former ice margin in late Otiran times. Exposures of all kinds of ice-marginal sedimentary deposits and till are still to be seen in recent slips along the main highway through South Westland, although these exposures are not as impressive as they were a few years ago when the road was being reconstructed. From the air, or from elevated outlook points like Alex Knob above Franz Josef Glacier, the festoons of moraine ridges, dotted with tarns and small lakes, form a memorable spectacle. Most of what can be seen reflects the work of the latest ice advances. Coastal cliffs at Okarito and Omoeroa Bluff (south of the Waiho River mouth) give some idea of the extremely coarse texture of Westland till deposits, which here include massive, angular blocks of schist up to 7 m through.

Most of the evidence for standard time-sequences of glaciation can be

WAIMAKARIRI TERRACES

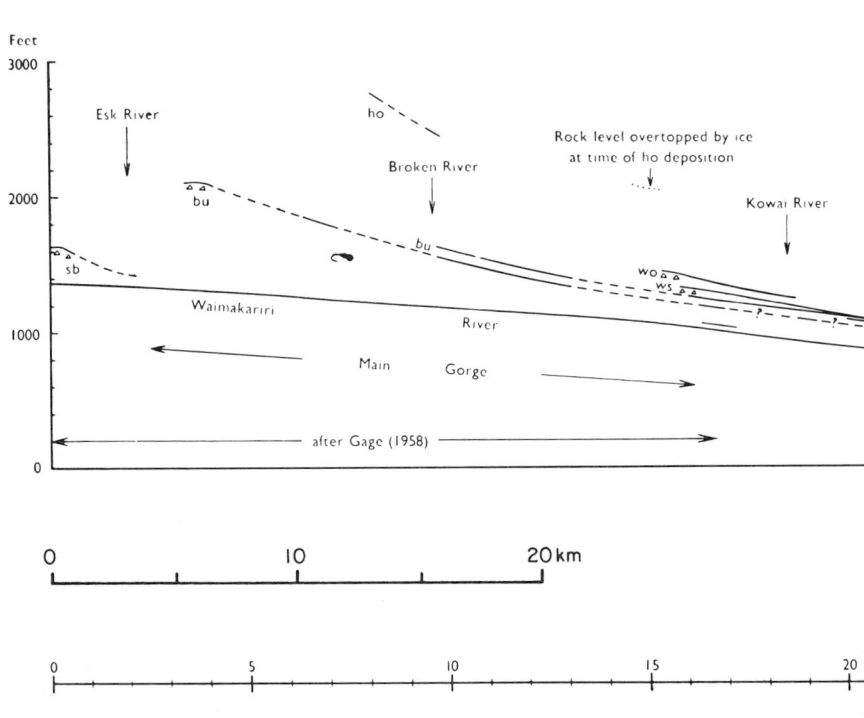

seen in North Westland. The vegetation cover here obscures much that is plainer to see in the open country east of the Alps, but with the help of the map with N.Z. Geological Survey *Bulletin 77* a great deal of the evidence for the Westland succession can be inspected at roadside exposures near Kumara and Hokitika.

A fine sequence of moraines and outwash surfaces due to successively younger ice advances is passed through on the road up the north side of the Rakaia Valley to Lake Coleridge power station, and several moraines can be seen from Highway 73 (Arthurs Pass Route) and Highway 7 (Lewis Pass Route). I could add further examples, but must leave it for interested readers to make use of the sources mentioned above.

Glaciation in the North Island

Small glaciers are barely surviving on the summit of Ruapehu. As noted earlier, extravagant claims were once made about the extent of Pleistocene glaciation in the North Island, but convincing evidence for it is limited to the Tongariro volcanoes and the Central Tararua Range. G. L. Adkin in 1912 published the first account of genuine glacial features, around the headwaters of the Otaki River.

Non-glacial Cold-climate Effects

At the beginning of this chapter it was noted that the 'ice age' was not everywhere a 'glacial period'. Extended glaciers and ice caps could

Fig. 9.20
Profile of moraines and outwash from successive late Pleistocene ice advances down the Waimakariri valley, Canterbury. (Suggate, 1965)

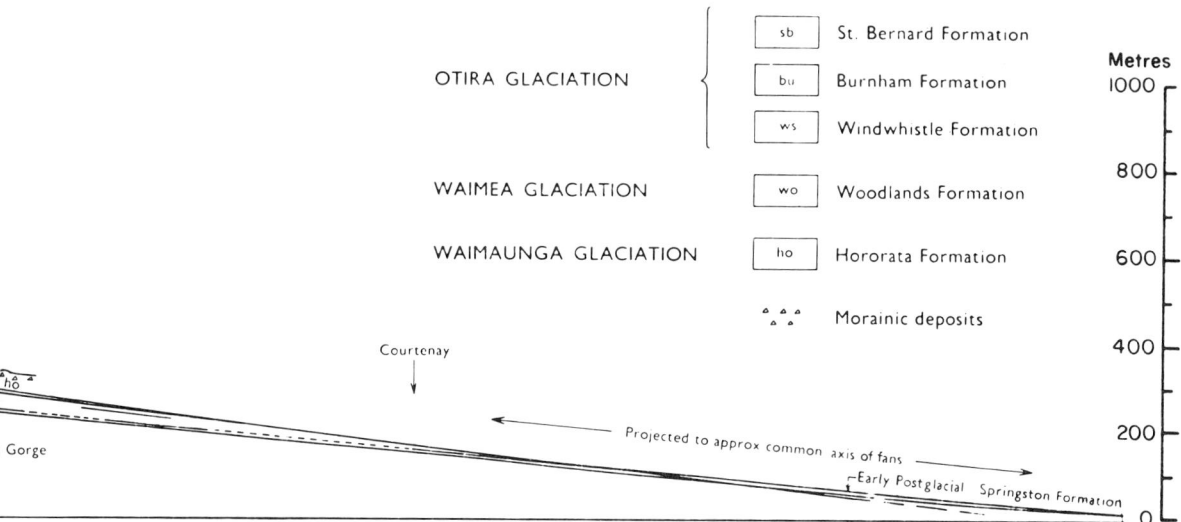

OTIRA GLACIATION

sb	St. Bernard Formation
bu	Burnham Formation
ws	Windwhistle Formation

WAIMEA GLACIATION

wo	Woodlands Formation

WAIMAUNGA GLACIATION

ho	Hororata Formation

Morainic deposits

Metres
1000
800
600
400
200
0

Courtenay

Projected to approx common axis of fans

Early Postglacial Springston Formation

ho

Gorge

25   30   35   40   45   50   Miles

Fig. 9.21
Three different kinds of
non-glacial cold-climate
deposits exposed in an old
gold mining claim at
Livingstone, North Otago.
The gravel at the bottom
was deposited by the
Maerewhenua River
during the Waimaunga
Glaciation, and covered
during the Otira Glacia-
tion by solifluction and
slope-wash deposits and
then capped by loess.
Buried soil horizons (not
well shown in this photo-
graph) marked the warmer
interstadial intervals.
(Interpretation after D. J.
Young, 1964; photo-
graphed by M. Gage)

grow only where there was a sufficient supply of moisture from
unfrozen, evaporating ocean surfaces to provide an ample snowfall,
and where the climate was such that year after year, indeed, century
after century, more snow accumulated each winter than melted in
summer. The accruing surplus became glacier ice. In regions where
snowfall is deficient though the climate is very cold, continual low
temperatures and lack of protective vegetation produce distinctive
effects. Where the mean annual temperature is lower than zero Celsius,
the subsoil becomes permanently frozen (permafrost) and the water
yielded by shallow summer thawing percolates down only to re-freeze
and to be added to continuous masses of solid ice below ground.
Meanwhile, surface saturation and poor soil drainage promote a variety
of kinds of mass flowage and earth creep summed up by the expression
'solifluction'. These phenomena are marked by distinctive superficial
deposits and surface forms. Such conditions prevail today in the
sub-polar latitudes.

Areas in which permafrost conditions, or even merely prolonged,
intense frost seasons occur are sometimes referred to as 'periglacial'.
This means simply 'surrounding glacial areas', which is not always the
case. On the one hand, the severest frost climates are experienced far
away from any glaciers, and on the other, conditions near existing
glaciers are sometimes surprisingly mild. Heavily forested areas
adjacent to the Westland glaciers today are literally 'periglacial', but
were not severely frigid even during the Pleistocene advances, perhaps
because the temperatures were moderated by a prevalent cloud cover.
Frigid non-glacial (but not permafrost) conditions did, however, affect
most other parts of the South Island except those at low levels in west
Otago, and are recognisable in certain North Island areas too.

The most extensive tracts of country showing clear evidence of
severely frigid conditions are in the inland parts of the South Island,
especially Central Otago, the inland basins and alpine foothills of
Canterbury and inland Marlborough. Coastal Southland was affected
also. Some of the most spectacular effects have been described from the
Old Man Range in Otago (Chapter Eleven, p. 322). The most abundant
sign elsewhere is a rubbly mantle of solifluction debris exposed near the
tops of countless road cuttings in hard-rock areas, in the form of
crudely stratified sheets that accumulated layer upon layer during
successive spells of cold. Buried soil horizons with peaty silt beds are
sometimes found between solifluction layers, and are seen as evidence
for intervening spells of warmth. Similarly, 'fossil gullies' now filled
with solifluction rubble are ascribed to warm-climate stream activity
followed by a change to frigidity. Many fine examples of this kind of
thing have been brought to light during highway construction near
Wellington.

In steeper country, roadworks and erosional gullying often have
exposed stratified layers of scree that have obviously long been under a
cover of soil and vegetation. In these places it has to be inferred that
the scree accumulated in former times when vegetation was absent
from the area and frost action more severe. The Porters Pass area,
Canterbury, traversed by a main highway and roaded also for access to

Fig. 9.22
Underlying a prominent terrace in the Awatere valley at Seddon, Marlborough, these gravels accumulated when physical weathering was intense in the Marlborough hinterland during an early cold phase of the Otira Glaciation. Glaciers did exist on the Kaikoura ranges but they were not extensive and the gravels are not regarded as outwash deposits. The capping of wind-blown loess silt also reflects cold conditions, but the upper part of it at least may be more recent. A small fault, possibly due to gravitational slumping towards the Awatere River (to the left) during an earthquake, no longer appears to displace the soil. This might suggest that the fault is not very recent, however the small scarp may have been obliterated by ploughing before the road cutting was formed.

skifields, shows good examples. It is one of the areas in which these cold-climate features have been the subject of some research, but much remains to be found out about their history and origin.

Oddly, perhaps, effects of solifluction in New Zealand were noticed first, not in the southern districts, but in the Wellington area whence they were described by C. A. Cotton, M. Te Punga and G. R. Stevens in a series of papers from 1955 onwards. I have seen similar features near the crests of the Ruahine and Raukumara ranges.

Although often recognised far away from glaciated areas, 'loess' is generally considered among periglacial phenomena. This is the name given to the evenly fine-grained, light grey or yellowish-grey silt forming the upper subsoil and mantling the slopes in many places in New Zealand from the Rangitikei and Manawatu southwards. It has received attention chiefly from soil scientists. The origin of the silt is attributed to the work of wind in bringing out the finer products of rock-grinding by glacier ice and shattering by frost, and indeed loess mantles are most conspicuous in regions lying down-wind (in terms of the prevailing westerly airflows of the Pleistocene Period) from outwash plains of formerly glaciated regions.

Loess deposits in southern North Island areas could have been derived from glaciated parts of Nelson. The thickest accumulations are found in eastern South Island districts, and are up to 10 m or more in road cuttings on Banks Peninsula and coastal cliffs at Timaru and Oamaru. Darker bands are interpreted as old soils developed while warm climate cut down the silt supply and favoured soil development under the loess surface of the time.

The origin of the silt need not always have been in glaciated mountains. Some is known to have come from seawards, uplifted presumably by on-shore winds from the beds of extended rivers during glacial phases when sea level was lower than now, on the evidence of

Fig. 9.23
Less than a century ago the Franz Josef glacier trough was occupied by ice as far down as the dark, scrub-covered mounds in the centre of the picture. Eleven thousand years ago the glacier extended across the Waiho flats as far as the prominent bush-covered moraine-arc (Waiho Loop Moraine). The canoe-shaped hill in the foreground is crested by a loop of moraine formed towards the end of the last major late Otiran advance while piedmont ice still occupied the lowlands of South Westland. Judging from a radiocarbon date from a peat deposit near the mouth of Waiho River that was overwhelmed by this advance, the date was around 14,000 years ago. In this view Mount Cook (the sharp one) and Mount Tasman stand to the right of the Franz Josef trough. (Photo: N.Z. Geological Survey)

fine needles of silica within the loess which are identified as part of the skeletons of marine organisms—sponges, in fact. A seawards source seems necessary for the loess deposits capping river terraces of Otiran age in Westland (e.g. at Taylorville in the Grey Valley and inland from Hokitika), where strong winds of the Pleistocene would have been dominantly westerly.

When did the Pleistocene Ice Age End?
By 13,000 years ago the great ice advances of the Otira Glaciation were over, and at least in some valleys the ice must have disappeared rapidly. About 11,000 years ago or a little later, glaciers thrust well down the main valleys once more, remained long enough to build small terminal moraines, and retreated finally to positions probably farther back in the mountains than where Europeans first saw them in the mid-nineteenth century.

The best known representative of this last, significant ice advance is the Waiho Loop moraine. As a prominent bush-covered ridge, it is crossed by Highway 6 a few kilometres north of Franz Josef Glacier village. From high viewpoints it is seen as an almost perfectly circular arc, breached by the Waiho and Tatare rivers. Also readily seen is the Birch Hill Moraine in the Tasman Valley, followed by the main road for some distance about 10 km south of The Hermitage. Sibbalds Island Moraine protruding from the bed of Godley River beyond the head of Lake Tekapo and the most prominent moraine in the upper Otira valley near Arthurs Pass were both probably formed during the same episode.

Indications are world-wide for a rise of sea level between 10,000 and

Fig. 9.24
The steeply-dipping gravels were deposited layer by layer on the front of a delta (hence, 'fore-set beds') which grew into the head of Lake Kanieri in Westland at a period when the lake was held at a higher level because a decaying glacier still blocked lower outlets. These gravels appear to have been brought in from the Styx or Kokatahi valleys to the south by an overflow of glacial meltwater.

Fig. 9.25
The sea has risen from about 100 m below the present level since the Otira Glaciation. The solid line in this graph averages a number of minor oscillations superimposed upon the general rise. According to some authors, however, there were significant oscillations up to a few metres above the present level within the last few thousand years, as indicated by the dashed line. (Adapted from graphs by Schofield and others)

4700 years ago. The event has long been known as the 'Flandrian Transgression', and it is attributed to the return to the ocean of water that had been locked up in Pleistocene ice caps. It is probably the best of all indications of when the Ice Age really ended. The sea level rise is amply documented in New Zealand, and particularly clearly in some of the water-wells of the Christchurch City area.

Minor readvances occurred from time to time later than 10,000 years ago, and in New Zealand there is evidence also that small moraines were built in some valleys at different times between 2600 and 1100 years ago. These need have been no more than temporary advances, not necessarily experienced by all the glaciers at the same time, and were not of world-wide climatic significance.

In 1958 R. P. Suggate showed how clearly the Flandrian Transgression is documented in the records of drilling for water at Christchurch. The southern shore of Pegasus Bay advanced and receded, perhaps not varying far from its present position, as sea level rose, according to whether the rivers were bringing down enough sediment to make up the continual slow submergence of the land. By about 4700 years ago the sea had reached its present position, or perhaps a metre or so above, since when only minor oscillations can have occurred. Deposits laid down during the transgression provide the basis for the final subdivision of the Hawera Series, the Aranuian Stage.

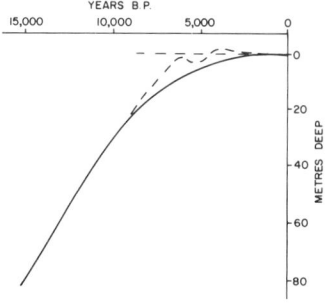

The 'Little Ice Age'

We move forward again to consider happenings in a period which is historic in most other lands, and indeed within the span of Maori occupation of New Zealand.

Fig. 9.26
Mount Evans (2600 m, right of centre) and the upper Whitcombe valley, showing shrunken remnants of the Wilkinson Glacier. Like many other smaller alpine glaciers, this one has been nourished since the end of the Otira Glaciation not directly by outflow of ice from the valley head but by avalanching of ice from névés perched high above the head of the trough. Early in this century the snout of the lower, reconstructed glacier filled the space between the two lateral moraine ridges at lower left. (Photo: Mannering and Associates, Christchurch)

European historians are well aware that, after a few centuries of genial conditions, the climate deteriorated sharply from about A.D. 1350 onwards. Frozen seas then isolated Iceland and the ancient Norse colonies across the Atlantic while alpine glaciers advanced over former forests and pastures and overwhelmed villages. The ice termini fluctuated during the next 400 years, but after about 1750 the losses from each recession were no longer made up by the next surge ahead. The episode has attracted the informal title of 'Little Ice Age', and it is considered as separate from the Pleistocene Ice Age.

A general recession of New Zealand glaciers has been going on, with minor readvances, since the middle of the nineteenth century. It is now known that these events were part of the recovery from the Little Ice Age. The ages of trees growing among small morainic ridges and on young outwash terraces in Westland alpine valleys have been determined mainly by counting annual growth rings. From the results, D. B. Lawrence, Peter Wardle and others have been able to show that all the glaciers studied were farther advanced at times in the

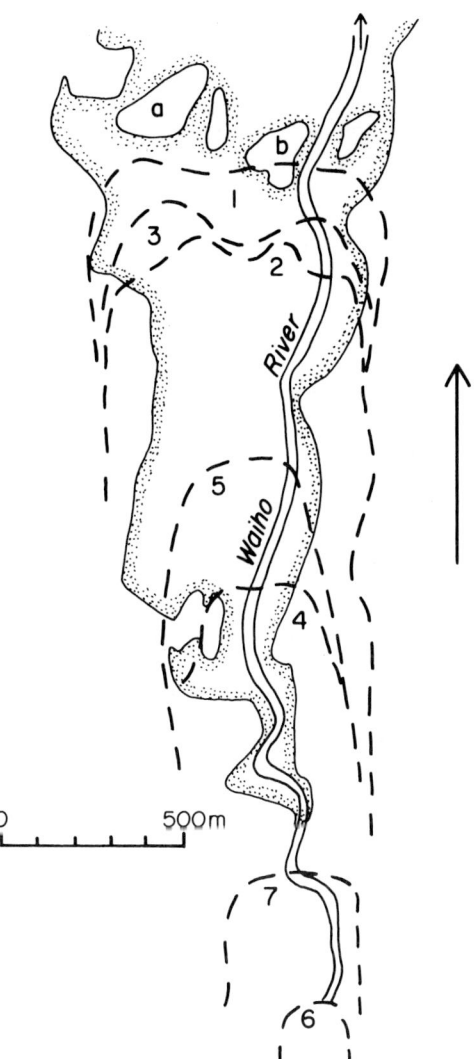

Fig. 9.27 When first seen by Europeans the Franz Josef Glacier was probably in retreat from an early nineteenth-century maximum. At the beginning of the twentieth century it still reached to the rocky bastions of Park Rock and Sentinel Rock. The limits of subsequent advances and retreats are shown in this sketch map. 1: 1894, 2: 1921, 3: 1934, 4: 1946, 5: 1951, 6: 1965, 7: 1967; a: Sentinel Rock, b: Park Rock. At present (April 1978) the snout has receded to approximately its early-1965 position. Note: The Waiho River channel is continually shifting. (Adapted from a map by W. A. Sara, 1973)

seventeenth, eighteenth and early nineteenth centuries than when first seen at close quarters by Europeans in 1861. Using similar information together with evidence from the rate of growth of certain kinds of lichen on morainic boulders, C. J. Burrows and others have claimed that the Tasman and neighbouring glaciers had been farther advanced in recent centuries than when first seen by the pakeha. Similar suggestions have been made about the Cameron Glacier in the Arrowsmith Range, inland mid-Canterbury.

Historians of Polynesia have thought that a deteriorating climate with greater storminess in the Pacific Ocean during the Little Ice Age period, accompanied perhaps by changes in prevailing wind belts, restricted the possibilities for safe ocean travel in open boats. Climate may thus have been responsible for the eventual isolation of the Maoris, who reached New Zealand while conditions were still favourable, from any further contacts with the brotherhood of Hawaiiki until the time of Captain Cook. It would make an interesting parallel with the experiences of the Norse colonies in the same period.

Chapter Ten

THE FRAMEWORK

Structural Outline and Crustal Setting

A Pattern with a Meaning?

The shape of New Zealand, especially on a relief map, could hardly be described as a formless blob. Neither the coastal outline nor the inland relief gives the impression of being random or accidental; rather, each suggests some underlying controls. Looking first at the more prominent relief features, the South Island has a clearly defined chief mountain axis which begins at Foveaux Strait with the Cameron Mountains, then swings into the north-easterly alignment of the Southern Alps proper, the Spensers and the Saint Arnaud Range. The same direction is taken by the Inland and Seaward Kaikouras and then continued north of Cook Strait by the Rimutaka, Tararua, Ruahine, Huiarau and Raukumara ranges to Hicks Bay. Roughly the same trend is shown by other important mountain chains, the Central Otago mountains from Taieri Ridge to the Livingstones, the Torlesse, Puketeraki and Lowry Peak ranges in Canterbury, the Paparoas in Westland, the Bryant, Pikikiruna and Wakamarama ranges in Nelson, the Aorangi Mountains in east Wellington and the Kaimanawas.

The north-easterly alignment seems to represent the true axial spine of that part of the New Zealand Platform that now stands above sea level, but not of the New Zealand Platform as a whole. Even above water there are other obvious alignments. A strongly contrasting north-westerly trend is expressed by the eastern Southland ranges, the Hokonui Hills, and the mountains of North Otago from the Horse Range and Kakanui Range across to the plexus of mountains west of Lake Ohau where the main ridges swing into line with the Alps. Again in the North Island, although there seems to be a lack of strong direction in the centre and west because of a cover of young sedimentary and volcanic rocks (the latter producing their own local patterns of radial ridges and valleys), farther north there emerges another strong north-westerly trend shown by the main topographic elements of the northern Auckland region and emphasised by the Coromandel Ranges and Great Barrier Island.

So far we have been looking only at topography. Looking now at geological maps, we can see that the rocks in most places show structural trends in general alignment with the chief topographic features while in others the grain of the rocks is transverse. Strong patterns there certainly are, and their interpretation has to be in terms of regional differences in tectonic history.

Now turning to an atlas map of the South Pacific, we may ask, 'Is there any special reason *why* New Zealand is *where* it is?' Few will be content with the universal answer to all such questions offered by Dr Pangloss in Voltaire's *Candide:* '. . . it could not be anywhere else. For it is impossible for things not to be where they are, because everything is for the best.' There must be some explanation, other than pure chance, for the position of New Zealand and the orientation of its main elongation relative to, say, Australia. Atlas charts showing only land relief offer few clues as to how this country fits into the framework of the south-west Pacific, but from any modern bathymetric map showing relief details of the ocean floors it is clear that, far from being structurally isolated, New Zealand is quite strongly linked by submarine ridges and troughs not with Australia but with other lands and island chains to the north, north-west and south. The aim of this chapter is to investigate how the major events of our diastrophic history influenced the structural and topographic grain, and how the main tectonic features of New Zealand fit into the crustal structure of the south-west Pacific region.

Thickness of the Crust in the New Zealand Region

Considering that direct observation of the earth's deep interior is impossible, it is remarkable how detailed, precise and seemingly reliable a picture of concentric shells with differing physical properties can now be drawn. It has been put together by combining indirect evidence from a variety of sources, including surface measurements of the gravitational field, astronomical evidence of the earth's motion in space, and especially from earthquake wave-paths through the interior. The chief remaining uncertainty is whether a central zone of the inner core has physical qualities to which we would apply the term 'liquid' under surface conditions of pressure and temperature. Under the unfamiliar, almost inconceivably high pressures within the core, it remains a question whether the familiar states of matter (solid, liquid, gas) have the same distinctions under those extreme conditions.

Travel-times of different types of earthquake waves over varying distances around and through the earth show peculiarities best explained by inferring that the physical properties of internal materials must change abruptly at several distinct depths. The levels at which abrupt changes must occur are called 'discontinuities'. The first of these to be located with fair certainty was named after the seismologist who determined its depth. The Mohorovičić Discontinuity ('Moho' for short) is now universally accepted as marking the boundary between the crust and the 'mantle' which enwraps the 'core'.

The distance down to the Mohorovičić Discontinuity varies around the world, being less under ocean floors than under continental land masses. In New Zealand the average depth is about 33 km. This is consistent with the world average for continental regions, and one of the strong reasons for regarding New Zealand not just as a group of oceanic islands but as a truly continental piece of the crust. The crustal thickness varies in different parts of the country from about 30 km to 40 km, greater under the main mountain belt of the South Island and rather less under most of the North Island. Beneath submerged parts of the New Zealand Platform it is intermediate between continental and oceanic thicknesses: 17 to 23 kilometres under the Campbell Plateau

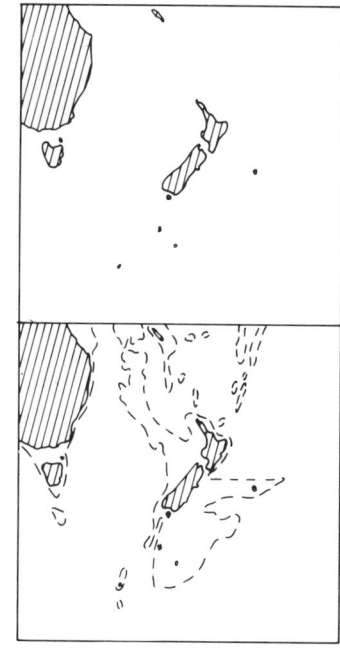

Fig. 10.1
Present-day surface out-lines (upper) suggest no obvious relationship bet-ween the Australian and New Zealand landmasses. (New Caledonia is just inside the upper map bor-der.) In the lower map the New Zealand Platform as outlined roughly by the −2000 m bathymetric contour line, emphasises the important, older con-nection with continental tracts north-east of Aus-tralia, and the newer, north-easterly alignments of the elevated land with the true eastern border of the Pacific Basin. (Adapted from map by C. P. Summerhayes)

and 20 to 25 kilometres under the broad rises and narrow ridges of the sea floor to the north-west and north. Under the deeper ocean troughs and basins it ranges from 5 to 11 kilometres.

What Determined the Pattern?

As we shall see in Chapter Eleven the raised surface features of the earth in detail are the result not of an indefinite number of local uplifts of the crust, but rather of the uneven effects of erosion having lowered the surface by greater amounts in some places than in others. Where raised features owe their prominence directly to vertical or tilting movements of the crust we speak of 'tectonic relief'; where relief has been determined by differences of erodibility within rocks affected by folding and faulting in earlier periods of crustal movement we are dealing with 'structural landforms'. The distinction is important for the purpose of this chapter, but it must be remembered that in either case the pattern of present landscape relief is expressing tectonic events, though in the case of structural relief the connection may be indirect and more ancient. Tectonic deformation was responsible for the internal grain of the underlying rock, but it does not follow that the internal rock structure will always be evident in landscape relief, which may be dominated instead by later block displacements oriented perhaps in different directions. Bold relief can also result of course from volcanic action.

The structural grain in different parts of New Zealand, expressed now in the general topographic plan, is a compound of the effects of the three main orogenic episodes in our geological history, as outlined in Chapters Three, Four and Seven. The late Paleozoic Tuhua Orogeny, to which is ascribed folding and faulting of the Greenland Group and in part the structure of the early Paleozoic rocks of north-west Nelson, cannot be claimed to have much influence on the present-day topographic pattern. Even in places where structures in the oldest rocks seem to be reflected in topographic ridges and valleys it is often hard to be sure that these features are not controlled by the effects of later movements following the older structural grain. Rocks that have once been deformed into folds and faults tend to yield again where possible along lines of weakness determined by the earlier movements. Cross-folding, new structures superimposed transversely or obliquely upon an older set, are recognised however, but in some cases the evidence has come from very detailed studies of structure on the microscopic scale. Examples of superimposed, transverse deformation have been recognised in North Auckland and in the metamorphic rocks of Otago. Needless to say, no influence of Tuhua movements upon present patterns can be claimed except in areas where Devonian and older rocks are still present.

Features determined by effects of the Rangitata Orogeny are more easily identified although the topographic expression is essentially secondary, the guidance of later erosion by folds and crushed rock along faults, while there also have been renewed movements during the Kaikoura Orogeny along earlier lines. It seems highly unlikely that relief forms due directly to Rangitata movements could have survived

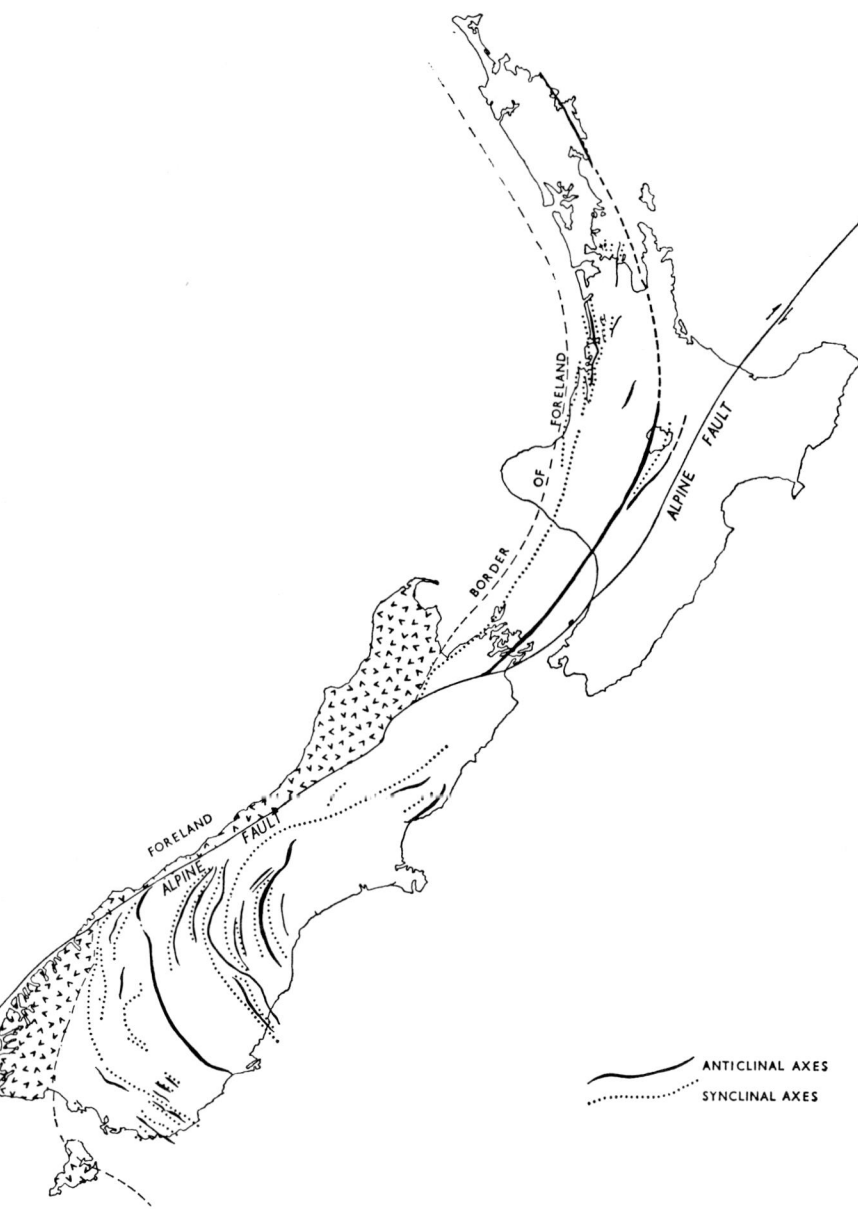

Fig. 10.2
*Axes of folding in sedi-
ments of the New Zealand
Geosyncline attributed to
deformation during the
Rangitata Orogeny. (From
C. A. Fleming, 1970)*

————— ANTICLINAL AXES
················· SYNCLINAL AXES

the erosion which resulted in the Late Cretaceous Peneplain. In so far
as one can 'filter out' superimposed effects of the Kaikoura Orogeny,
the trends of major structures attributed to the Rangitata Orogeny are
aligned north-east/south-west from Canterbury to eastern Otago.
Across Otago, the trend swings with the axis of the main belt of Haast
Schist into a south-westerly direction, while in North Auckland M. H.
Battey considered that the east-west trend of New Zealand Geosyncline
rocks there reflects the true Rangitata trend, whereas the elongation of
the Auckland Peninsula marks a younger Kaikoura axis of uplift. But
as we shall see later in this chapter there has also been the view that the

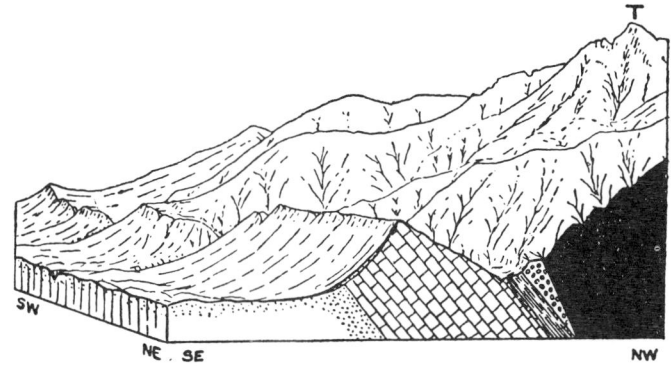

Fig. 10.3
One of the diagrams which C. A. Cotton used in an overseas publication half a century ago to emphasise the geological youthfulness of mountainous tectonic relief in New Zealand. It was intended to show undermass rocks pushed up along a fault which also involved rocks of Tertiary age during the upheaval of the inland Kaikoura Range. T = Mount Tapuaenuku, 2900 m. (From C. A. Cotton, 1925)

directions of Rangitata folds were later distorted during a 'twisting' of the axis of New Zealand.

The Kaikoura Orogeny is recent enough for its effects to be dominant over most of the country, not only in the folding and faulting of the younger covering strata, but directly in tectonic relief. In the west of Nelson, where few traces remain of strata equivalent in age to the New Zealand Geosyncline, very little information can be gleaned about the effects of the Rangitata Orogeny except by indirect inferences from the dates of when certain granites and metamorphic rocks were last at high temperatures (Chapters Five and Six). One thing that can be done is to assume that the lines along which remnants of a once much more continuous cover of Tertiary and early Pleistocene strata are now involved, with older rocks, in faulting were in fact directions determined during the Rangitata Orogeny.

Changing Ideas about New Zealand Structure

Interpretations of the main topographical features and those of the dominant rock structures of New Zealand have developed together, reflecting changes in the theory of crustal deformation and advances in the knowledge of regional geology over more than a century. The pioneering geologists noticed that the directions of North Auckland Peninsula, Cook Strait, Foveaux Strait and the Otago Schist belt differed from most other trends. Early interpretations were in crude terms of simple, anticlinal up-archings and block displacements of the crust, naturally enough, at that stage, with little sense of timing.

Commenting on the unsymmetrical cross-profile of the Southern Alps, its western slopes descending abruptly to western lowlands, Hochstetter suggested in 1863 that only the eastern flank now remains of a formerly broader mountain chain. The western flank was supposed either to have subsided below the Tasman Sea or to have been eroded away owing to its greater exposure to the violence of westerly storms. This view seems to have been accepted by both Haast and Hutton at the time, and was still being quoted by Marshall more than half a century later.

At the beginning of the nineteenth century a remarkable global synthesis of the earth's structural features was put together in Austria by E. Suess. It was to have a powerful influence in broader geological thinking for two generations. Knowing about the contrasting topographical and structural trends in New Zealand, Suess suggested that two major crustal fold trends merged here, making a 'syntaxis'. This word has such an impressive sound that it was used for many years by authors who may or may not have comprehended fully what it implied. Folds have indeed been crowded one upon another, but the notion of a single fold system splitting, or of separate folds merging into one, has some mechanical difficulties.

Early authors were also puzzled about what happens to the main axis of the Southern Alps to the north of Cook Strait, although they could see that the southern Wellington ranges continued the trend of the Kaikouras. The most popular speculation was that a continuation of the alpine chain had subsided before being buried under younger sediments and volcanic products in western and central North Island districts. It was also noticed that the structural axis of the 'core' of the supposed New Zealand 'anticline' lay westwards of the alpine axis and curved across Otago as the axis of the schist belt. This arrangement had much to do with the whole 'anticlinal' concept, because it was still conventionally assumed at the time that metamorphic rocks and granite would have to be the oldest rocks of all.

Patrick Marshall publicised the foregoing ideas in his *Geography of New Zealand* (1905) and again in *Geology of New Zealand* (1912). Incidentally, he was one of the first to see clearly that this country belongs structurally with the eastern rim of the Pacific Ocean basin. Survey Director P. G. Morgan also had misgivings about the simple anticlinal concept. During that same period, mapping in Westland for the newly reorganised Survey, he had been unable to see how greywackes of the almost unmetamorphosed Greenland 'Series' (as they were then labelled) and granites could possibly make the core of an anticline underneath metamorphic schists. Having observed also that an important fault runs along the western base of the Alps (the 'Alpine Fault' was not to receive its title until many years later), Morgan concluded that the alpine mass had been thrust westwards over the less metamorphosed rocks of Westland. The climate of geological thought was then appropriate for such ideas, for the existence of enormous overthrust sheets (refolded and overthrust again) had been confirmed in the western European Alps during the building of several long railway tunnels, but Morgan did not pursue the idea here. For one thing, the forward margins of great overthrust masses are usually quite irregular, whereas the Alpine Fault is remarkably straight.

The next phase was an obsession with explanations involving faults, promoted in particular by Morgan's successor as head of the Survey, J. Henderson. Until about 1937 New Zealand structure was being represented on Geological Survey maps as a mosaic of raised or tilted, rigid blocks, having either moved vertically or rotated about horizontal axes so as to present arcuate fault-boundaries resembling the curved cracks developed by rigid materials under stress. There was little field evidence to support either version.

Within the next few years a great revolt was to develop. First came a suggestion from H. W. Wellman and R. W. Willett in 1942 after a Haast-style flying reconnaissance of South Westland geology that the Alpine Fault had moved laterally as well as vertically. Then, in the

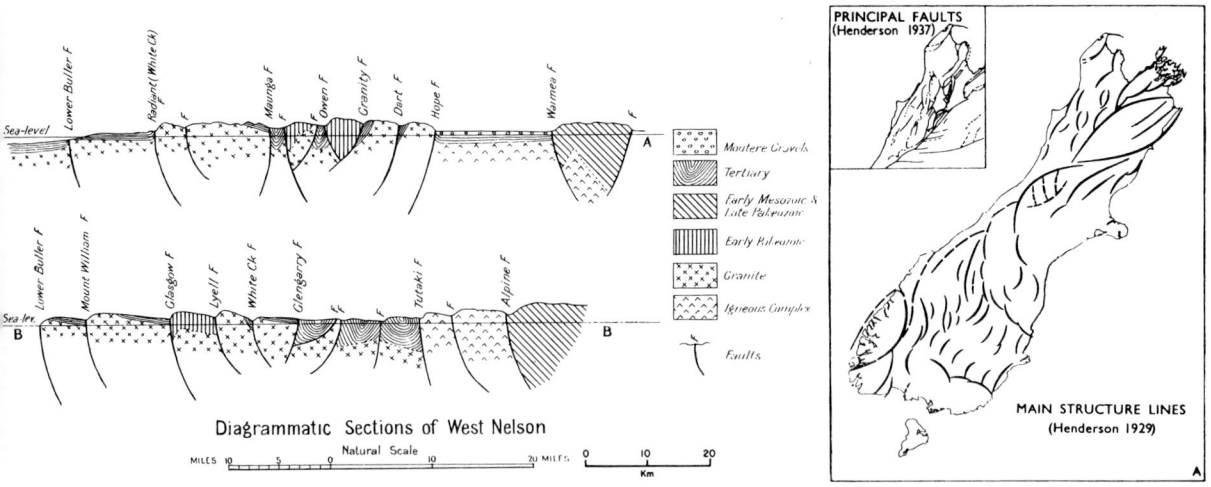

Diagrammatic Sections of West Nelson

PRINCIPAL FAULTS
(Henderson 1937)

MAIN STRUCTURE LINES
(Henderson 1929)

Fig. 10.4 (above)
There was little real evidence to support Henderson's interpretation of faulting in terms of horizontally rotating crustal blocks, or his picture of arcuate structural axes. An inspection of modern geological maps will soon show that no such idealised schemes could possibly accommodate the complexities of rock structure we now know to exist. (Henderson, 1937; R. P. Suggate, 1963, after Henderson)

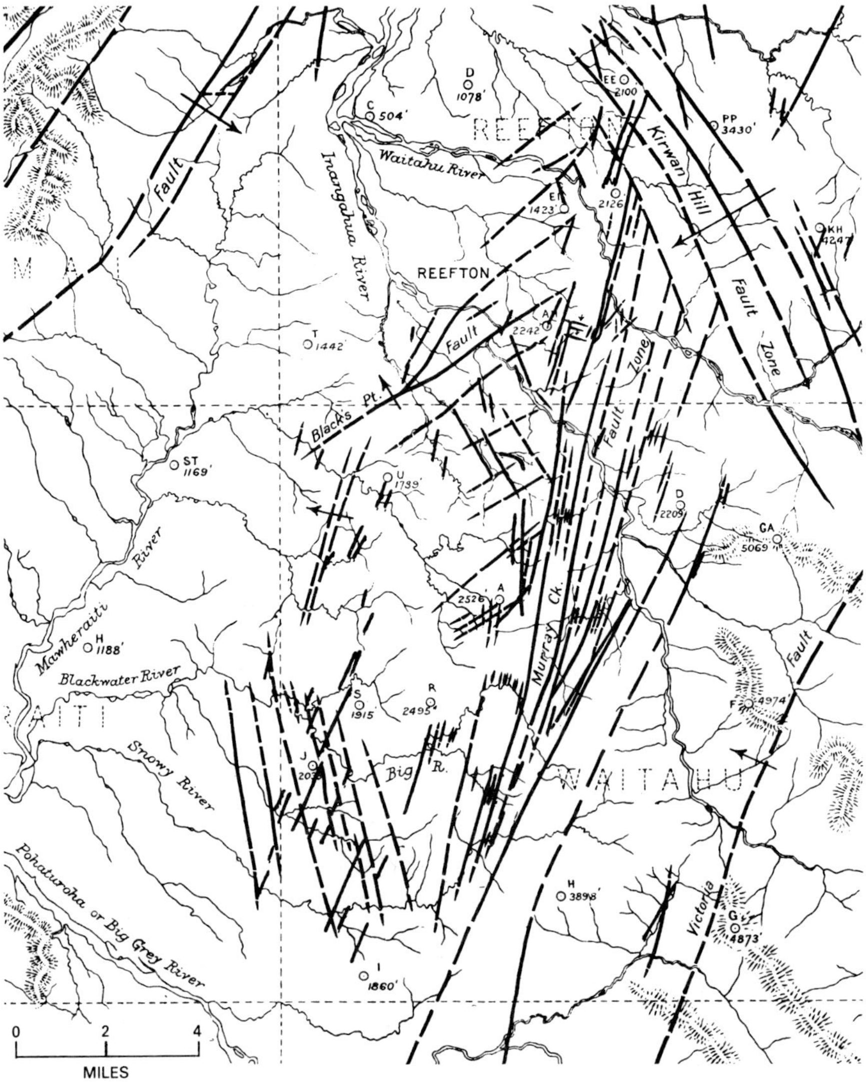

Fig. 10.5
This portion of a map of the Reefton district by Henderson in the World War I era illustrates the earlier fashion of interpreting the structural relationships of the rocks in terms of faults, largely hypothetical. Compare with Fig. 10.6. (Adapted from Map 9, with N.Z. Geological Survey Bulletin 18, 1917)

1940s came a number of papers and Survey *Bulletins* giving greater importance to folding, both in the basement and covering strata. For instance, the results of a revision of Reefton geology completed before the Second World War in connection with gold-quartz prospecting and compiled by the author (who also did some of the remapping), but not published until 1948, replaced Henderson's previous unrealistic picture of a fault mosaic with a very different pattern of well-authenticated fold structures in the Greenland rocks. In 1946 an exciting, new synthesis of New Zealand structural information appeared under the authorship of E. O. Macpherson, with major emphasis upon folding. The breakaway from old, rigid ideas was complete.

Macpherson was recognised as a man of ideas, an independent thinker with a vast fund of knowledge about New Zealand stratigraphy and structure gathered in the course of long experience in regional mapping, mining and petroleum exploration. Though folding was all-important in his scheme, the collapse of overturned folds, with shearing-out along the axial plane and overthrusting of the upper flank, figured largely in his interpretations. He claimed to recognise important overthrusts in the East Cape region and along the line of the Alpine Fault (he did not use that name either). For the first time, the linear arrangement of Cretaceous and younger volcanoes was related to the trends of crustal folding on a basis of field relations rather than theory.

Macpherson dismissed as of secondary importance the northwards continuation of North Island structures from East Cape into the Pacific Basin, which Marshall had emphasised. Instead, he united the two contrasting north-westerly and north-easterly trends into one system by supposing that a single major axis of crustal folding had later been doubly-bent into a reversed-S curve convex to the east in the North Island and to the west in the south of the South Island. Although vague on many crucial points, blind to some implications of superimposed, transverse foldings, and rather less than objective as regards other contrary evidence, the Macpherson scheme was a refreshing change. Though in some ways a return to the philosophy of Suess (to whom he refers as 'the master'), it foresaw what is now the essence of modern interpretations of distortion and stretching-out of the New Zealand Geosyncline during the history of the Alpine Fault displacements. It was the source of stimulus and provocation to investigate New Zealand structure until 1952, when Wellman proclaimed another revolutionary notion which he had aired tentatively a few years earlier. This was to the effect that the small amount of lateral shift on the Alpine Fault which he and Willett had reported in 1942 was merely the most recent increment of a total displacement of nearly 500 km since the Jurassic Period. Wellman's Alpine Fault findings coincided with recognition of other similar 'transcurrent' or 'wrench' faults of large displacement in other parts of the world, notably in California.

Various authors have since added detail to what is known about the structure of the older rocks, pondered about the true meaning of the two trends, and applied an increasing amount of evidence from micro-structures, cleavage and other smaller-scale deformational

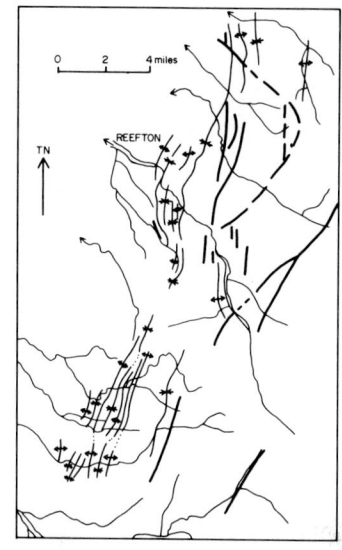

Fig. 10.6
Roughly the same area as in Fig. 10.5, as re-mapped in 1936–7, structural relationships now in terms of folding in the older rocks, and faults represented much more conservatively. (Adapted from maps with N.Z. Geological Survey Bulletin 42, 1948)

features towards a better understanding of the nature and directions of stresses within the crust of this region at different times in the past. Confirmation of great lateral movement on the Alpine Fault marked the end of an era when broad structural interpretations were based on reconnaissance and scattered information from earlier surveys. Beginning with the 1:250,000-scale mapping programme in 1959, much old work was revised, new areas mapped for the first time, and structural views brought into line with modern thought on tectonic matters. A further revolution, and a re-evaluation of the way in which the pre-Cretaceous rocks in particular have acquired their structures, has followed the appearance of the concept of moving crustal plates and sea-floor spreading (see later). This greatly affects ideas about our place in the Pacific, and if correct, would limit the possibilities as regards the directions and origins of the crustal stresses which folded the rocks, and elevated the 'New Zealand Anticline'; we have come a long way since Hochstetter's time.

The New Zealand Orogen—Long-lived Crustal Mobility

When early geologists described the general structure of New Zealand as merely an arching of the crust, one side of the arch having subsided or otherwise vanished, it must have been apparent even then that this was an oversimplified picture. The term 'anticlinorium' has been available for more than a century to denote rock folds grouped together so as to constitute a major arch structure, with the tacit implication that the subsidiary folds run parallel with the trend of the major fold. It has been applied occasionally to the New Zealand ridge as a whole, but never achieved general acceptance probably because it conveys no idea of the degree of structural complexity and diversity here, and is inappropriate where the trends of subsidiary folds by no means consistently run parallel with the major axis. From another page of the geological dictionary, 'geanticline' to begin with was almost synonymous with anticlinorium but later came to mean an upheaved belt supplying sediment to a neighbouring geosyncline, in which sense it is very near to 'foreland'. Despite its ambiguity, 'geanticline' has been applied by New Zealand geologists in fairly recent times, by some to denote a general axis of upheaval (J. T. Kingma, for example), and by others in the 'foreland' sense.

How, then, are we going to describe this crustal welt? Structural geology has developed a deplorably large vocabulary with many subtle shades of meaning, some pertaining only to particular theories. Luckily, there is one widely understood, relatively uncomplicated term, neutral as regards competing theories of diastrophism. The word 'orogen' denotes a belt of rocks, deformed and usually metamorphosed, making up a mountain system. It is convenient when dealing with a region where structures and metamorphic effects have resulted from a long history of crustal mobility, expressed by a succession of orogenic cycles. Such is the history of the New Zealand region.

In 1974 the Geological Society of London published a world-wide review of information about the histories of crustal movements in different countries. G. W. Grindley contributed a section about New Zealand, in which he used 'orogen' in the sense indicated above for the entire system of superimposed foldings, igneous intrusions and metamorphic belts making up the structural basis of New Zealand. He summed up its complexity and long history in these words: 'Because of its unique situation on the mobile boundary of the Pacific, successive orogenies have been superimposed through Phanerozoic [= Cambrian and later] time giving a complexity of structure that has made this a testing ground for geologists and a graveyard for tectonic theories. Even now, with modern plate tectonic theories carrying all before them, difficulties remain in applying these theories rigorously to the New Zealand orogen.'*

It will be useful here to bring together a summary of the successive diastrophic episodes described in earlier chapters of this book. The earliest history of the New Zealand orogen is veiled by the mists of the past, but we know that in late Precambrian times an ancient terrain composed of granite and other continental kinds of rock must have existed in this sector of the ancestral Pacific. Sediment eroded from it, together with material erupted by contemporary volcanoes, went into the formation of the Haupiri, Greenland and other rock groups of the Buller Geosyncline.

The well-recorded history of the orogen as a belt of deformed rocks begins with the start of the Tuhua Orogeny about 400 million years ago, when the Buller Geosyncline deposits were folded, mildly metamorphosed and upheaved, according to Grindley's interpretation. Perhaps it should be thought of as beginning a little earlier, as a curtain-raiser to the main Tuhua event when deposits in the eastern part of the geosyncline were compressed sufficiently to be squeezed out of the trench and caused to slide gravitationally as overthrust sheets across the western part of the geosyncline. Following a lull in early Devonian time, during which the sea flooded over part of the orogen, the second and main phase of the Tuhua movements then produced sharp folds with a north-south trend in the older strata and refolded the thrust-sheets. The Karamea Granite was generated at the same time.

This was not the first attempt to apply to New Zealand the kind of tectonic ideas evolved in Europe after a century of investigation of the western Alps. The concept of squeezed-out and detached thrust-sheets has been tested in the East Coast region on several occasions since the mid-1920s but for various reasons has had to be rejected. Grindley was indeed bold to advance ideas of re-folded thrust sheets as a way of explaining the complexities of Paleozoic geology which he had discovered in the bush-covered mountains of west Nelson. At the time it enabled him to make a consistent and credible (if surprising) story out of his field observations, but recent new fossil discoveries by R. A. Cooper may necessitate substantial modification of Grindley's views. This is the way that every active science must progress.

---

* Introduction to 'New Zealand' in: 'Data for Orogenic Studies, Mesozoic-Cenozoic Belts', A. S. Spence, ed., *Geological Society of London Special Publ. 3*, 1974, p. 387.

Fig. 10.7
Sketch maps illustrating
Suggate's view of how and
when the lateral movement
on the Alpine Fault began.
(Adapted from Suggate,
1963)

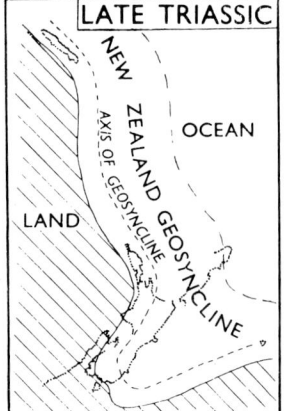

LATE JURASSIC

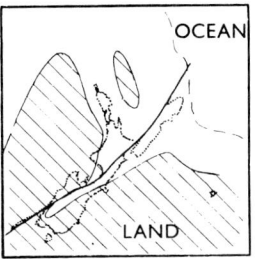

MID-CRETACEOUS

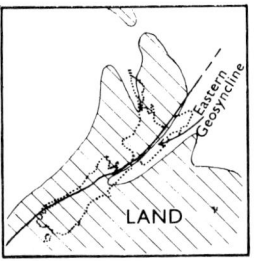

END CRETACEOUS

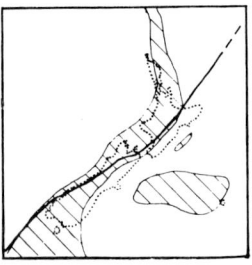

The episode to follow was one of prime importance in the history of the New Zealand orogen. The rocks which accumulated in the New Zealand Geosyncline were destined to become, directly and indirectly, the major source of raw material from which most of New Zealand has since evolved—in substance and in form.

It was mentioned in Chapter Four that the developing phase of the New Zealand Geosyncline was not without interruptions. Compression, folding and some upheaval of the older geosynclinal contents happened about the middle of the Triassic Period. Whether this should be considered as heralding the start of the Rangitata Orogeny depends upon one's choice of orogenic theory. Certainly there is strong evidence that the trough was more than adequately supplied with sediment from early Jurassic time onwards, because a large proportion of the sedimentary rocks dating from then is non-marine in origin. By the end of the Jurassic Period the main axial zone of the geosyncline, marked now by the main belt of Haast Schist, began to rise and the climax of the Rangitata movements accompanied by the raising of a mountainous terrain was imminent.

Complete agreement is lacking as to the timing of events, but in one view, the distortion which twisted an originally straight orogen into the reversed-S shape took place in early Cretaceous time and became more pronounced until rupture along the axis initiated the lateral, strike-slip movement along the Alpine Fault. In the same period, about 130 million years ago, metamorphism and deforming of the geosynclinal rocks was approaching its peak. There is no doubt that the crescendo of Rangitata movements is a complicated story of folding, shearing, metamorphism, dyke-injection and granite emplacement spread over at least 50 million years. The details of timing and sequence of events have been pieced together with the help of three-dimensional microscopic study of slices made from pieces of rock of which the orientation in the outcrop has been recorded, along with detailed field measurements of the attitudes of cleavages and joints, and occasional datings by radiometric means. Chemical analyses of individual mineral components and of whole-rock samples help to recreate a picture of the chemical and physical environment in the orogenic 'cooking pot'.

The Rangitata climax is recorded by breccia and conglomerate deposits in Buller, Westland and Otago, marking a phase wherein vigorous erosion was being outpaced for a time by mountain upheaval; the decline is marked by profound weathering and erosion overtaking orogenic upheaval and producing the Late Cretaceous Peneplain.

With invasion of the land by the ocean again before the end of the Cretaceous Period the final cycle had begun. New troughs of sediment accumulation were not all aligned parallel with the structural grain imposed by the Tuhua and Rangitata orogenies, and some of them subsided sufficiently to accommodate exceptionally thick sequences of late Cretaceous and early Tertiary sedimentary rock in Westland, southern Nelson and the North Island East Coast. Following a lull in the Oligocene Period when, according to some authorities, the whole New Zealand region was broadly up-warped (and to a few others, also disrupted by large-scale fault movements) the whole tenor of events in

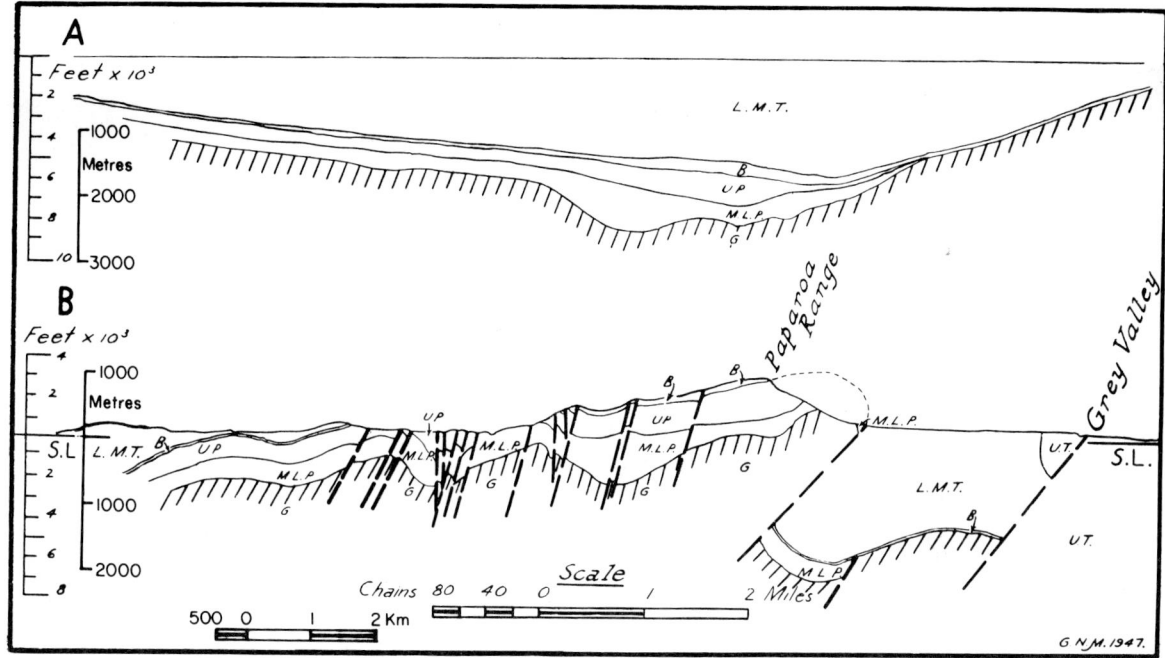

Fig. 10.8
The upper half of the diagram is a hypothetical cross-section of the Paparoa tectonic trough in Oligocene times, at the end of its long period of subsidence, and at the maximum stage of sub-mergence under the sea. The lower half depicts the southern Paparoa Range at the present time. The range rose during the late Tertiary and early Pleis-tocene along the site of the former trough with ever-sion of its contents and the initiation of a new trough on the eastern side, the Grey-Inangahua Depres-sion. The boundary has been mapped as the Roa Fault, and is part of the 'Paparoa Tectonic Zone'.

the New Zealand orogen suffered a dramatic change. In Westland, for example, troughs which had been subsiding since even before the Late Cretaceous Peneplain was formed began instead to be upheaved and everted, their sedimentary contents eroded and redeposited in new, adjoining troughs. Trenches became ridges. The directions of movement on some major faults reversed at this time, so obviously there were important changes in the crustal stress systems affecting the orogen. The effects are also well marked in North Auckland where limestone and mudstone deposited in troughs on the eastern side in early Tertiary time were squeezed out bodily on the sea floor, sliding as detached sheets ('olistostromes') and fragmented but coherent masses ('chaos-breccia') towards the western side. The confusion of the stratigraphy of North Auckland from these events baffled everyone until the last few years. All these dramatic happenings seemed to come to a head in the Miocene Period, when there was also widespread volcanic activity in many parts of the country. It is widely held that they mark the onset of the Kaikoura Orogeny, the climax of which in Pliocene and early Pleistocene time was to be responsible for shaping the present mountainous relief of New Zealand.

This condensed account gives an impression that the history of the New Zealand orogen was a perpetually busy affair, and indeed it was by comparison with what was happening over the same span of time in some other regions of the world. It is worth remembering, though, that we deal with events spread through some 600 million years or more. Even the dramatic 'finale' (not the right word, perhaps—the Kaikoura movements have not ceased) has taken 30 million years to develop. After all, consistent displacement averaging only a few millimetres a year can amount to kilometres in a million years.

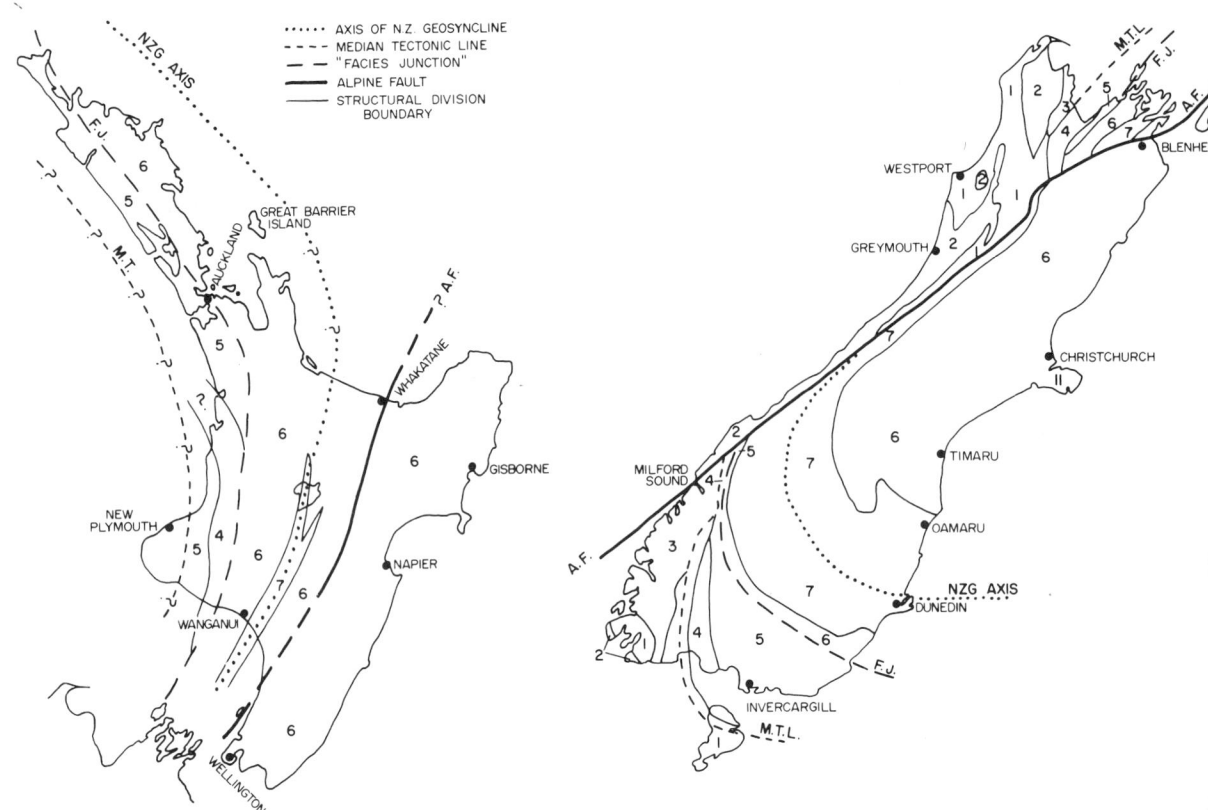

LEGEND:
- ······ AXIS OF N.Z. GEOSYNCLINE
- – – – MEDIAN TECTONIC LINE
- – · – "FACIES JUNCTION"
- ——— ALPINE FAULT
- ——— STRUCTURAL DIVISION BOUNDARY

Fig. 10.9
Major structural features of New Zealand arising from the Rangitata Orogeny and earlier tectonic events. (Figures refer to text, pp. 279–80.) (Compiled from various sources)

Structural Divisions of New Zealand

In 1956 in a small book called *The Structural Outline of New Zealand* H. W. Wellman suggested a way in which the country could be divided into 'structural regions', twenty in all, which he described as natural divisions distinguished by structural differences that originated before the Cretaceous Period. It was not possible to carry this out consistently, however, and in the North Island especially, where so much of the earlier geology is hidden under Cretaceous and Tertiary deposits, he had to rely on the way in which the structure and distribution of the younger strata *appear* to reflect conditions underneath. The basis on which he separated the regions in fact varied considerably. The publication was welcomed by most New Zealand geologists as a timely survey of current knowledge of the stratigraphy of different regions as well as a structural analysis. It was, incidentally, the vehicle by which the concept of the New Zealand Geosyncline came to the notice of many people, although Wellman did not invent the term. Other structural classifications have appeared from time to time, but in relation to some specific topic such as earthquake distribution, grades of metamorphism, etc. The most recent, by G. W. Grindley in 1974, was devised especially to illustrate his historical review of the New Zealand orogen as discussed above.

The following paragraphs present yet another scheme of structural divisions to suit the requirements of this book. It is essential to make a

primary distinction between structural divisions determined chiefly by
events prior to the Rangitata Orogeny and those defined by later
events.

Divisions Determined by Rangitata or Earlier Events

(1) *Older granite and gneissic-granite areas* of western Nelson and
Westland which, though now discontinuous, may originally have been
parts of a continuous pluton in depth. The granites were emplaced
during the Tuhua Orogeny around 300 million years ago, generally
follow the structural grain of the early Paleozoic rocks and became part
of the foreland for the New Zealand Geosyncline. The division must
include the areas of Karamea and Paparoa granites, and probably some
Fiordland granites as well.

(2) The area covered by sedimentary strata of the *Buller Geosyncline*,
comprising the Greenland, Haupiri, Aorere, Golden Bay, Mount
Arthur, and Baton River groups of marine and volcanic sediments and
contemporary lava rocks, makes a well-defined unit. Though in a
different setting, the Reefton Group must be included. The structural
trend is generally northerly or north-westerly in Nelson and North
Westland, but is complicated by the north-easterly trend of these rocks
nearer to the Alpine Fault (which, incidentally, led to the separation of
Waiuta Group from Greenland Group).

(3) The *high-temperature/low-stress metamorphic and associated igneous
rocks* (for which C. A. Landis and D. S. Coombs regrettably chose the
already overworked prefix 'Tasman' in 1966) constitute the schist
terrains of north-west Nelson, the Tasman and Cobb intrusive rocks
and the gneissic areas of Fiordland.

(4) *Younger granites and gneisses* of the Separation Point batholith in
Nelson, the western granites of Fiordland and probably the Rakehua
Granite of Stewart Island, all emplaced around 100 million years ago,
deserve to be distinguished from division (1). In Nelson the trend is
mainly north-north-easterly, and in the far south it is south-easterly in
conformity with the curvature of the whole orogenic belt.

(5) The areas of the *Hokonui Facies* of New Zealand Geosyncline
sediments and associated igneous rocks make an easily definable unit
in Southland and Nelson, and in western Auckland. Its presence under
younger rock cover in Taranaki and under Cook Strait has been
detected geophysically because of magnetic properties due to the
abundant mafic igneous detritus it contains. There is some question as
to how the tracts of Rotoroa Igneous Complex rocks in Nelson and the
Darran Diorite in west Otago should be treated, but I have decided to
include them with the Hokonui unit. The pattern of folding in the
Hokonui Facies belts, dating from the Rangitata Orogeny, is
comparatively simple. The Southland Syncline, Nelson Syncline and
Kawhia Syncline are broad, relatively simple structures, modified by
longitudinal faults and minor superimposed folds, some of which are
probably later effects of the Kaikoura Orogeny. Trends, however, are
essentially those of the re-curved arc of the New Zealand orogen.

(6) The *Alpine Facies* of the New Zealand Geosyncline, represented
most widely by the Torlesse Supergroup rocks, forms the basement

under eastern parts of the country for most of the way from North Otago to North Cape. Internally, structures are more complicated on the whole than in the Hokonui Facies, including overturned folds, steeply plunging folds, thrusts and innumerable faults. The strike of minor structures by no means consistently follows the general trend of the orogen, and in many places it can be shown that the deformation occurred soon after deposition, before the sediment was compacted to rigidity. Early attempts to resolve the structures were hindered by the scarcity of time-marking fossils.

From about 1950 onwards, and especially during the 1:250,000 mapping programme, an understanding of the structural style of the Torlesse rocks was built up with the help of field mapping techniques designed to overcome the difficulties due to lack of fossils and the limited extent of distinctive, individual beds. (The structure of the central Southern Alps is discussed separately later.)

An important structural boundary is provided by the 'Facies Junction' between structural units (5) and (6). (Chapter Four, p. 94.)

(7) The *Haast Schist Belt of low-temperature/high-stress metamorphism* completes the list of major structural units in the basement rock. It too follows the curvature of the New Zealand orogen, and the axial line for the New Zealand Geosyncline has sometimes been shown as lying somewhere within the Haast Schist Belt. Offset by the Alpine Fault at Lake Rotoroa, it continues north-eastwards again along the Richmond Range to the Marlborough Sounds and Cook Strait, and reappears in the southern end of the Kaimanawa Mountains. Presumably it runs beneath the intervening area of younger rock cover.

The *Median tectonic line* is a significant, if ill-defined structural boundary between the zones of high-stress and low-stress metamorphism. It is believed to represent a narrow zone across which the geothermal gradient (the rate of temperature-increase with increasing depth below ground) changed markedly in the past. According to Landis and Coombs it is indicated by a fault in western Otago—the Skippers-Skelmorlie Fault.

Divisions Determined by Post-Rangitata Events

It is harder to make generalisations about these. I shall not attempt a complete coverage, but will present examples of the important distinctions reflecting the more diversified events leading up to the climax of the Kaikoura Orogeny. Crustal movements and different patterns of sediment distribution were more localised than in earlier times.

(8) *Sedimentary Basins.* Structurally, these are rarely if ever true basins, but usually furrows or fault-angle depressions which continued to subside long enough for substantial thicknesses of sediment to be trapped in them. Separate 'basins' developed at various times between the late Cretaceous and early Pleistocene Periods. Examples notable for the thickness of sediment that accumulated in them are the early Tertiary Paparoa Trough and the late Tertiary Grey Valley Trough alongside it. The former everted in Miocene and later times to become the site of the present southern Paparoa Range; the Murchison Basin

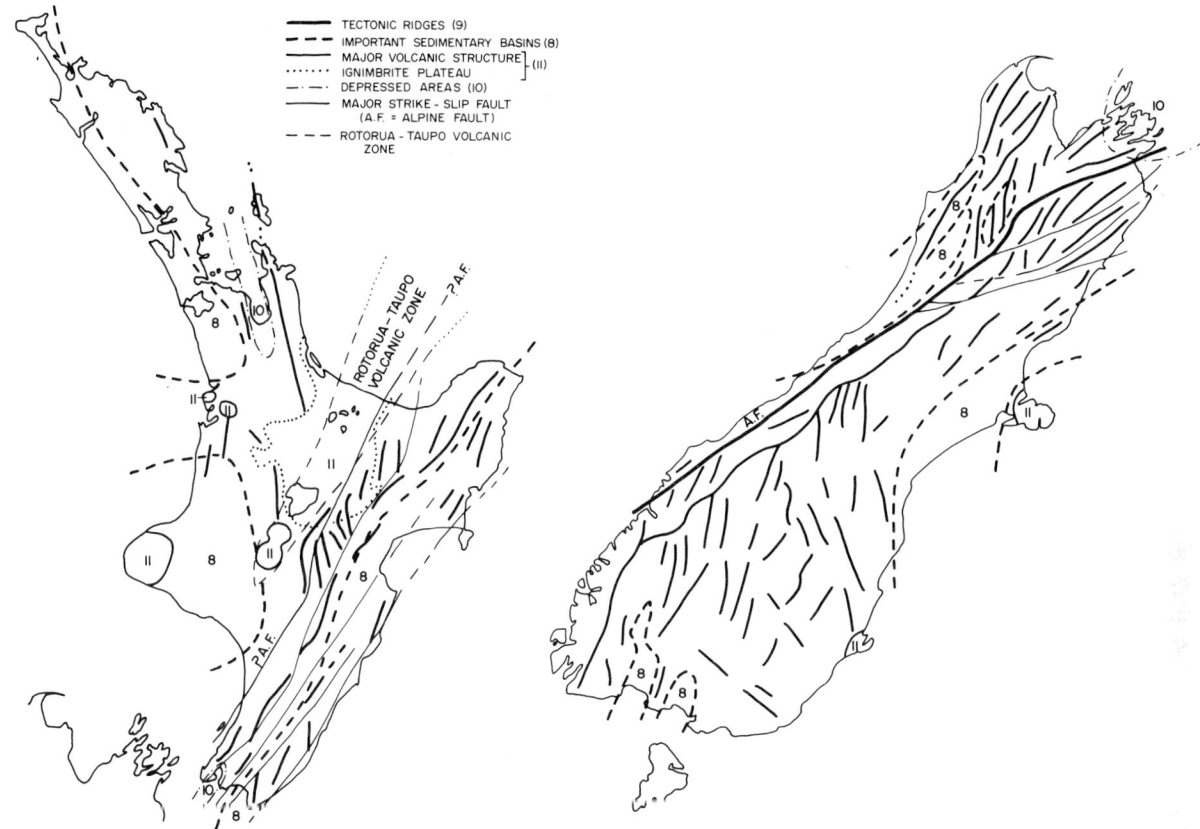

TECTONIC RIDGES (9)
IMPORTANT SEDIMENTARY BASINS (8)
MAJOR VOLCANIC STRUCTURE
IGNIMBRITE PLATEAU   (11)
DEPRESSED AREAS (10)
MAJOR STRIKE - SLIP FAULT
  (A.F. = ALPINE FAULT)
ROTORUA - TAUPO VOLCANIC
  ZONE

Fig. 10.10
Major structural divisions arising from post-Rangitata events. (Figures refer to text, pages 280–83)

containing several formations of varying thickness up to more than 3000 m; the Waiau Basin in Southland, with sediments up to 2600 m thick; the Taranaki Basin incorporating several smaller basins, gentle fold structures and small faults, and containing as much as 5000 m thickness of mainly late Tertiary sediments; and the North Island East Coast belt of Cretaceous and Tertiary sediments, structurally extremely complex in places and containing up to at least 7000 m of sediment in the Hawkes Bay and Poverty Bay regions. Oil prospecting and geophysical surveys in recent years have indicated that surprisingly thick Cretaceous and Tertiary rocks underlie western Northland, the Nelson depression and Tasman Bay. All of these basins are economically important in connection with our oil, gas and coal resources. Offshore oil exploration shows that similar structures and thick Tertiary sequences continue beyond the present coastlines. Being essentially the result of the movements preliminary to the Kaikoura Orogeny, these structures nearly all have a north-easterly trend.

(9) *Tectonic ridges* (for want of a better term) including some symmetrical fault-bounded horsts but mainly tilted crustal blocks fault-bounded on one side only, provided the basis of most of the individual mountain ranges of both islands. In some cases it can be shown with virtual certainty that a former cover of Cretaceous or Tertiary strata has been stripped away to expose basement rocks in the core of the present ranges during the Kaikoura upheaval. To mention

Fig. 10.11
The axial plane of this fold in Cretaceous strata in the Gisborne district appears to lie roughly horizontally, making it a 'recumbent' fold. It is not always possible to decide whether the complex folding shown by Cretaceous rocks in this region is the result of crustal compression movements connected with the Kaikoura Orogeny, of large-scale disturbances of the sediments before consolidation, or of gravitational slumping and collapse of the strata after the Kaikoura uplift of the region, in late Pleistocene times. (Photo: R. D. Black)

Fig. 10.12
Looking northwards along the western margin of Castle Hill Basin, inland Canterbury, across the headwaters of Thomas and Broken rivers with the Craigieburn Range rising abruptly at the left. The margin is marked by a system of west-dipping reverse faults. Structures within the Basin itself are complex, involving strong folding and faulting of the late Cretaceous and Tertiary covering strata and faulting of the underlying Torlesse greywacke basement. A south-westwards-plunging anticline towards the upper right of the picture is shown up by a curving ridge of outwards-dipping limestone and a low spur of greywacke exposed in the core of the fold. These structures are attributed to crustal compression during the Kaikoura Orogeny. (Photo: V. C. Browne)

a few examples, this has been demonstrated for the Central Otago mountain blocks, the Paparoa Range in Westland, the Tararua-Ruahine Range near Manawatu Gorge, and the Kaikoura Ranges themselves. The Hokonui Hills in Southland and the North Auckland regions provide the main exceptions to the rule that these features tend to be aligned in north-easterly directions. Similar basement ridges, not yet uncovered, are known to underlie Tertiary strata in Westland and probably also the Canterbury Plains gravels.

(10) *Depressed areas* attributed to the continuation of Kaikoura movements into late Pleistocene and recent times include the Marlborough Sounds sunkland, the Taupo Graben (affected perhaps also by the amount of material that was emitted by volcanoes in the

Fig. 10.13
Crushed and sheared
Paparoa Granite thrust
over Oligocene limestone
by the View Hill Fault on
the Westport-Karamea
road a few kilometres
east of Mokihinui. The
compressive phase of the
Kaikoura Orogeny pro-
duced many such over-
thrusts and steeper reverse
faults in the West Coast
region.

Fig. 10.14
A reverse fault (see p.
393), dipping westwards
at a moderate angle, dis-
places glacial gravels and
loess deposited early in the
Otira Glaciation. The
faulting reflects a
continuation into late
Pleistocene times of the
compressive Kaikoura
movements responsible for
the uplift of the foothills
ranges of Canterbury. The
locality is in a ballast pit
alongside the Midland
Railway in the
Waimakariri valley close
to a possible northwards
extension of the crush-zone
of the Porters Pass Fault.

Rotorua-Taupo Volcanic Zone), the Port Nicholson (Wellington-Hutt
Valley) depression; and the Hauraki trough occupied by the Ngatea
Plains and the Hauraki Gulf. Some of the Tertiary sedimentary basins
(8) (above) continued to subside into late Pleistocene time.

(11) *Major volcanic structures* built up by constructive types of
eruptions since Miocene time cover substantial areas of both islands.
The Tongariro volcanoes, for example, have substantially built up the
level of at least 1000 square kilometres of the North Island central
highlands with volcanic products since early Pleistocene time; Egmont
has built up a comparable area in the same period. The area originally
overwhelmed by ignimbrite sheets would be hard to determine. In
1935 P. Marshall, who first determined the true character of the

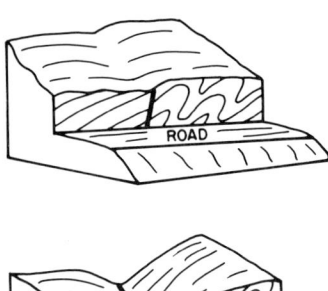

Fig. 10.15
A fault directly observed
from outcrop (upper
diagram), and a fault
inferred as being the most
likely and simplest
interpretation of the
obscured boundary
between two unrelated rock
formations (lower
diagram). It is also
inferred that the alignment
of the stream valley has
been determined by erosion
of weakened rock exposed
along the trace of the
fault.

ignimbrite plateau, gave an estimate of 10,000 square miles (say, 25,000 square kilometres) which is not far short of one quarter of the North Island area. Even if this estimate probably errs on the high side, there is no question about the importance of the ignimbrite plateau as a structural entity. Banks Peninsula volcanoes, Dunedin volcano, Pirongia and Karioi in south-west Auckland probably complete the list of substantial, geologically young volcanic features.

The foregoing classification does not take into account the larger alluvial plain areas (Canterbury Plain, Southland Plain, Takapau Plain) as structural features. They are, however, important physiographic units.

Fault Lines on the Maps—What they Can Mean

The term 'fault' applies properly to any dislocation in rock in which the material on one side has moved relative to that on the other. There are no restrictions as to the amount or direction of movement in this definition, but a rather complicated language for describing faults has had to be devised to provide names for all conceivable relationships of movement-direction, attitude of the fault surface and attitude of the stratification or other layering in the dislocated rock. Luckily, it is not necessary for the purpose of this book to become involved in the terminology of faults to any extent. (See Appendix 1.) It would be as well, however, to draw attention to the fact that, as a consequence of changing ideas and policies through the years, the lines denoting faults on geological maps drawn at different periods and for different institutions do not always mean the same thing. The map legend should be an adequate guide, but this is not always the case.

Depending upon the scale of the mapping, there must always be a lower limit to the magnitudes of faults than can be shown clearly without cluttering the map. The width of the finest line that can be printed usually represents a belt of country many times wider than that affected by the fault in the ground. Fault lines find their way on to geological maps in different ways. A fault separating rocks of different type and age actually observed in a mine or roadside outcrop, for example, may be considered important enough to show on the map, and the line showing it may have to be exaggerated in length far beyond the observed extent of the fault in order to show at all. Separate exposures of what the geologist believes to be the same fault may be joined up by a line across the map, drawn (alas, too rarely) with due regard for probable deviations from straightness because the fault surface is inclined and the topography hilly. Surface features due to differences of rock on either side or due to recent displacement of the ground may help to trace its course. Faults thus mapped should be geological realities and reasonably accurately located.

Faults appear on maps also on much less real evidence. Particularly in reconnaissance, faults are often inferred to be present as the most reasonable way of explaining the relative positions or attitudes of different formations, though no actual fault is visible. Usually such inferred faults are indicated by a broken line, which, however, does not always show the distinction between faults that are assumed to exist (for good reasons) and those of which the existence is not in doubt but the location is uncertain.

Faults are very numerous in New Zealand, especially in the older rocks, and it would be impracticable to show them all on any reasonable map scale. Some selection is necessary, but it is often difficult or quite impossible to estimate the amount of relative displacement ('slip') on faults visible in an outcrop. Many visible ones are of minor importance, while major ones of large displacement tend to be obscured in natural outcrops by slumping debris from crushed rock.

The policies of geological mappers in this country have not been very consistent in these matters. The early explorers were commendably cautious and conservative about showing faults, being well aware of the inadequacy of their knowledge. On the other hand, many maps produced by the Geological Survey, especially during the times of Morgan and Henderson as directors, are criss-crossed by fault lines for which evidence is slim or absent. Moreover, the lines as printed take little account of how the topography should affect the surface trace of any fault which is not vertical. The situation improved rapidly from 1940 onwards, largely because topographical maps showing heights by contouring became available, together with aerial photography, to assist the geologist in the field.

Compilers of the 27 different sheets making up the 1:250,000 scale geological map series were not over-regimented. Some co-ordination was essential, but there are nevertheless some differences from sheet to sheet reflecting not only real variations in the abundance and style of faulting on the ground, but also the exigencies of scale and differences in authors' philosophies as to how best to give an impression of regional structural patterns. The scale is, of course, a reconnaissance one, and only a minute fraction of the faults actually observed could be shown on the final maps.

The contrast between Sheet 6 (East Cape) and Sheet 8 (Taupo) as regards the pattern and closeness of spacing of faults illustrates the point. It would be hard to confirm the existence of many individual faults as shown on Sheet 6, but it was the author's aim to bring out the general pattern of deformation which in his opinion showed up the kinds of tectonic process that had been dominant in that region.

A majority of the faults shown on our maps were active at some stage during the Kaikoura Orogeny, including in the older rock areas some that had histories of movement dating from the Rangitata Orogeny or earlier. Those that have moved historically or since the present landscape evolved are distinguished by colour (red) on the 1:250,000 series, but not consistently on older maps.

Slices of New Zealand—the Great Strike-slip Faults

A system of horizontally-offsetting dislocations, including the Alpine Fault, dominates the structural picture in central districts. Altogether they constitute one of the more distinctive aspects of New Zealand structural geology.

As noted earlier in this chapter, the fashion of representing New Zealand structurally as a mosaic of earth-blocks differentially raised in

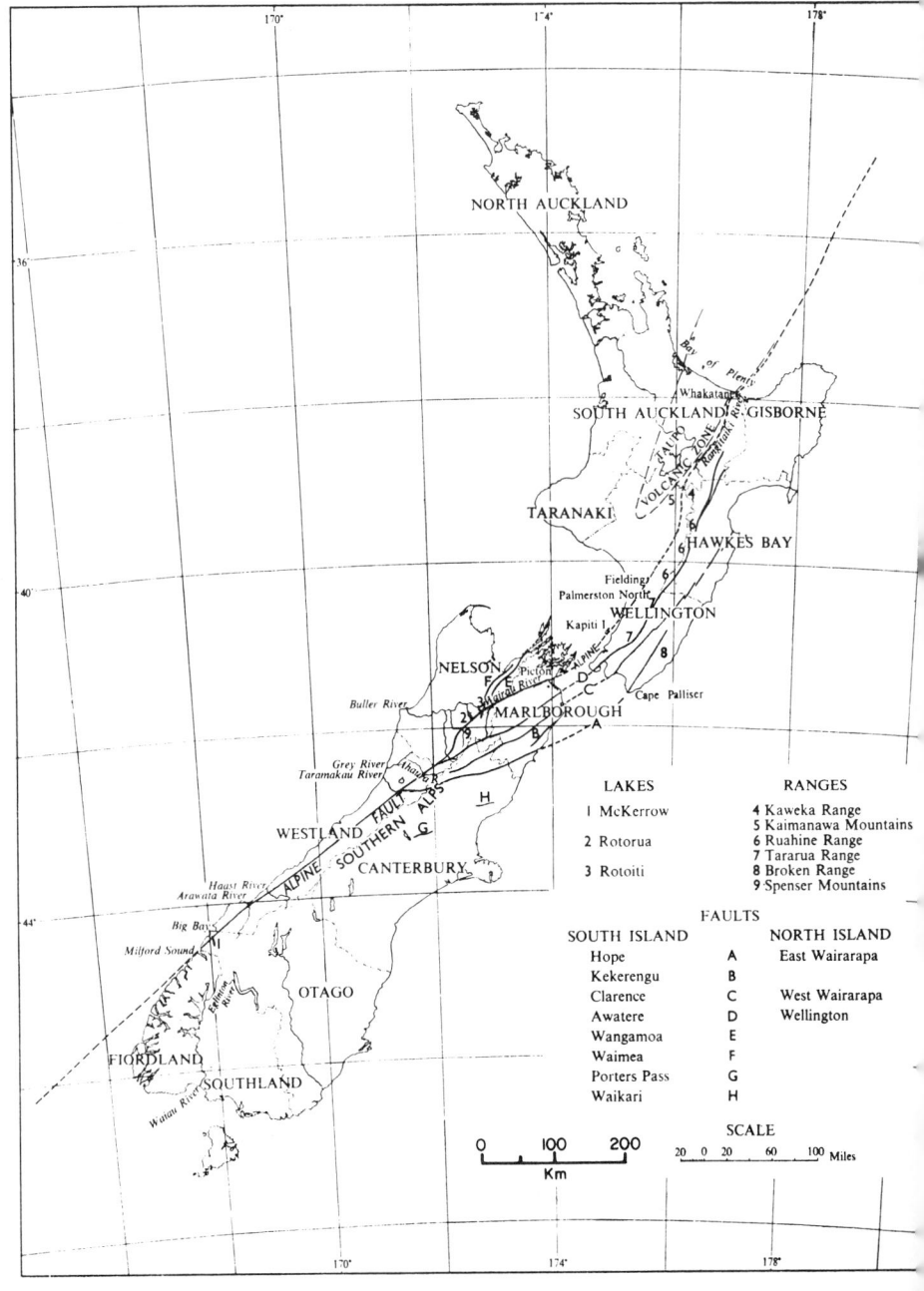

Fig. 10.16
The Alpine Fault and associated north-east-striking faults in Marlborough and the North Island East Coast regions, on most of which some amount of strike-slip displacement can be detected. (From R. P. Suggate, 1963)

the Kaikoura Orogeny originated in a paper by C. A. Cotton in 1916. This dogma profoundly influenced the way in which faults were shown on Geological Survey maps during the Morgan-Henderson regime, and was followed by university geologists too. The ensuing revolt, however, was so strong that faulting tended to be underrated for a time, and it was envisaged by some that folds in the covering strata were reflected by similar mass-distortion of the undermass rocks beneath, rather than by faulting.

Considering the amount of publicity given to horizontally offset fences and ground features due to horizontal movement on the Hope Fault in the 1888 North Canterbury earthquake, it is surprising that the importance of components of displacement in directions other than vertical (or rather, in the direction of dip of the fault plane, i.e. 'dip-slip') was neglected for so long. It is now known that many faults in New Zealand have had components of horizontal ('strike-slip') motion. Interest in strike-slip faulting really began with the report by Wellman and Willett in 1942 that stream courses and other features intersected by the Alpine Fault had been offset by as much as a mile (1.6 km) during the Pleistocene. In 1953 Wellman published a catalogue of active faults, noting the few cases where strike-slip components had been recorded up to that time. Within a remarkably few years after that it had been established that a system of major, steep or vertically dipping strike-slip faults exists between South Westland and East Cape. Suggate now proposes to call the whole assemblage the 'Axial Fault System of New Zealand'.

The Alpine Fault

It may be wondered why nearly a decade lapsed between the first recognition of a substantial amount of strike-slip movement along the Alpine Fault and the startling discovery that the movement probably amounted to nearly 500 km. The main reason for this delay was that until G. W. Grindley and B. L. Wood had carried out detailed mapping in remote western parts of Otago, it was not fully appreciated how closely the sequence and structure of Permian and Triassic strata in that area resembled those of the Nelson belt that is cut off to the south by the Alpine Fault (or rather by its northwards continuation, the Wairau Fault).

A few geologists, notably J. T. Kingma, remained hard to convince, not only about the amount of horizontal displacement but also about the supposed steepness of the fault plane in depth. Admittedly, at many exposures of the fault the way in which Haast Schist rocks from the east seem to override the rocks west of the fault, including Pleistocene deposits, suggested overthrusting from east to west, but eventually all were satisfied that this effect was the result of gravitational slumping of the weak and fault-crushed rock exposed along the steep fault scarp. Alternatives to accepting a large amount of strike-slip were suggested, but none was as credible or as well supported by observations, and there has been little debate on this aspect in recent years.

There has been some doubt about which fault, if any, truly represents a continuation of the Alpine Fault northwards from Lake Rotoiti to Cook Strait and beyond, and much discussion of the tempo of movement, in both horizontal and vertical senses. At first the Waimea Fault, east of the Nelson depression, was assumed to be the extension, but it is now accepted that the greater part of the horizontal movement is taken up by the Wairau Fault, despite mechanical problems raised by the implied bend near the Matakitaki River. Let us now look a little more closely into these questions.

Fig. 10.17
With this simplified map of the basement rocks of the South Island, accompanied by a sketch of the displacement of Holocene river terraces in the Maruia valley near Springs Junction, Wellman in 1952 promoted the view that recent small horizontal offsets of geologically young features by the Alpine Fault were merely the continuation of strike-slip movement which since Jurassic times had amounted to almost 500 km. (From Wellman, 1952)

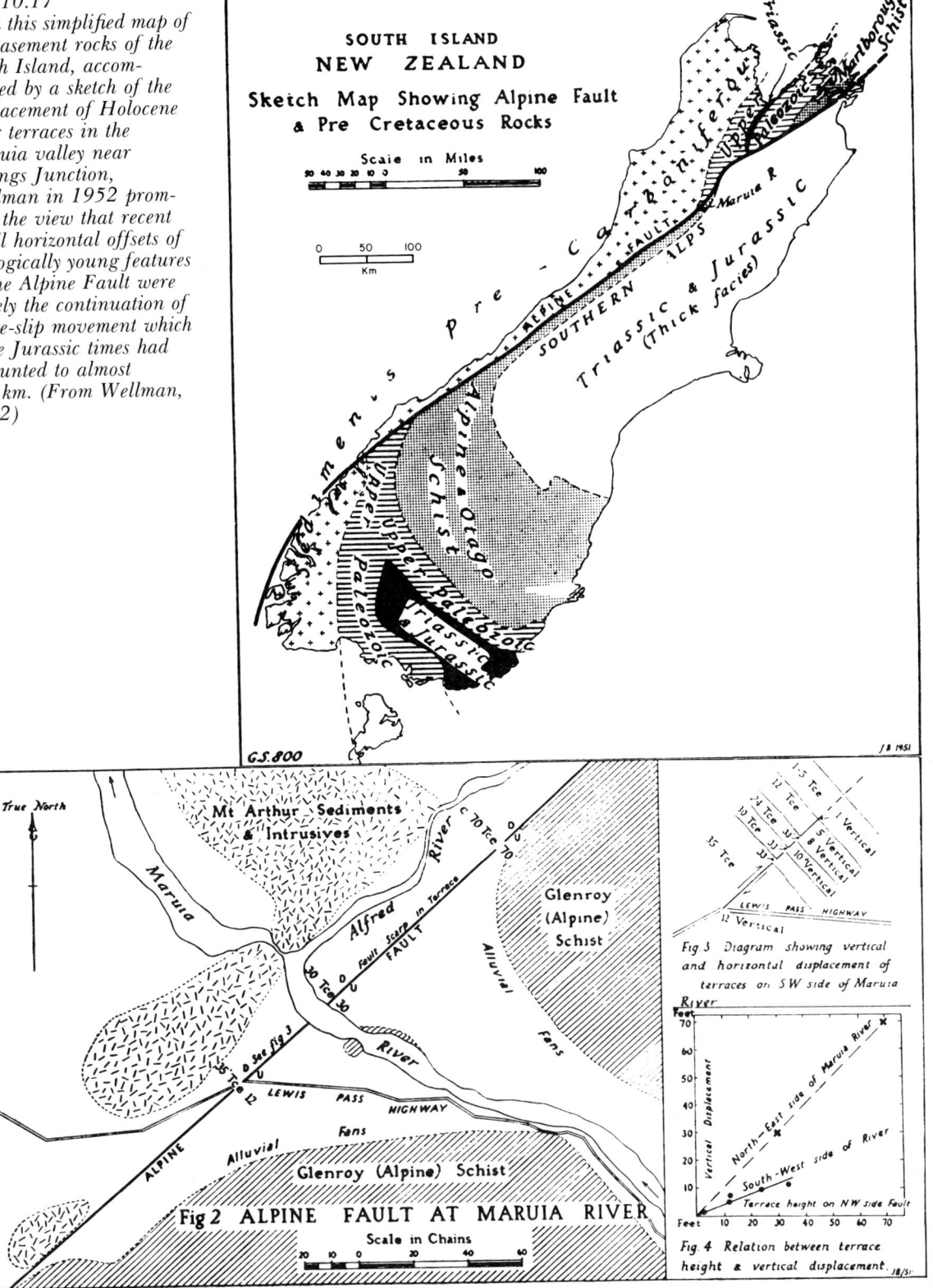

To begin with, Wellman had supposed that strike-slip movement began in the Jurassic Period, which amounted to linking its beginning with the 'first stirrings of the Rangitata Orogeny', as Suggate expressed it later. There were other possibilities: it could have been moving at a uniform rate since the Jurassic Period; it could have started then, ceased for a time perhaps during the orogenic lull when the Late Cretaceous Peneplain was formed, and resumed movement with the build-up of Kaikoura movements; or, if a faster rate of average movement is accepted, it may not have begun until the inception of Kaikoura movements in mid-Tertiary time.

Wellman has come to favouring the latter view, on the grounds of the amount of deformation that has affected Tertiary rocks close to the line of the fault on both sides. It also has the support of evidence for continuing distortion of a tract of country across the Wairau Fault at a rate of about 13 mm per year, according to re-triangulation of survey points. If maintained continuously since the Miocene Period, this amount of movement could account for the entire 500 km of displacement. Majority opinion, however, favours the idea that about two-thirds of the strike slip occurred during the Rangitata Orogeny, followed by a static period, and about one-third during the Kaikoura Orogeny.

The amount and timing of the vertical component have also been questioned. Using geological arguments to do with the question of when schist debris first became abundant in conglomerates (that is, in

Fig. 10.18
River gravels of post-glacial age resting on till formed in a late phase of the Otira Glaciation, here seen displaced upwards to the east on the line of the Alpine Fault where it crosses Crooked River, North Westland. The date of movement is uncertain, but must have been within the last 10,000 years.

Fig. 10.19
Looking north-eastwards
along a section of the
fault-line erosional trench
of the Alpine Fault in
central Westland. (Photo:
N.Z. Geological Survey)

Fig. 10.20
Exposures at several places
along the Alpine Fault
show Haast Schist, sheared
and crushed, over-riding
Pleistocene glacial and
alluvial beds apparently
along a gently east-dipping
thrust plane. But this
is not compatible with
other strong evidence for
great sideways ('strike-
slip') displacement, for the
surface trace of the fault
should then be a straight
line, which it is not.
Wellman, however,
showed that the apparent
over-thrust of schist was
really a gravitational col-
lapse of the steep western
mountain front as it was
heaved up by a vertical
component of movement
on a predominantly
strike-slip fault.
(Wellman, 1955)

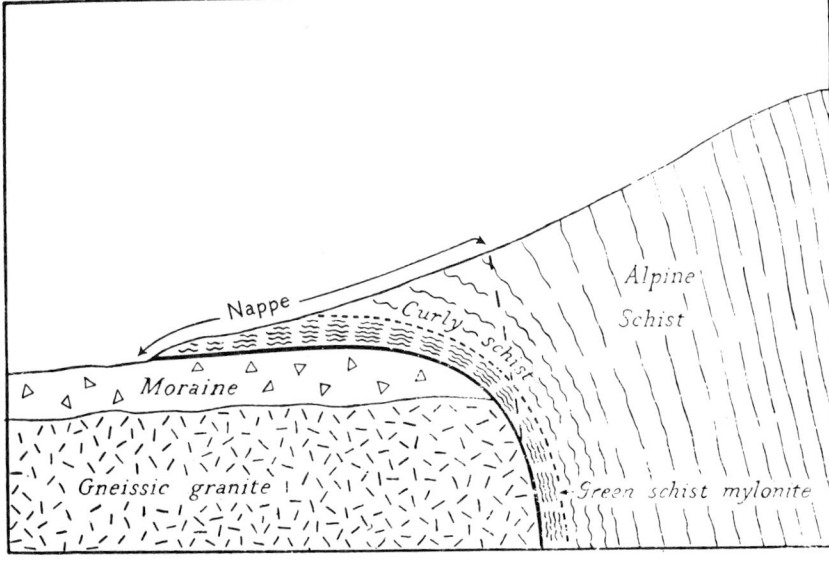

late Tertiary times) together with extrapolation from radiocarbon datings of Pleistocene materials displaced by the fault in South Westland, Suggate inferred that most of the vertical movement occurred in later times, perhaps as much as 18 km of it during the Pleistocene Period. Additional information has been contributed recently by geochemists working with minerals in the schists exposed along the fault scarp. Radiometric determinations of when these minerals could last have been at high temperatures suggest that as much as 10 km of the uplift occurred in mid-Cretaceous times, between 2 km and 6 km late in the Tertiary Era, and the remainder since about the beginning of the Pleistocene Period or a little earlier. Two main phases of uplift are thus related with the climaxes of the Rangitata and Kaikoura orogenies.

Connections across Cook Strait remain rather an open question. Perhaps the Alpine Fault loses its identity northwards, the movement being taken up by other faults of the same system. In the absence of good geological reference-points on either side, one of the difficulties is how to decide which of them have experienced large amounts of strike-slip movement. Besides the Alpine Fault and the Wairau Fault, also to be considered are the North Island connections (if any) of the Awatere, Clarence, Kekerangu and Hope Faults, in relation to the Wellington Fault, the Owhariu and Pukerua Faults and the Eastern and Western Wairarapa Faults. The interpretation shown by Graeme Stevens in *Rugged Landscape* (p.42) identifies the Wairau Fault with a major fault on the east side of Kapiti Island, and joins the Wellington Fault with the Awatere Fault. Considering that the total amount of gross, lateral displacement on the Alpine Fault may in fact be shared by the whole Axial Fault System, there is little point in trying to show that one particular fault in the North Island is *the* continuation of the Alpine Fault. On the other hand, there is much to commend Suggate's suggestion that the North Island continuation of the Axial Fault System should be known by a separate name, such as 'Marlborough-East Coast Shear Belt'.

To mention briefly one other fault suspected to have a large amount of strike slip, M. G. Laird has shown that an important zone of faults bounding the southern Paparoa Range east of the lower Grey Valley, known as the 'Paparoa Tectonic Zone', when followed northwards appears to cross diagonally over the topographic axis of the range. It eventually becomes the western boundary of the elevated tract south of the Buller River ('Papahaua Block') upon which the big coal mines are located. The strike-slip movement may help to explain how north-west and north-east trending areas of Greenland Group strata are now opposite one another in Westland. The dip-slip component is in the reverse direction, scissors-fashion, at opposite ends about a 'pivot' in the central Paparoa Range.

The relative sense of strike-slip movement is consistently 'dextral' (clockwise, or right-handed) for faults striking in the north-east quadrant of the compass, and 'sinistral' for the few active ones in the south-east quadrant. Inferences may be drawn as to the directions of the crustal stresses from these orientations, which also involve

DEXTRAL

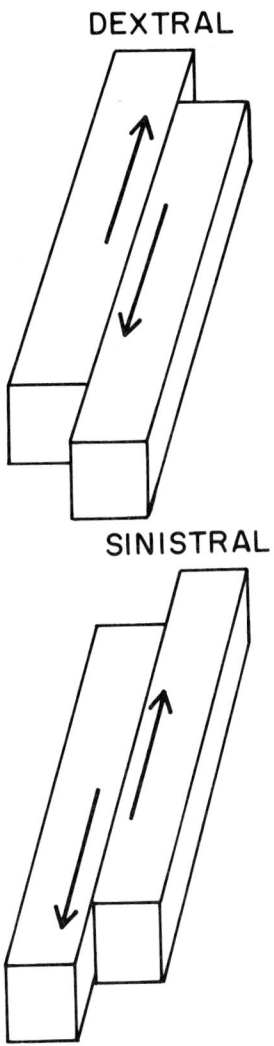

SINISTRAL

Fig. 10.21
The two 'senses' of strike-slip fault displacement; sometimes described respectively as 'right-lateral' or 'clockwise' and 'left-lateral' or 'anticlockwise'.

differences in the steepness and directions of dip of the fault surfaces. Many examples seem to fit the theory of shear-failure of materials under stress, but in application to the rocks of the crust, the directions of later movements are likely to be influenced by the existence already of some surfaces (bedding planes, older faults, etc.). The outer crustal rocks are far from being ideal, homogenous material.

Structure of the New Zealand Alps

The greater part of the alpine chain of the South Island as far south as Copland Pass is composed of sedimentary strata of the Torlesse Supergroup, in which many glimpses can be seen of folds, faults and belts of crushed rock in separate, bare areas. Farther south along the axial range and in other mountains that are composed of schists, structures can also be made out from a distance, but the surfaces which show them up are cleavage and schist-foliation planes rather than original stratification, reflecting distortion suffered by the rock since it has been metamorphosed. Older structures may have been obliterated, though they can usually be detected still by studies under the petrological microscope. In neither case is it easy to discover in the field how the structural features of one rock exposure may be joined up with those of another area.

Classical studies in the European Alps early in this century showed how much could be learnt from patient and detailed field mapping despite a high degree of structural complexity. Naturally, New Zealand geologists interested in structure hoped to achieve similar results in the Southern Alps, but for a long time they were thwarted by the general lack of useful fossils and the uniform nature of so much Torlesse rock. Notable progress has been made over the past few decades by geologist-mountaineers, beginning with reconnaissances by A. R. Lillie, B. H. Mason and B. M. Gunn in the early 1950s, and extended since in greater detail by D. G. Bishop, J. D. Bradshaw and others. Attention has been paid not only to the attitude of stratified sedimentary rock, but to other dimensional properties of deformed rocks, including joints, cleavage, and stretching and distorting of conglomerate pebbles. Folds occur in all possible orientations, several tracts of overturned, folded strata being detected.

Some (not excluding geologists) find it difficult to free themselves from mental hang-ups when trying to visualise entire rock-fold systems having been refolded, rotated in any direction and overturned, so that anticlines from an early phase of folding now appear as 'synclines' (in this case to be called 'synforms'), or else plunging steeply into the crust. It is essential to be able to picture the whole system three-dimensionally, and to be able to imagine the effects of each phase of deformation being, as it were, 'undone' in turn. Not many can do this easily, if at all. To get one's eye in, structural complications of this kind on a small scale can be studied on the coastal rock platforms of Cook Strait west of Ohiro Bay and around Kaikoura Peninsula.

The style of folding typical of the Southern Alps rocks includes a great deal of complexity of these varieties. What seemed to be simple areas have mostly proven to be structurally complex when examined

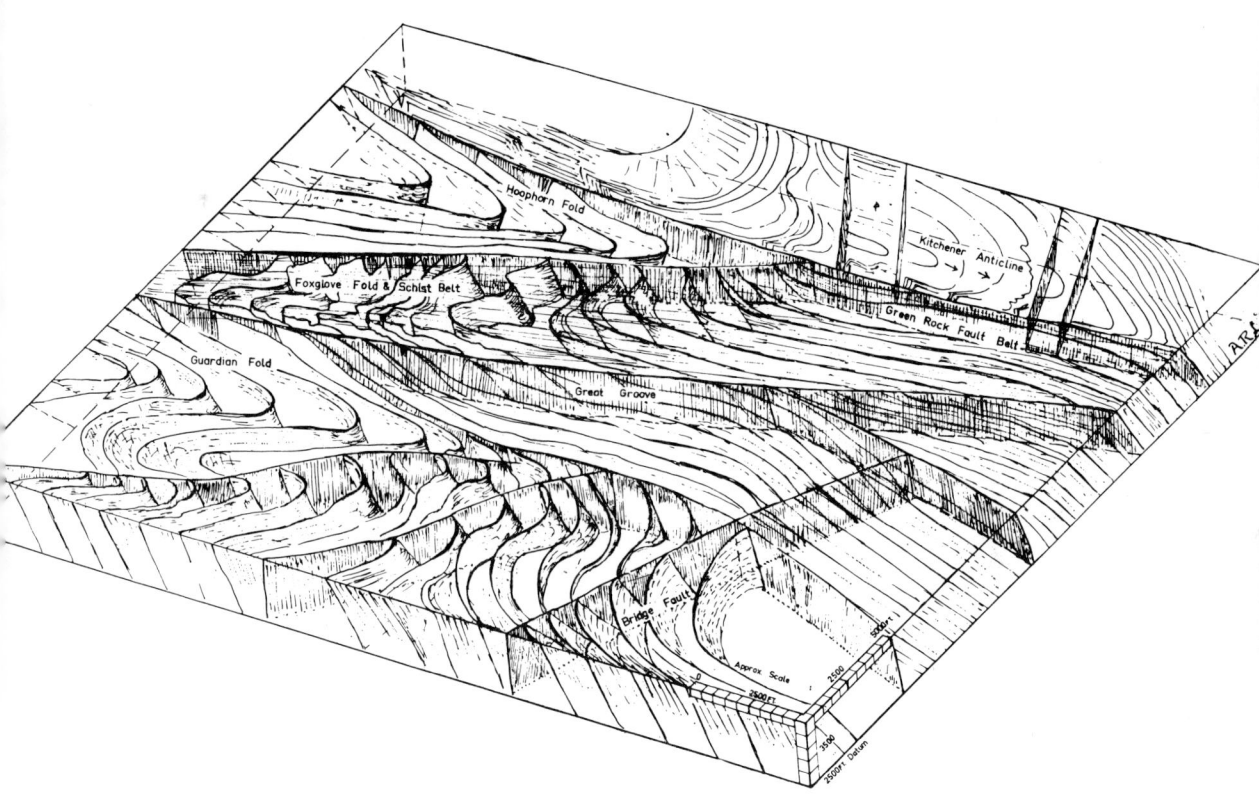

Fig. 10.22
A stereo-diagram of part
of the central Southern
Alps showing structures in
the Torlesse Supergroup
greywacke rocks in which
folds plunge steeply into
the ground. (From A. R.
Lillie and B. M. Gunn,
1964)

more closely. It is now established that our Alps do not closely resemble
the European Alps structurally. Some impression of the style of folding
can be gathered from the kind of stereo-models used by Professor
Lillie, as reproduced here in Fig. 10.22.

At the present stage a number of separate areas have been structually
investigated in detail and the results presented in a variety of
geometrical ways that bring out the style and important trends of the
deformation. A complete structural analysis of the Alpine chain,
integrating the results from separate studies, is still some way off.
Indeed, because of the limited assistance that paleontology is able to
give, it may never be possible to approach the detail and precision that
has been achieved in the European Alps, but considering the relatively
small number of geologists participating here, the progress that has
been made within a generation is highly creditable.

To comprehend the reason for the degree of complexity presented
by the structure of a single suite of rocks in the New Zealand Alps one
has to remember that the initial stages of folding began while sediment
was still accumulating in the New Zealand Geosyncline, that more
occurred in the depths of the geosyncline at the stage when slaty and
schistose cleavages were first introduced; that still more folding and
faulting accompanied the climax of the Rangitata Orogeny; and finally
that large displacements on faults and rotation of blocks of country
occurred *en masse* during the Kaikoura Orogeny.

Fig. 10.23
Global crustal structure is now seen as an assemblage of independent 'plates', usually named as follows: 1: Eurasian. 2: American. 3: Pacific. 4: African. 5: Indian. 6: Antarctic. A smaller, unnamed one takes in the very mobile Caribbean region. Mid-oceanic plate boundaries (e.g. between plates 2 and 4, 4–5 and 6) are also ridges margining volcanic rifts which emit sub-crustal mafic material. For the most part they are sub-oceanic, and as the emitted lava solidifies it pushes apart the adjacent oceanic plates, causing 'sea-floor spreading' and the under-riding of the ocean floor beneath some continental margins.

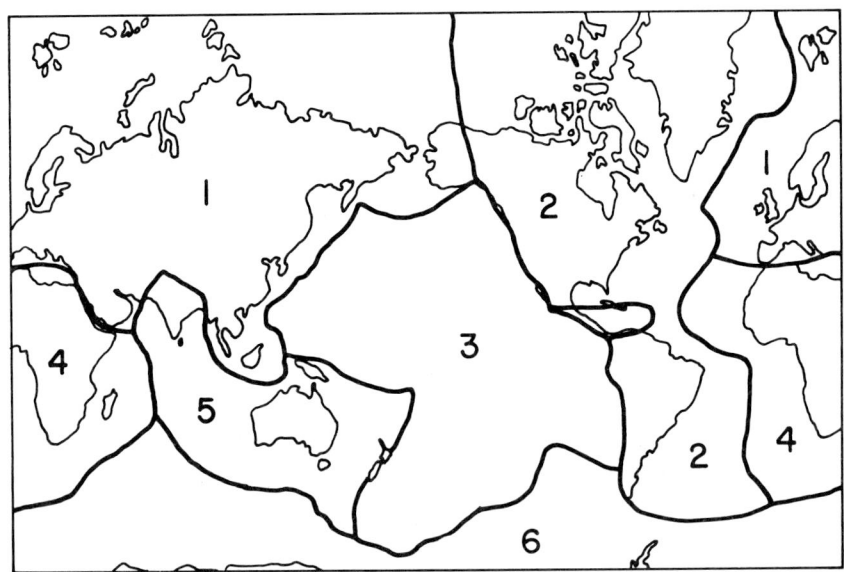

Structural Plan of the South-west Pacific

An atlas map shows the islands of New Zealand standing alone and aloof in the south-west Pacific, looking rather askance at Australia but seeming to beckon towards New Caledonia and Melanesia. As our under-sea connections became better known a closer kinship with regions to the north-west was indeed confirmed. Ocean depth data around the New Zealand shores became much more abundant from World War II times onwards, thanks to the availability of results from sonic soundings from naval survey vessels, oceanographic research ships of several nations and merchant ships. The picture clarified of an interesting sea-floor topography between New Zealand, New Hebrides and the Coral Sea, in distinct contrast with the lack of strong connecting features between here and the Australian continent.

The 'spine' or tectonic axis of New Zealand continues south-westwards in a 'tail' made up of a belt of ridges, furrows and several seamounts, one of which rises above the ocean surface to form Macquarie Island. The tail then swings to east-of-south and almost joins up with the broad mid-ocean rise that separates the Antarctic Ocean from the South Pacific and the Tasman Sea. A submerged plateau of sub-continental size, a part of the New Zealand Platform known as the Campbell Plateau, rises to within 500m of the surface over three broad areas between the Campbell, Bounty and Antipodes islands; it is separated by the Bounty Basin from the Chatham Rise which juts out eastwards from Canterbury for 750km. (Figs 2.1, 10.1.)

The prominent axial line of the New Zealand landmass seems to continue north-north-eastwards along the Kermadec-Tonga and Colville ridges as far as Samoa, whereas the topographical axis of the Auckland Peninsula, heading north-westwards, links obviously with a system of ridges and rises (including Norfolk Ridge and Lord Howe Rise) and curves northwards towards New Caledonia and the New

Hebrides. It was noted in Chapter Two and elsewhere that the Kermadec Trench marks the true western boundary of the deep Pacific Ocean floor, and the eastern limit of Asian crust with continental properties and also of andesite volcanoes. Lines of active volcanoes, above water and undersea, are aligned along the Kermadec-Tonga and Colville ridges. In contrast with the more commonly curved 'island arc' pattern of ocean margin volcanism, this alignment is straight.

Public interest has been attracted to the ocean-floor features south of New Zealand by the offshore drilling ventures in the deepest and stormiest seas yet to be tackled by the oil-men. Geophysical evidence has indicated that the Solander Trough which appears to continue the axis of the Waiau Syncline southwards from Te Waewae Bay should contain favourable strata and structures.

The prominent undersea structures are not all of the same geological age. Most likely, the north-westerly trend including the Lord Howe Rise and the structures assumed to underlie the Campbell Plateau date from Paleozoic time and were part of the early history of the New Zealand Geosyncline, whereas the Chatham Rise, Norfolk Island and New Caledonia are ridge features generated by the Rangitata Orogeny. The north-easterly trend which continues the main New Zealand tectonic axis northwards towards the Kermadecs and southwards to Macquarie is a late Tertiary development.

In terms of the theory of sea-floor spreading outlined very briefly in Chapter Two, the New Zealand tectonic axis and its continuations mark the position of a subduction zone, a compressed belt of crust beneath which material of the spreading Pacific floor is descending into the mantle. The New Zealand Geosyncline, before the orogen was twisted into its present 'Z' shape in plan, marked the direction of an earlier subduction zone. The change in direction is connected with a tectonic revolution which took place with the inception of drift and rotation of the New Zealand continental platform away from Australia, and of both from Antarctica during the Cretaceous Period. Relative rotation of the crustal plate containing New Zealand relative to the southern Pacific Ocean plate is expressed in a great strike-slip system of faults which since mid-Mesozoic time has produced more than 1000km of displacement. In New Zealand it is represented by the Alpine Fault system.

While it was believed that movements on the Alpine Fault were dominantly horizontal until the climax of the Kaikoura Orogeny, there was also a need to postulate some major change in the field of crustal stresses in late Tertiary time. Recent results from the datings of rock exposed by uplift along the Alpine Fault, as noted above, indicate that the ratio of horizontal to vertical components of movement need not have changed so much during the history of the fault. However, some explanation is still needed as to why the whole pattern of tectonic activity seems to have changed after the Oligocene Period. This question, and indeed also the reason for the Cretaceous revolution, remain very much matters of speculation.

Chapter Eleven

SHAPING THE LANDSCAPE

New Zealand Geomorphology

The Genealogy of Landscapes

Those who not only enjoy scenery but go on to become interested in the origins of landforms soon learn that landscape beauty is more than skin deep. Individual features that catch the traveller's eye and, on a broader scale, distinctive landscape textures and stream patterns visible from the air must have their explanations. Sometimes the explanations are simple and obvious, more often complex and rather obscure. It is also soon learnt that every landscape has a past, a pedigree.

I expect it will have dawned upon many readers already that the expression 'everlasting hills' is a fairly extreme example of poetic licence, the hills and valleys and the sea-shore being anything but static and unchanging. We are familiar enough with the idea that today's landscape is an outcome of interactions between the subsurface materials and the surface environments of the earth, and that it very probably looked different yesterday, geologically speaking. Traces of its ancestry are usually to be found, for although present conditions and events of the immediate geological past tend to dominate the picture, few landscapes present no hint of yesterday's situation to the discerning eye.

Seldom can one properly comprehend the meaning of landscape features or describe them usefully without reference to the local geology. It is therefore not only convenient but logical as well for this chapter to come after those describing New Zealand rocks, structure and geological history.

More than half a century ago the late Sir Charles Cotton set out to write an explanatory account of the landforms of New Zealand but found that he had first to produce a preliminary book showing how the theoretical principles of landscape evolution then being developed mainly by American geographers and geologists could be applied with great success here. This was his immensely popular *Geomorphology of New Zealand–Part 1: Systematic* (1922; second edition 1926), succeeded in 1940 by *Geomorphology* and in that revised form reprinted many times in this country and abroad. For a generation the original book remained the only satisfactory geomorphic textbook in English. The promised sequel did not eventuate, and although he wrote many scientific papers on the subject, no comprehensive regional description of New Zealand geomorphology has appeared from Cotton's or any other pen. There have been descriptions of separate regions, as in *The Face Of Otago* (B. J. Garner, ed., 1948), *The Natural History of Canterbury* (G. A. Knox, ed., 1969) and most notably in Graeme Stevens's *Rugged Landscape* which explains the geological background to scenery on both sides of Cook Strait. Many more geomorphic articles and scientific papers have appeared in *New Zealand Geographer* and other journals.

In Chapter One of this book the most important geological processes acting at or near the earth's surface were briefly described, and in Chapter Two an attempt was made to instil a sense of proportion, as to the magnitude of surface features relative to the whole earth, and of the time-spans of geology. Chapter Two considered how the shape of New Zealand in plan has evolved, and the question was taken further in Chapter Ten. The following paragraphs are aimed at helping readers to discover meaning in landscapes, and perhaps thereby to gain more interest and pleasure from travelling.

Pushed up or Carved out?

Some may remember Will Hay's comic schoolroom sketches entitled *The Fourth Form at St Michaels*. One of these produced the following dubious definition: 'What is a mountain?' 'A piece of land standing higher than the surrounding ground—on account of the forces that pushed it up there.' The intended humour no doubt lay in the peculiar logic of the teacher's answer to his own question, under pressure from an aggressive pupil, but it serves also to draw attention to a once common misconception about the origins of terrestrial relief. Only in exceptional cases have hills and ridges been 'pushed up' individually. True, it was pointed out in Chapter One that any relief the primeval earth's surface might have had would long ago have faded, and the Geological Cycle stalled, had it not been continually regenerated through uplifts of segments of the crust and also through constructional activities like the piling-up of volcanic products upon the surface. In a way, therefore, Will Hay's answer was geomorphically sound in that whole mountain tracts have thus grown up independently, but it fails by neglecting to add that individual valleys have been excavated by erosion and that the ridges and hills between them are merely what has not yet been eroded. This is an important corollary to the 150-year-old principle known as Playfair's Law which stresses that valleys in general are the work of the rivers that flow along their floors.

The infinite variety of ways in which that work is done by streams and their tributaries, influenced by climate and by conditions in the underlying rock, shows up in the fact that no two catchments are quite identical. Moreover, the picture is a dynamic one of continually changing and evolving landforms. The situation, instantaneously at any moment of time (such as the present) can be represented as a kind of equation, thus: 'Regardless of how the relief originated, the amount surviving equals the algebraic sum of the amount of uplift (or upbuilding) *minus* the amount so far eroded away.'

Different Ways of Describing Landforms

It would be infinitely tedious and practically useless to write detailed, descriptive word-pictures of landscapes. Far better to represent them by means of maps on which the outlines in plan are shown by streams and coastline, and the relief indicated by shading, or, where there is enough survey data, by contour lines. Special kinds of mapping enable the steepness of slope to be read off directly at any place, while various statistical methods are available to condense and generalise altitudes, steepness, stream-spacing and other numerical data for purposes of comparison between landscape and landscape. This rather arid approach, however, is unsatisfying as an end in itself. Most people prefer to find descriptions which bring out the origins of landforms, as far as is known. The science and the language of geomorphology arose in reponse to this preference, enabling whole landscapes to be described genetically, and individual landforms to be classified and named according to their supposed histories and origins.

Some ideas, more or less rational, about the origins of landscape features appear in very ancient writings, but not until the later half of the nineteenth century did an organised body of knowledge on the subject begin to be assembled. Engineers, geologists and geographers then evolved a number of basic principles, mostly arising out of Playfair's Law, of which the most important ones recognised, first, that landscapes eventually tend to be reduced by erosion to a general flatness, and second, that the effectiveness of erosion below sea level is strictly limited. Sea level was thus seen as providing a base-level down to which landscapes may be worn by erosion. Other empirical laws and deductions took into account the relationships between stream flow, sediment supply, gradients and land slopes, the influence of underlying rock conditions, and the consequences of vertical shifts of base-level.

In order to be able to write about landscapes in these terms, the American geographer W. M. Davis and others had, by the early part of the present century, developed a whole lexicon of terms and definitions for landform types, embodying the diagnostic signs of which kinds of physical process, what typical series of events, had been important in their generation. This language of 'explanatory description', developed largely by Davis between about 1893 and 1930, was introduced to New Zealand by C. A. Cotton from 1911 onwards. Despite much criticism, and with many modifications and refinements, some of the Davis geomorphic terminology is still widely used and understood.

Landscape-modelling Processes

The agents which have shaped the landscape are essentially the agents of physical geology noted in Chapter One—solar radiation, gravity, the atmosphere, rain and running water, snow and ice—weathering the rocks and transferring material from place to place, here lowering ('degrading') the surface and there building it up with deposits. Constructive effects of volcanic eruptions on land have already been mentioned (Chapter Eight), and the raising of the land directly by crustal movements, but it must be remembered that either of these processes can also lower the surface. Mainly, the shaping of the surface, as in the arts of sculpture and modelling, has been achieved by removal of rock material.

Earth-modelling processes can be observed in action today, sometimes at measurable rates, so applying what is known as the Uniformitarian Principle, which declares that 'the present holds the key to interpreting the past'. We recognise in the landforms examples of where earth-modelling processes have operated similarly in the past. Thus, far away from any present-day glaciers, we can identify distinctively rounded forms modelled in solid rock with similar features revealed by the drastic shrinkage of alpine glaciers in historic times. They provide contributary evidence (not proof on their own) of more extensive glacier action in the past. Again, it is often useful to be able to recognise a kind of peculiar, hummocky, stepped and broken slope as having been due to a former downhill slump and flow of weak surface materials under the direct influence of gravity, even if erect, mature trees or an undisturbed mantle of volcanic ash shows that no movement has occurred for hundreds, or tens of thousands of years. (Useful, for example, if you are thinking of having a house built there.) Textures and structures in the soils and subsoil materials may yield corroborative evidence.

The results from different modelling processes, however, are not always as distinctive as we would like. I have been involved in arguments, for example, as to whether mounds of rock debris on certain valley floors were ancient morainic deposits or the result of a rockfall from adjacent steep slopes. The older the landscape, the greater the variety of conditions which have governed its formation, the greater the chance that changes of level or slope have intervened, and the more complicated the explanatory description must be.

Crustal Uplift and Landscape Evolution

Progressive lowering of relief by erosion ('base-levelling') having been accepted, a beginning to every episode has to be visualised in the form of a raising of the land relative to the sea, and also, ideally, an end to a completed geomorphic cycle when the land is worn down to an almost flat peneplain. W. M. Davis taught that a full understanding of an erosional landscape involved knowing the character of the underlying rocks (which he summed up as 'structure'), the environmental conditions, especially climate ('process') and how far the wearing-down had progressed ('stage'). He recognised successive stages of 'youth', 'maturity' and 'old age'. These deduced, conceptual models, imaginary landscapes, if you like, have been found useful, along with Davis's names for the various elements in them, by several generations of geomorphologists. The models are readily enough matched in actual landscapes, but it is no longer a matter of dogma that the 'sequential forms' of the geomorphic cycle have necessarily followed one another in nature in the standard order, nor is it believed that complete cycles have often had the chance to be completed without changes of climate (affecting 'process') or interruption by further changes of base-level. But does that invalidate the whole idea? This and other debatable aspects have never been clarified to the satisfaction of all. Alternative theories,

Fig. 11.1
The steep-walled, narrow, rock-floored condition of the upper Kakanui River valley in Otago exemplifies a stream valley in the condition of 'youth' as envisaged by geomorphologists of the Davis school.

Fig. 11.2 (centre)
An erosional landscape of subdued relief, the result of prolonged erosion in an area of relatively undisturbed early Tertiary strata. The relief features are due mainly to harder limestone beds and volcanic rocks, and the landscape would have been described in the language of the Davis school of geomorphologists as 'late mature'. The locality is Ngapara, North Otago.

giving more importance to the question of how valley-side slopes and forms evolved, were developed in Europe (where Davis was largely ignored) and later popularised in English by L. C. King. Mathematical models representing the working of both kinds of theory also have been devised with the aid of computers, and it seems that the different approaches are complementary rather than in conflict.

Fortunately the basic tenets of geomorphology that concern the lowering of land by erosion stand independent of the Davis cycle scheme, although some were in fact born with it. That mountains have been eroded away to expose deep-seated rocks and structures is not in doubt. Irregularities both in the side slopes and longitudinal gradients of valleys become smoothed out in time, the hollows and gentler slopes being built up with sediment deposits while steeper segments are eroded down at a faster-than-averate rate. Gradients generally adjust to a steepness governed by the nature of the rock underneath, and by the dominant erosional processes, which in turn are governed largely by climate, and in the long term tend to merge into a continuous, even curve ('graded profile') adjusted to the base-level provided in most cases by sea level—so long as the latter remains stable.

The capacity of a stream to erode its bed and banks is reduced when there is a heavy load of sediment to be transported. Instead of eroding, in these conditions it is more likely to build up ('aggrade') its bed. It may lose vigour because of reduced rainfall, blocking by landslips or dams, back-tilting by crustal movements or loss of water by infiltration, with a similar result. Conversely, any change which reduces the sediment load or increases the vigour of a stream improves its capacity to degrade. As a general rule where running water is the important agent, bold relief means vigorous erosion, and an abundant supply of sediment for deposition elsewhere.

Oversimplified, no doubt, yet the last few sentences embody the chief factors governing how stream gradients evolve in a wide range of climatic regions.

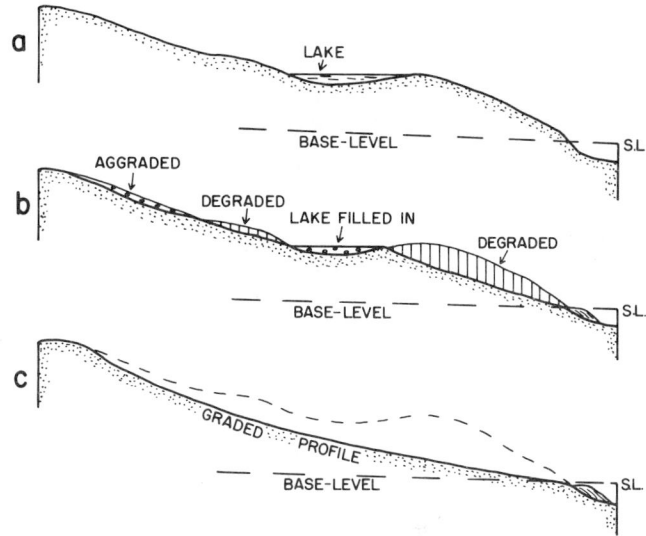

Predominantly Stream-modelled Landscapes

The importance of climate in determining which type of weathering processes will play the dominant role has already been stressed. In New Zealand the present-day climate is moist and temperate in most places except in the high mountains and in a few exceptionally dry localities. This was generally true through most of the period in which the present landscape has been evolving. Although glaciers, frost and snow were important in the past over much wider areas than now, running water and the gravitational mass-movements of waste between them modelled the scenery of most of the country. Summits and ridge-crests tend to be sharp in most areas, but in some the upper slopes are subdued and convexly rounded because mass-movements due to frost played an unusually important part.

Influence of Rock Structure—the Bones Showing Through

Structures in the underlying rock often show quite strongly in both the relief forms and the pattern of streams. It happens in the following way. Naturally, all kinds of physical erosion degrade surfaces of softer, less resistant rock more rapidly than harder, more durable kinds such as hard sandstone and limestone. In areas of stratified sedimentary rocks, exposures of weaker mudstone, for instance, are likely to erode more rapidly than hard sandstone areas, which will be left standing as ridges. Eroding streams tend to develop their valleys along the strike, following the outcrop of weaker rocks. Adjustment of the stream pattern to folded structures is most marked where the folding is simple and dips mainly gentle, and less obvious where strata are tightly or complexly folded. Besides folding, erosion picks out other lines of weakness such as belts of crushed rock along faults and prominent joints. A distinctively gridded pattern of streams has developed in eastern Otago south of the Clutha River because of a system of faults and joints transverse to the trend of broad folds in the Mesozoic strata.

Fig. 11.3
Base-level, and successive stages in the re-grading of an irregular landscape profile by the work of streams. a: An initial condition with a lake occupying an original, undrained hollow.
b: Re-grading progresses through a combination of stream-down-cutting, where slopes are steep, and in-filling of hollows and flat area. c: Re-grading completed to the extent that streams have a continuously graded profile related to base-level, which is assumed not to have shifted either upwards or downwards while regrading was in progress. (After Cotton)

Fig. 11.4
Intensified erosion of the upper slopes of the Ruahine Ranges caused excessive amounts of rock waste to be discharged into the river headwaters. Aggradation and the rapid building of alluvial fans followed, as in this example in the upper reaches of the Waipawa River. Discharge of waste must recently have abated in this gully, allowing the stream to entrench itself below the fan surface.

Some very prominent relief has been generated by differential erosion in folded rock areas, as in northern Wairarapa where the Puketoi Range east of the Manawatu Gorge, for example, is supported by a ridge of limestone along a tilted block. The dramatic, rocky scenery of Castle Hill, Canterbury, owes its character to the rather complex folding of Tertiary strata as expressed by the sinuous curves of hard limestone ridges. On the other hand, the stream pattern of inland Taranaki exemplifies the opposite state of affairs in which gently-dipping, mainly soft rocks present no strong, structural grain to direct the erosional activity of streams. (Compare Figs 11.7, 11.11.)

Patterns of relief in detail have been determined by the ramifications of rivers and their tributaries of all magnitudes down to the smallest gullies. In geomorphic language most of our landscape is described as 'maturely dissected' because the run-off of rain is discharged by way of a network of permanent stream valleys, reaching in most places to the divides. The spacing of streams and tributaries, which determines the texture of dissection, tends to be wider (i.e. coarser texture) in areas of homogeneous hard rocks such as granite, and closer (i.e. finer) in soft rocks. Fine texture develops also where rocks are much affected by closely spaced joints, fractures and bedding planes, as in many Torlesse greywacke and Haast Schist areas.

The very thick, homogeneous and rather structureless conglomerates of late Tertiary and early Pleistocene age in the South Island (Moutere, Kowai, Old Man and Maori Bottom formations) develop a distinctive arrangement of minor streams and gullies, which tend to be straight, parallel, uniformly spaced, open and separated by subdued ridge-crests, giving an unmistakable pinnate or herringbone pattern, especially when viewed from above. Owing to the porous character of the conglomerates, the gullies are often dry. Good examples are to be seen from the roads across the Moutere Hills and Spooner Range in Nelson, and in the Loburn Downs near Rangiora in Canterbury.

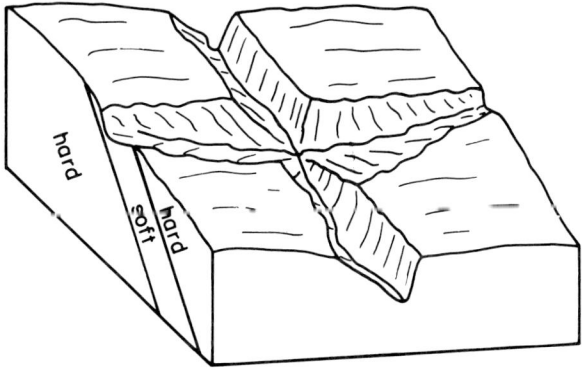

Fig. 11.5
Channels scoured by the Awatere River at an earlier level while cutting its valley in soft, Tertiary mudstone. The old bed is now a terrace 15 m above present river level, underlain by a thin layer of irregular, coarse gravel typical of deposits on river-cut erosional terraces. The capping of fine material may be river silt covered by wind-blown loess and soil.

Fig. 11.6
One way in which a stream pattern develops 'adjustment' to underlying rock structures that are reasonably simple. Tributaries can more readily extend their valleys headwards along belts of relatively weak rock. (After a diagram by Cotton)

Faults find expression in various ways. First, where displacements have occurred since the present landscape took form, the surface trace of an active fault may make an obvious fault scarp, as at the White Creek Fault in the Buller Gorge, which moved 5m in 1929. More ancient faults are likely to be indicated in the landscape by fault-line scarps that have developed in the course of erosion because the rock exposed on one side of the fault has been eroded more rapidly than on the other. Examples well known from their illustrations in Professor Cotton's textbooks include the Wairoa Fault in the Hunua Hills south of Auckland, and the Hunter Fault near Waimate in South Canterbury. Fault-line valleys are the work of streams favoured in their excavating efforts by the presence of severely crushed and sheared rock close to the plane of a major fault, as in the case of the upper Kaiwharawhara and other straight valleys of western Wellington Peninsula that are not guided directly by recent fault movements.

As mentioned in Chapter Seven (footnote, p. 164), C. A. Cotton introduced to New Zealand the distinction between 'undermass' and 'covering strata'. This is of unusual geomorphic importance in this country over the wide areas in which the tempo of erosion is much

Fig. 11.7
The stream pattern of south-east Otago inland from Nugget Point shows a strong influence of NW-SE folds and NE-SW faults and joint systems in Triassic and Jurassic strata. Such rectangular drainage patterns are described as 'trellised'. (From Marshall, 1912)

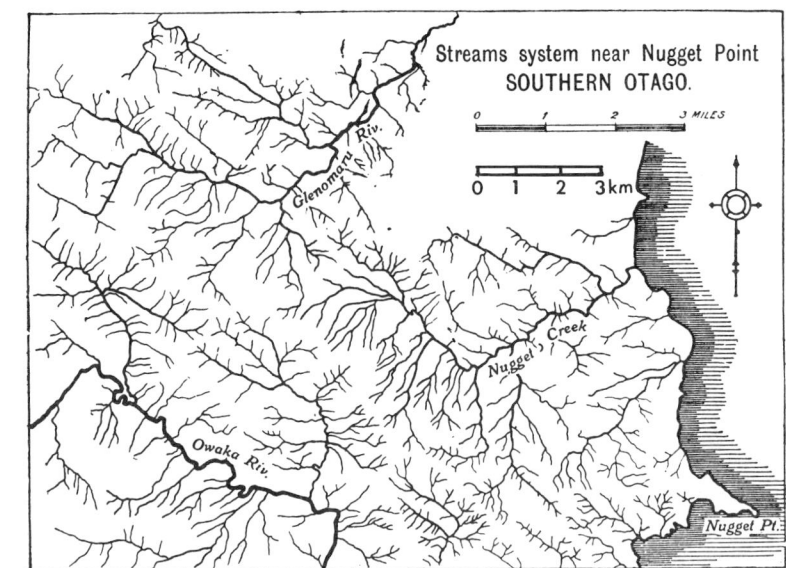

Fig. 11.8
Gently dipping layers of alternately softer and harder rock favour a strong expression of structure in an erosional landscape. 'Cuesta' ridges, capped on one flank by 'dip-slopes' due to the protective effects of harder layers, emphasise the strike-direction of the strata.

Fig. 11.9
Structural control of erosion produced these plateau forms in a region of flat-lying Tertiary strata near Tokarahi, North Otago. On the right, the plateau expresses the protective effect of hard Otekaike Limestone (Oligocene) upon softer rock beneath, and its greater resistance to erosion compared with higher, soft beds that have been removed. Farther away, a similar plateau surface demonstrates the durability of a thick basaltic sill that was intruded into Eocene strata, now stripped from above it.

faster in weaker rocks above the Late Cretaceous Peneplain unconformity than in harder rocks below it, causing the uplifted and deformed old erosion surface to be re-exposed. Some Central Otago landscapes owe their special character to this stratigraphical-structural situation. During the Kaikoura movements, the schist and greywacke basement rock broke up into a jumble of differentially elevated and depressed and warped blocks, from the higher elements of which erosion has since stripped away younger covering strata to re-expose the peneplain. In structural troughs and fault-angle depressions the younger beds survive, though largely buried under young river gravels. One or both margins of elevated blocks in many cases are faults.

Faulting is not necessarily represented obviously or at all in the landscape unless marked by a belt of crushed and sheared rock; unless it separates tracts of rock of different hardness; or unless the fault has moved recently enough to have displaced the surface or offset streams. Very many geologically important faults have little or no surface expression.

Excellent examples of landscape features directly due to geologically recent fault displacements occur in the Cook Strait region, and of these the Wellington Fault is probably the best known. Many fine illustrations and descriptions of fault features in that area are given in Graeme Stevens's book *Rugged Landscape*. Of all the fault features in New Zealand the most spectacular is the long, straight western boundary of the Southern Alps, determined along hundreds of kilometres by the Alpine Fault. In parts it is a true fault scarp, while in others its broad belt of crushed rock has determined the sites of fault-line valleys, including that followed by the Taramakau River between Inchbonnie and Wainihinihi.

Faulting has blocked out major landforms in the South Auckland district. The depressed trough of the Hauraki depression is a 'graben' bounded by the Thames and Hauraki faults. The Drury and Wairoa faults make prominent escarpments separating the upland areas of Torlesse rocks from the Manukau Lowland and Whangamarino Swamp areas, respectively.

Crustal Movements and the River Pattern

The web of major rivers and largest tributaries which provides the framework for finer physiographic detail, when simplified on small-scale maps, shows where are the chief 'crustal bulges' from which erosion has sculptured the landscape. By looking only at the major river pattern we obtain some impression of the distribution of the basic relief, whether due to piling-up of volcanic masses (as at Banks Peninsula) or to arching-up or thrusting-up of the crust regionally in the most recent orogeny (as in the Kaikoura Ranges). Omitting the minor elements of drainage which bring out secondary effects of adjustment to the internal structure of major blocks helps to clarify the picture of regional crustal upheaval.

The picture in north-eastern Marlborough, for example, is one of complete domination by the geologically recent uplift, tilt and rotation of adjacent blocks, among which the main rivers still find their way to the sea along depressions between an assemblage of upheaved masses. The Clarence River in particular follows a most tortuous route. Other eastern rivers farther south have maintained direct courses from the Main Divide, cutting across the grain of differentially elevated blocks and depressions, through series of gorges (see later, p. 312). From the Waimakariri River southwards to the Rangitata, present courses appear now to be more direct than those which preceded them, having been straightened and shortened during repeated occupations of their catchments by large glaciers in late Pleistocene times.

When viewed from heights of between one and two kilometres, the crests of mountain ranges, as far as the eye can see, appear to rise to a uniform level. The impression is very strong over the ranges of the South Island, except in the north-east. Among the explanations suggested for this 'summit accordance', the most popular for a long time was that it indicates the former presence of an arched-up peneplain, now totally dissected by the modern valley systems. However, for various reasons this interpretation is no longer acceptable. Indeed, it seems likely that summit accordance needs no special explanation, being a normal result of dissection of any uplifted mass by regularly spaced streams.

More than thirty years ago H. W. Wellman experimented with contour maps that generalised summit heights only, ignoring the valleys below, with some intriguing results. In some areas the resulting contours seemed to express a simplified doming or 'anticlinal' upheaval of mountain ranges as a whole, but in others no simple picture emerged. Major rivers in some cases are aligned roughly perpendicularly to the contours drawn from summit heights, as though their courses had originated simply as part of a primitive drainage developed in the early stages of the last orogenic upheaval. Such an impression was given, for example, by contouring the mountain tops of north-west Nelson and the central Southern Alps, whereas the rivers in the Tararua and Rimutaka ranges appear to be more strongly governed by structures within the Torlesse Supergroup greywackes (dating from the Rangitata Orogeny) than by the more recent Kaikoura upheaval of these ranges. Knowledge of regional geology and orogenic history has advanced considerably, but the full meaning of these contrasts remains uncertain.

Rivers that Flow through Mountain Ranges

Streams traversing alternate tracts of harder and softer rocks are restrained from widening their valley floors where the rock is hard but can develop broader, continuous valley plains and widely curving courses where the rock is softer. Narrow, gorge-like reaches alternate with wider reaches for this reason in many New Zealand valleys, but there are also some deep gorges through which rivers surprisingly flow from one side of a mountain range to the other with apparent disregard for seemingly easier and more logical routes to the sea in some other direction. As the normal habit of streams is to run down-hill, this odd behaviour has long been a puzzle and gave rise to fanciful theories, of which one of the more popular held that the 'back' of the

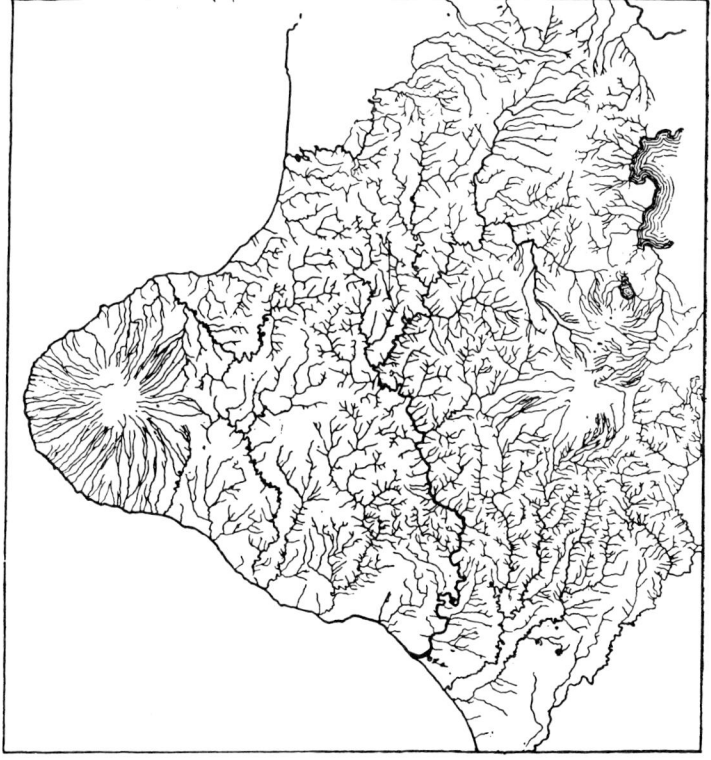

Fig. 11.10
Erosional forms in the
Waipara Valley, North
Canterbury. In the centre,
a cuesta developed on dipping Weka Pass Limestone
(Oligocene), its escarpment
facing west and its dip-
slope facing right towards
a higher escarpment of
Miocene limestone.
Nearer, terraces carved by
the river while lowering its
channel during late
Pleistocene times.

Fig. 11.11
The multiple-branching
streams of central
Taranaki and western
Wellington show little evidence of guidance by rock
structure. Note also the
strongly radial pattern of
drainage down the slopes
of Mount Egmont (L) and
the Tongariro volcanoes
(R). (From Marshall,
1912)

Fig. 11.12
The texture of stream dissection in areas of mainly soft, rather uniform Tertiary strata, as here in inland Taranaki, is characteristically fine. (Photo: N.Z. Geological Survey)

Fig. 11.13
Serial block-diagram by C. A. Cotton to show (from right to left) the Davis concept of the invasion of a newly uplifted area by erosional stream valleys which cut into it until none of the original surface remains unmodified. A, young stage; B, dissection approaching maturity; C, mature stage. (C. A. Cotton, 1942)

Fig. 11.14
The western slopes of Makara Valley, Wellington, exemplify moderately fine dissection by small streams and gullies. Ridge crests were rounded by down-slope flowage of weathering products under frigid conditions during the Otira Glaciation. The underlying Torlesse greywacke and associated rocks behaved like a homogeneous material, their structure apparently not having influenced the course of erosion here.

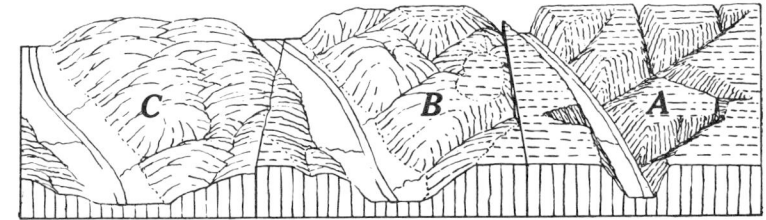

Fig. 11.15
From the air, the Moutere Hills, Nelson, present this distinctive pattern of regularly spaced tributary gullies and herringbone ridges to be seen also in most other areas where the underlying rocks are thick, uniform late Tertiary-early Pleistocene conglomerate. (P. 190.) (Photo: N.Z. Geological Survey)

Fig. 11.16
a: Fault-scarp, the result of dislocation of the ground surface by faulting. b: Fault-line scarp, caused by the more rapid erosional lowering of the land on the soft-rock side of the fault. c: Fault-line valley, developed by stream erosion which has been favoured by a belt of crushed and sheared rock along the line of a fault.

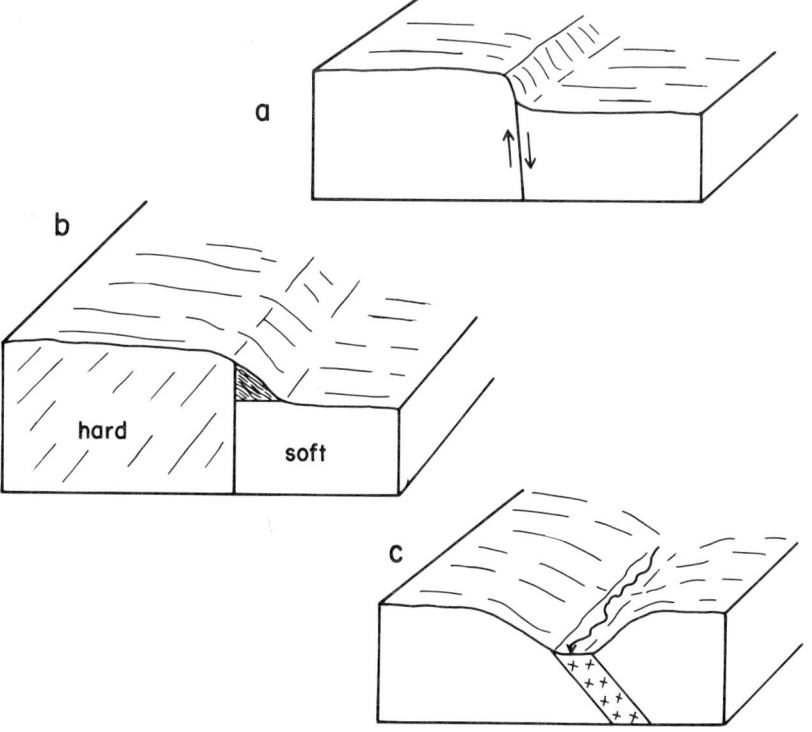

Fig. 11.17
The steep eastern wall of the Takaka Valley, northwest Nelson, separates a fault-angle depression from the high-standing Pikikiruna Range block; it is a tectonic scarp which originated during the Kaikoura Orogeny.

Fig. 11.18
From Harper Pass, looking down the long, straight fault-line valley excavated by the Taramakau River along a broad belt of crushed and sheared rock due to movements over a long period on the Hope-Taramakau Fault.

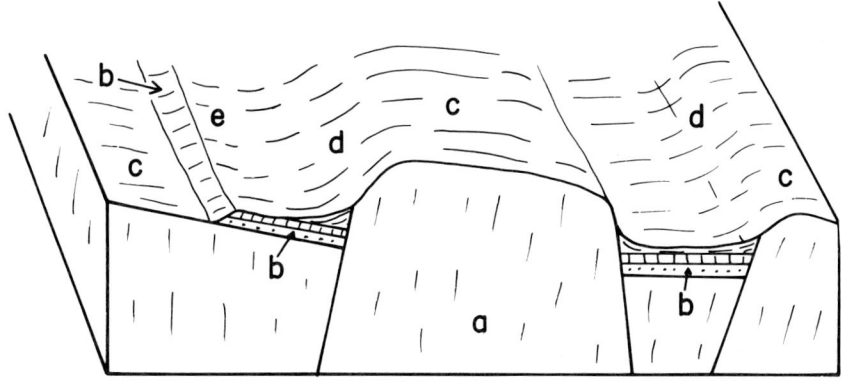

Fig. 11.19
Younger covering strata (b), resting unconformably on a more erosion-resistant undermass (a), which they once completely covered, have been stripped from uplifted and upwarped areas, re-exposing a 'fossil' erosion surface (c), bounded by fault-line scarps (d) and erosional escarpments (e). (See also Fig. 11.16)

Fig. 11.20
The Late Cretaceous Peneplain, here underlain by Haast Schist, was stripped of most of its former cover of Tertiary strata following uplift in the Kaikoura Orogeny and is now deeply dissected by Awamoko Stream and other tributaries of the Waitaki River, North Otago.

Fig 11.21
a. Horst (h) and graben (g) in the sense of geological structures, not expressed in landscape features. b. Horst and graben as landscape forms ('block mountains' and 'fault troughs') resulting directly from crustal movements. c. Fault-angle depression.

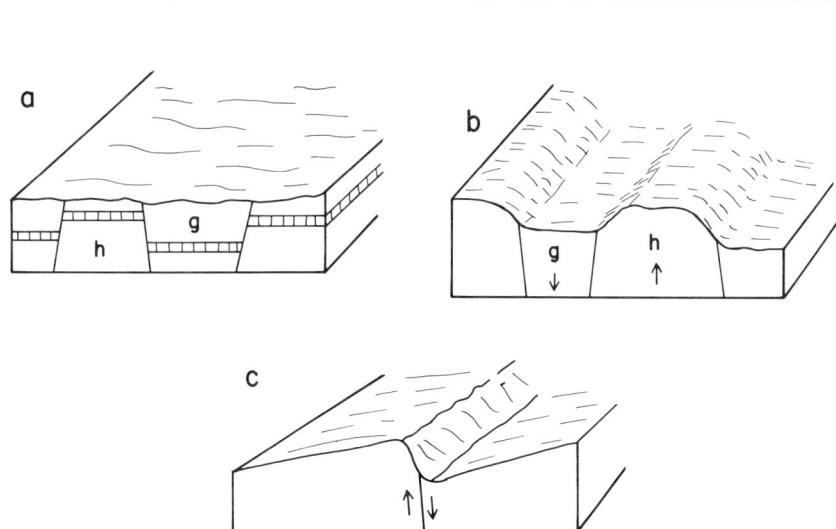

Fig. 11.22
The apparent accordance of mountain range summit levels is no illusion. Contours drawn on Tararua-Rimutaka summit heights suggest a simple up-doming, but the stream pattern is not simply related to this. The smooth contours may be the 'ghost' of a low relief erosion surface—a peneplain—since upheaved and dissected, but it has been noted that accordant summit heights must result where an up-lifted region is dissected by regularly spaced streams. (After Wellman, 1948)

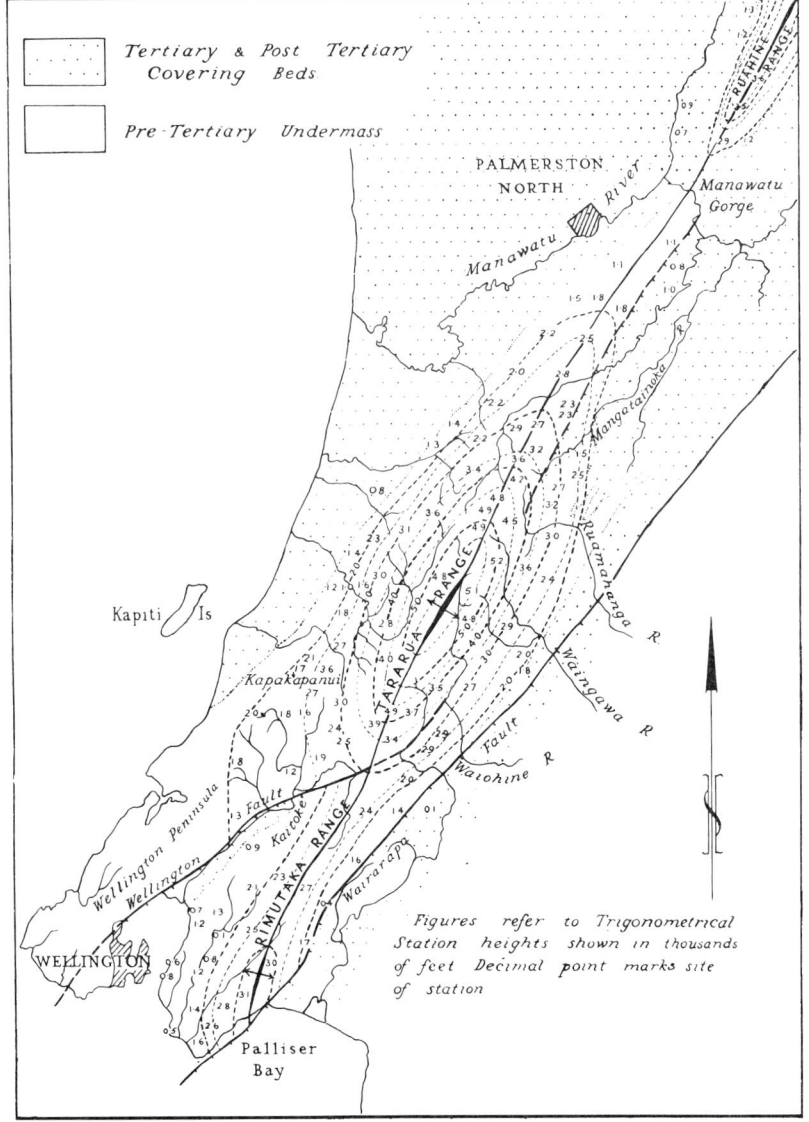

range had been 'broken' in some indescribable cataclysm. How it came about that the river was ready to dash through and where the drainage might have been going beforehand were not often considered.

The trans-mountain gorges of the Manawatu, the Buller, the Grey, the Hurunui and other rivers attracted the attention of geologists and have been debated since the beginning of the century. Leaving aside some special explanations, there were two general theories. According to one, the course of the river is supposed to have been well-established prior to the uprise of the range across its path, and by some aggradation upstream and vigorous down-cutting across the rising axis the original course was maintained while the range grew in height. Gorges of such origin are described as 'antecedent'. The alternative explanation requires that a transverse, older ridge of hard rock lay

buried unconformably beneath a cover of younger, more homogeneous soft rock or gravel. The obstacle is 'discovered' by the river, but being entrenched in the younger cover, is not diverted but keeps on cutting down across the hard-rock barrier as it emerges (by differential rates of erosion) during the general degrading of the region. The resulting gorge is then described as 'superposed'. This term is also applied to situations where the drainage pattern disregards all kinds of structural features and differences of hardness in the basement rocks of a region, and is then inferred to have had its ancient course 'stencilled' through former covering strata that have since been stripped away.

Evidence in support of antecedence comes from faulting, tilting and warping of geologically recent deposits in the vicinity of a gorge. The case for superposition, on the other hand, usually rests on the discovery of remnants of some formerly thick and widespread cover which is believed to have provided the 'stencil'. The notable gorges in fact seem to belong in neither class entirely, but partly in both.

Superposition through an earlier filling of the valley with gravels, usually glacial in origin, provides the best explanation of the numerous cases where rivers now cut through the nose of a bedrock spur in a short, rocky gorge where it would seem to have been more logical to excavate a course in gravels around the end of the spur. Visitors to Lake Sumner pass near two fine examples of this kind of gorge in the upper Hurunui gorge between the junctions with South Branch and Sisters Stream.

Fig. 11.23
Manawatu River flowing west across the young tectonic mountain axis of the southern North Island. Remnants of Pliocene and early Pleistocene strata near the gorge form an anticline over a core of Torlesse greywacke. If the river established this course before up-arching began, the gorge would be 'antecedent'. But the westward drainage possibly was not established until later, although probably before the hard rock core was exposed by erosion. (Photo: N.Z. Geological Survey)

Fig. 11.24
1. A vigorous river that can maintain a pre-existing course (a) by downcutting while a transverse ridge rises across its path (b) produces an 'antecedent' gorge. 2. If a river flowing over younger deposits that have buried the relief of a former landscape (a) can maintain its course while erosion re-exposes the old relief (b), a 'superposed' gorge will result. 3. A river course that disregards underlying rock structures may also result from superposition; established originally over a cover of younger strata (a), it was 'stencilled' on to the surface of the undermass (b) before all the younger cover had been eroded (c).

River Plains and Shingle Fans

Persistent aggradation leads to the development of broad valley-plains, beneath the floors of which, in contrast with plains carved in rock by lateral erosion by the river, substantial thicknesses of stream deposits should exist. The lower reaches of many valleys are occupied by river plains, showing that despite a falling-off in the supply of erosion products as the climate improved after the Otira Glaciation, sufficient sediment was still coming down in most cases to ensure that aggradation kept pace with the rise of sea level through the post-glacial period and so prevented invasion by the sea. River plains have since continued to be aggraded in inland areas, well beyond the influence of sea level rise, indicating that the post-glacial work of modifying the glaciated landforms in the upper valleys has continued, especially in the frost-vulnerable greywacke and schists of many mountain catchments, despite the return of vegetation up to altitudes of well above 1000 m. Extensive river plains in the central North Island region resulted also from the great quantities of volcanic debris which the rivers have had to export from eruptions of the Tongariro group and in the Rotorua-Taupo Volcanic Zone.

In marked contrast, some large rivers flowing from regions that have been uplifted continually through later Pleistocene time occupy narrow-floored valleys almost to their mouths, and lack valley plains of any extent or continuity. Notable examples are provided by the Mohaka and Clarence rivers, and by some of the smaller rivers of Wellington and Taranaki north of Wanganui.

A tendency to swing from side to side in unstable series of curves or 'meanders' is a confirmed habit of unconfined stream flow, as

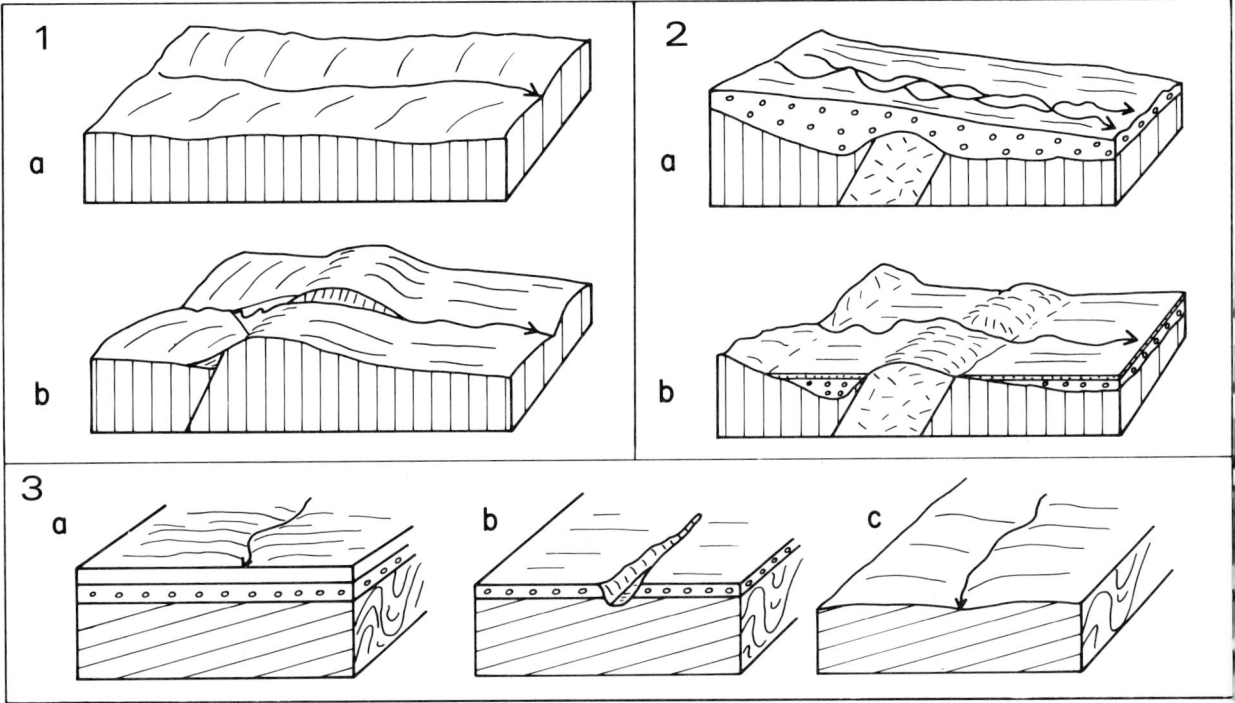

Fig. 11.25
The Orongorongo River, in the Rimutaka Range, eastern Wellington, has been building up its bed intermittently for at least half a century. The cause is uncertain, although instability of the mountain slopes following the 1855 earthquake is a possibility. Forest deterioration owing to interference by introduced animals is helping it to continue.

hydrological experiments will show. Natural stream meander loops on widened valley plains, especially where gradients are gentle, velocities normally low and sediment loads light, tend to expand in radius until adjacent loops break into one another. Short-circuited loops may remain as horseshoe-lakes or swamps. Traces of abandoned meander-channel loops can be made out on some river plains, when viewed from above.

Wide gravel-floored valleys of the major rivers in the South Island mountains seldom are submerged entirely even at flood times. Normally the surface flow is distributed over an ever-changing mesh of channels which diverge and rejoin to make a 'braided stream pattern'. This pattern is accepted as an indication that a river is still aggrading, or at least for the time being is in a condition of equilibrium. Streams that are lowering their beds tend to be confined in a single channel.

Usually because of a sudden loss of velocity upon reaching gentler slopes or a flat valley floor, mountain-side tributary streams construct 'alluvial fans', which are among the most striking features of New Zealand mountain scenery. Where still going on, fan-building takes place intermittently during exceptionally heavy rain or snowmelt in gully catchments. The characteristic flat-cone form and arcuate outlines of alluvial fans are attributed to the fact that the stream, though fixed in position at the apex of the fan, is free to shift its channel anywhere over the surface as the fan grows, more or less evenly occupying all possible radial positions at different times. Fans may be attacked or destroyed from time to time when the main river channel impinges upon their perimeters, but will be rebuilt after the channel swings away again. As a rule, the smaller the discharge from the gully and the coarser the debris, the steeper the slope of the fan. Many fans have been built by periodic flows of mud and debris rather than by stream torrents, and it is hard to draw a sharp distinction between alluvial fans and debris fans.

Many fans in glaciated regions grew to their present size very soon after ice vacated the valley, in some cases against the side of the wasting glacier. Alongside the Lewis Pass Highway one can see good examples of ice-margin fans, and also of fans built out as steep deltas into former ice-dammed or moraine-dammed lakes.

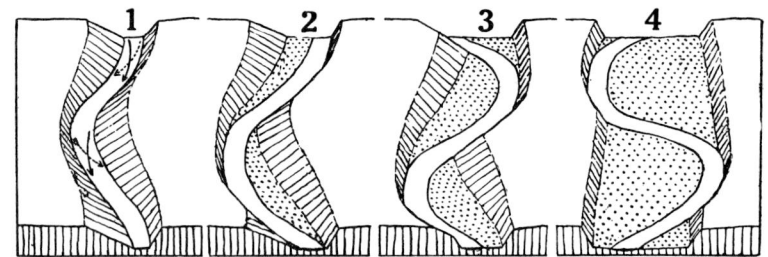

Fig. 11.26
Sideways-cutting by a stream at the foot of valley-side slopes ('lateral corrasion') widens the valley floor and leads to development of flood plains. (After Cotton)

Fig. 11.27
Whareama Stream, near Tinui in the eastern Wairarapa, swings in wide meander loops across a former flood plain surface beneath which it has now incised its bed by as much as 9 m. (Photo: N.Z. Geological Survey)

Fig. 11.28
Braided stream channels on the bed of the Rakaia River and on the terrace above it. The river is entrenched below the Canterbury Plain, which consists of a series of gently-sloping gravel fans built of glacial outwash brought out from the mountains by the Rakaia and other major rivers during the later ice advances. Valley sides at each terrace level were cut back into concave curves and cusps as the river widened its bed. (Photo: N.Z. Geological Survey)

Intervention of Cold Climate

The first part of this chapter emphasized the dominant role of rain and running water in the sculpture of most New Zealand scenery. The most important exceptions are in areas where cold-climate types of process intervened, once or more often, during the Pleistocene. Such interventions were called 'climatic accidents' by Davis, implying that modelling of the landscape had gone on for the most part under temperate conditions, the effects of the 'accidents' being to modify the earlier work of running water rather than to replace it entirely with the work of ice and frost. Such a view fits the New Zealand situation reasonably well, although there have been not one, but several episodes of cold-climate modification. Moreover, owing to the persistence of Kaikoura movements, little evidence remains of the landscape before the last two or three glaciations, so it is not simply a matter of glacial modification of a strictly 'pre-glacial' scene.

Where the Glaciers have been

Glacier ice is now confined to the Southern Alps, adjacent high ranges in mid-Canterbury, western Otago and Southland, and a small capping on Ruapehu. Prior to this century's devastating decline the alpine valleys indeed held more ice than now, but even when seen by the first explorers would have given little idea of how the whole mountainous region appeared during the glaciations. It now requires some mental effort to imagine the scene—sharp peaks and ridges here and there protruding as 'nunataks' above sweeping sheets of ice which coalesced between catchment and catchment. In all probability, there would have been little indication of how deep were the underlying troughs which conducted thick ice tongues far down the main valleys and in some areas well out on to the lowlands beyond the mountain fronts.

Our grandest mountain scenery is provided by the characteristically bold forms of ice-sculptured rock. Most spectacular of all, the imprint of deep glacier erosion in tough, crystalline granite and gneiss survives in Fiordland because these rocks were resistant to the attack of frost after the glaciers receded. Similarly bold forms can be seen on a reduced scale and over more limited areas in the granitic ranges of North Westland and west Nelson. Travellers on Highway 7 between Rahu Saddle and Reefton see some fine examples.

Generally wider, more open valleys were modelled by glaciers in regions of Torlesse greywacke and Haast Schist, but the post-glacial defacement of glacial forms has affected these rocks more strongly because the innumerable bedding planes, cleavage and joint surfaces and faults admit water superficially and promote the action of frost. Some fine, steep glacial scenery is to be found nevertheless in the headwards parts of most alpine valleys.

Glacial landforms of the kinds most widely seen in New Zealand are illustrated and explained diagrammatically in Figs 11.31–11.40. The constructional features due to glacier action, however, require some special comment.

Accumulations of debris at the snout of an active glacier, if released in amounts and in places such that the outflow of meltwater cannot

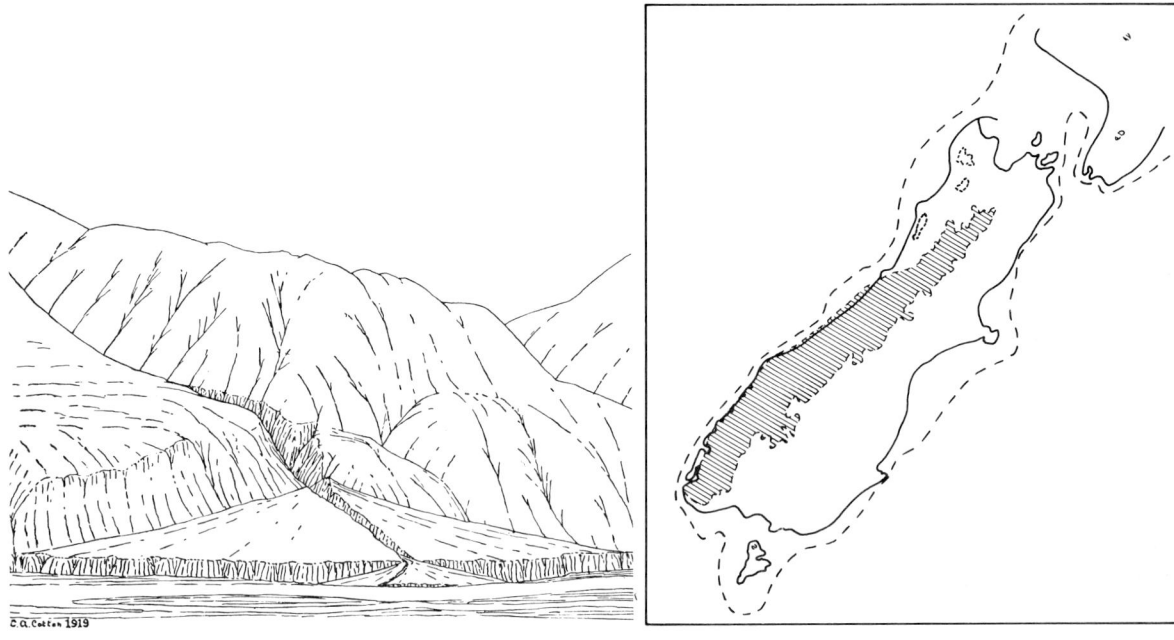

Fig. 11.29
An alluvial fan built on to the valley floor at the mouth of a tributary may later be cut back by meandering channels of the river, and then dissected by the stream that built it. (After Cotton)

Fig. 11.30 (right)
Estimated extent of maximum ice cover during the Otira Glaciation, and the estimated position of the contemporary coastline. (After Suggate)

remove it all, builds a 'terminal moraine'. It can be taken as a sign that the position of the snout remained unchanged for a time before receding. An advancing glacier tends to override materials in front of it rather than to push them forwards bulldozer-fashion, whereas a receding glacier dumps its burden of debris in uneven heaps ('ablation moraine'). 'Lateral moraines' and 'medial moraines', the latter formed by the merging of two lateral moraines where glacier tributaries join, are made up mostly of debris that has fallen on to the ice from rocky walls above. Deposits formed directly by the release of debris from the ice are called 'till'; strictly, the word 'moraine' refers to the surface features thus produced.

Deposits and landforms created by streams draining from glaciers are described as 'fluvioglacial' (or 'glacifluvial'). The bed of the Tasman River downstream from the moraines of the Tasman Glacier provides a modern example of an outwash plain. On a vastly extended scale, the Mackenzie Plains remain as an indication of the vast quantities of debris discharged by meltwater outlets through the Pleistocene moraines of the three biggest Upper Waitaki glaciers, named usually after the lakes Tekapo, Pukaki and Ohau which occupy their former sites. Full-sized rivers flowed alongside these glaciers; their deposits, stranded on the valley walls since the glaciers retreated, now form 'kame terraces'. These are common in the South Island glaciated valleys. One of the most impressive, built of gravels 70 m or more thick while the glacier was still advancing or thickening, now forms the south wall of the Hope valley upstream from Hope Bridge on the Lewis Pass highway for a distance of more than 10 km. Sloping terrace features flanking lakes Tekapo and Pukaki, formerly regarded as lateral moraines, are at least in part kame terraces.

Fig. 11.31
Shaped mainly by glacier erosion during the Pleistocene Period, gneissic rocks above Milford Sound have been sculptured into a bold, inhospitable mountain landscape. (Photo: N.Z. Geological Survey)

Fig. 11.32
Mount Aspiring, surrounded on all sides by steep cirque headwalls, is described as a horn, although it probably did not develop directly from a pre-glacial summit, as in Fig. 11.33. (C. A. Cotton, 1942)

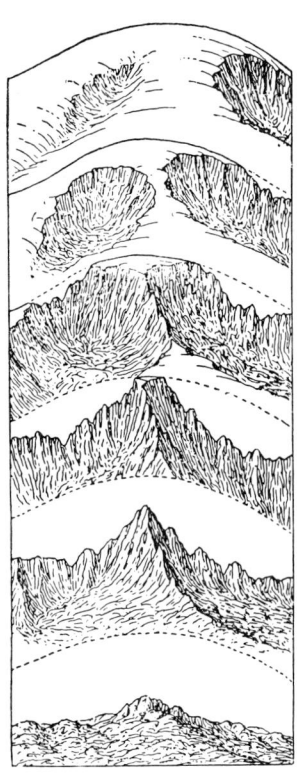

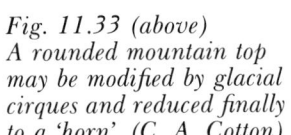

Fig. 11.33 (above)
A rounded mountain top may be modified by glacial cirques and reduced finally to a 'horn'. (C. A. Cotton)

Fig. 11.34 (upper right)
At the maximum of each ice advance the surfaces of tributary ice streams tended to merge with those of the main trunk glaciers. Carrying less ice, tributaries required smaller, shallower channels; when the ice melted the mouths of their valleys remained perched above the main trough floors as 'hanging valleys'. (C. A. Cotton)

Fig. 11.35 (right)
Hanging valleys over the Havelock River, Canterbury. (Photo: Robert Speight)

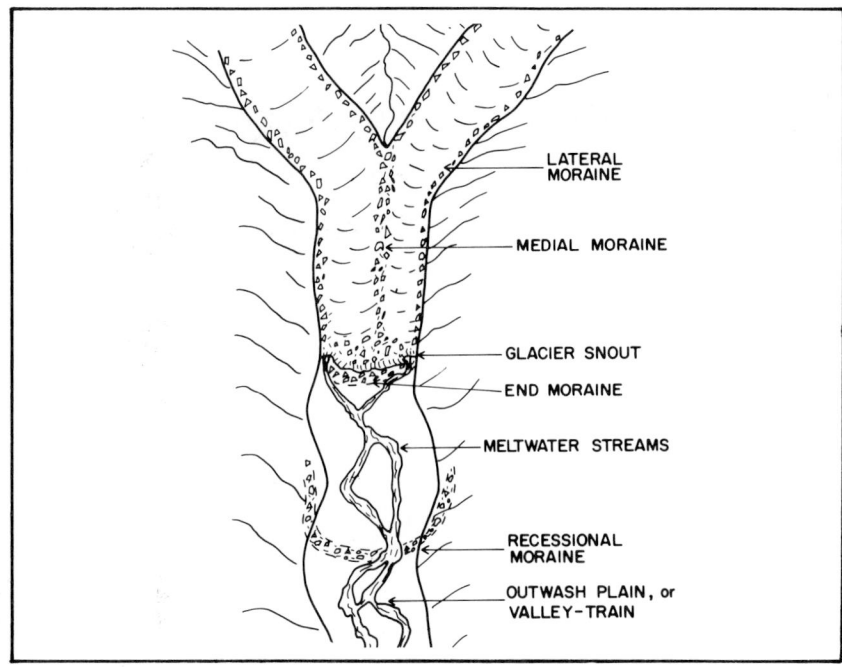

Fig. 11.36
Common terms for features of the lower reaches of a valley glacier.

Fig. 11.37
Types of deposits associated with snouts of stable valley glaciers.

Fig. 11.38
A bulky moraine in the Waimakariri valley opposite the mouth of Poulter River (centre) accumulated at the snout of the Waimakariri Glacier during the last major ice advance. Outwash gravel (right) is now 130 m above the river. When the glacier receded, moraine and buried ice impounded 18-km-long 'Glacial Lake Speight'. A stairway of lake shorelines, formed as the outlet was lowered in stages, is faintly visible in the foreground.

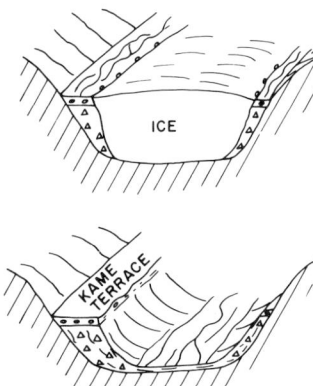

*Fig. 11.39 (above)
Streams flowing alongside
large valley glaciers leave
deposits which, after the
ice has melted, may remain
as 'kame terraces'.*

*Fig. 11.40 (right)
A prominent kame terrace
flanking the Hope Valley,
viewed from above the
Lewis Pass Highway. It
was built up by a powerful
stream flowing alongside
the glacier which last
occupied this part of the
valley in late Otiran times.
(p. 253) (Photo: A. Lush)*

Cold Climate without Glaciers

From the geomorphic point of view, periglacial conditions (Chapter Nine, p. 260) in New Zealand in the Pleistocene Period were important in several ways. Although the forests did not entirely disappear even from the South Island, vegetation cover was greatly reduced. Physical weathering was severe down to quite low altitudes, mantling hill slopes with scree and providing streams with heavy burdens which led to aggradation, as in the case of glacier-fed rivers. Although the ground may have been frozen for much of the winter in places as far north as the central North Island, the summer thaw and odd warm spells in winter produced melting down from the surface, loosening superficial debris, lubricating particles with water films, in all ways providing favourable conditions for down-slope movements. Deposits and forms ascribed to solifluction (page 260) during the Pleistocene are thus common from Wellington southwards, especially in the east, but no convincing evidence for permanently frozen ground (permafrost) has been found.

A variety of distinctive surface features due to disturbance of superficial debris by frost, summed up in the expression 'patterned ground', characterise the frigid sub-polar zones today, and extend into upper-middle latitudes in the Pleistocene. In New Zealand these are known only on a small scale. Even today, frost at high altitudes is severe and frequent enough to sort fine debris into 'stone-rings' and 'stone-polygons' on flat surfaces and 'stone-stripes' on slopes. Larger stone-polygons of probable Pleistocene age are reported from mountain crests in the south, as for example on the Kirkliston Range and Old Man Range at altitudes of around 2000 m. Large stone-stripes and solifluction lobes are also described from Central Otago.

Secondary features of Wellington's hilly landscape were recognised about twenty years ago by C. A. Cotton and M. Te Punga as being the result of Pleistocene periglacial conditions. Most of the credit for replacement of originally sharp ridge crests by more subdued rounded summits is now granted to solifluction and related processes. Developing the idea in subsequent years, Cotton considered that the complex profiles of Wellington hills showed that there had been alternating phases in which solifluction and normal stream-dissecting acted in turn, in response to the climatic oscillations which in the South Island were expressed by glacial and interglacial sequences.

Finally, we must include as a periglacial geomorphic effect the mantling of many landscapes, from eastern Wellington to the far south, with loess. Derived from barren, windswept mountain sides, outwash plains and dried-up lake beds, fine dust travelled great distances down-wind from source areas before being redeposited as a mantle of silt. Successive loess layers built up in a series of cold phases are often separated by fossil soils developed during intervals of more genial climate. The total thickness reaches more than 10 m in favourable sites for accumulation. Besides the rounding of slopes, loess has important effects on run-off and drainage, and its tendency to erode easily is widely known.

Terraces

Terraces are recognisable in river valleys in many parts of the world, but seem to be particularly obvious in New Zealand; indeed, they contribute one of the distinctive aspects of our scenery.

The term 'structural terrace' is applied where a wide ledge has developed on valley walls directly above the outcrop of a horizontal or gently-dipping, hard, resistant stratum because softer rock above has been eroded farther back. The classic example of this type of terrace, on a spectacular scale, is in the multi-stepped walls of the Grand Canyon of the Colorado River in Arizona. Ledges of similar origin, though not as wide, are common enough in areas of alternating sandstone and mudstone of Tertiary age in various parts of New Zealand. Rarely do they justify the designation of terraces.

The majority of terrace features in New Zealand valleys mark the survival of portions of earlier valley floors, formed when the stream flowed at a higher level. These valley-plain terraces fall into two categories. In the first, a stream which had been flowing on a valley floor which it had widened for itself by eroding laterally at the foot of the side slopes, later cuts its channel below the floor and carves an inner, narrower valley within the confines of the earlier one. A common reason for this behaviour is that the higher valley floor represents the earlier profile of the stream at a time when its gradient was adjusted to a base-level higher than the present one. Then either the land rose or sea level fell, making it necessary for the stream to regrade its channel down to the new, lower base-level. Such an event theoretically initiates a new geomorphic cycle leading in time, if there is no further change of level, to modification of all gradients and slopes throughout the region. Theoretically, down-valley tilting of the land

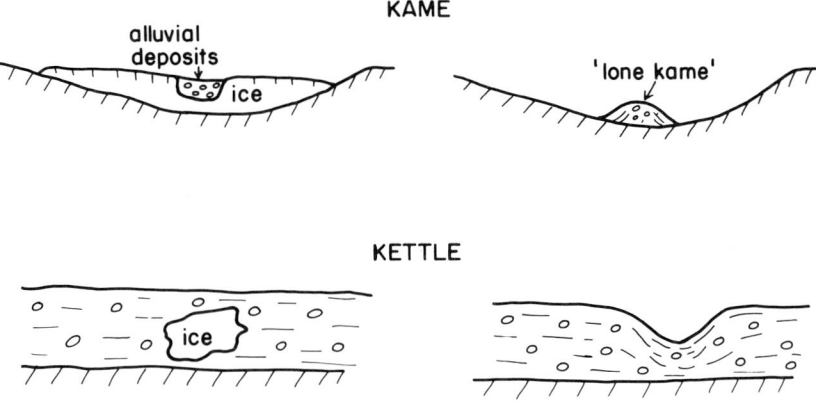

Fig. 11.41
Streams flowing on top of large glaciers leave gravel and sand deposits which are dumped when the ice melts to form 'kame' mounds. 'Kettles' are hollows in the surface of outwash gravel caused by collapse when buried blocks of ice have melted.

Fig. 11.42
a. Structural terraces of the 'Grand Canyon' type.
b. Valley-plain terraces, remnants of former valley-floor levels that survived as the river lowered its bed.
c. Terraces produced during re-excavation of an earlier filling of gravels. The filling resulted from stream aggradation which may have been due to variations in the intensity of physical weathering processes induced by climate oscillations; but can have other causes as well, e.g. an episode of severe earthquakes generating rockfalls and landslides.
d. Alternate phases of aggradation and downcutting can produce flights of terraces.

can have a similar result. If a bedrock surface is to be seen underneath a rather thin veneer of stream deposits it is quite likely that the terrace is one, perhaps of a series, which originated this way during pauses interrupting a long period of uplift. Valley plain terraces are certainly common enough in soft-rock areas of the North Island that have had just such a history during the later Pleistocene, and they are not hard to find in the South Island either.

Climatic fluctuations and periodic volcanic outbursts at different times in the Pleistocene Period have both been responsible for alternating episodes of sediment-filling and down-cutting which few valleys have escaped. The second class of terrace, resulting from the incision of a stream below the surface of the last gravel filling, is likely to be underlain by a substantial thickness of alluvium, and the chances that it will escape destruction by erosion or subsequent burial under an even deeper sediment filling are enhanced if base-level is continually being lowered.

The combination of continual though perhaps intermittent uplift of most of New Zealand through the Pleistocene Period, with many pulses of river-aggradation, is expressed by the presence of elegant flights of

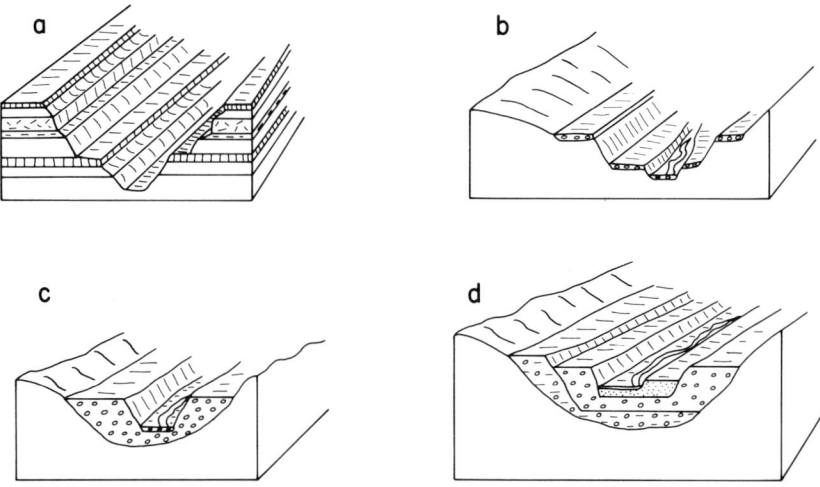

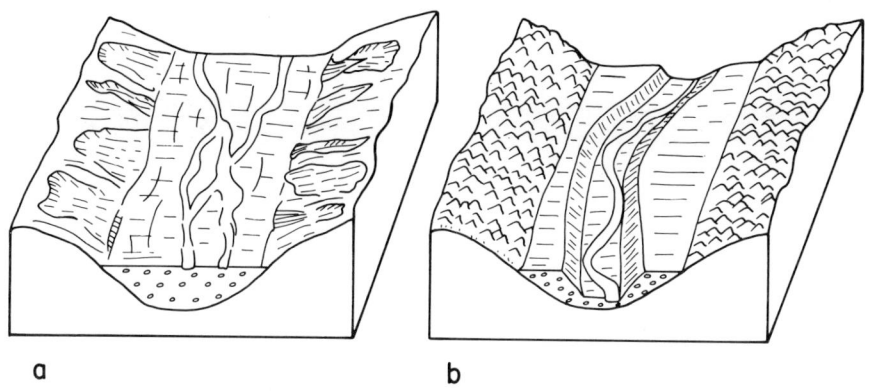

Fig. 11.43
Climatic terraces, caused by phases of aggradation due to intensified physical weathering and debris supply while vegetation cover is reduced in cold periods (a), alternating with downcutting when vegetation cover is restored and erosion reduced upon the return of warmer conditions (b).

terraces in the great majority of rivers throughout the country except the far north.

The mapping of 'climatic' terraces from highland areas towards the coast runs into a complication in middle and lower reaches because, while cold climate conditions are promoting heavy sediment loads and aggradation in the upper reaches, sea level has fallen, causing the river to cut down near its mouth; and conversely, when melting ice-caps are raising the ocean level, causing aggradation in the lower reaches, the streams of the hinterland are degrading. As a result of this out-of-phase relationship, there may be two sets of terrace profiles, alternating in age, which cross over at some intermediate point. Theoretically, that is; identifying and tracing them is another matter.

The cycles of terrace-formation we have been considering have involved time-spans of the order of thousands of years and more. It would be as well to mention that smaller-scale terraces have been seen to form and be destroyed by erosion over periods as brief as a few years or decades in response to cycles of weather, for example, in the Otira Gorge during the 1960s, and also the mobilizing of scree and hillside debris during earthquakes and the temporary overloading of rivers from this cause.

Landscapes on the Move

Soil creep, shallow slumps and intermittent mudflows are among the recognised mechanisms for transferring the products of weathering into the realm of stream transport, but they are seen as playing a subsidiary role. Over some thousands of square kilometres in the North Island East Coast region, however, it is a different story. Here, landscapes have been shaped almost entirely by mass-movement on a major scale. The areas concerned are underlain by late Cretaceous and early Tertiary marine strata, some of which contain swelling clay minerals of the bentonite group which lose their bulk strength when water gains access. In that condition, the rock is ready to creep and flow gravitationally down slopes with gradients of less than 20°.

The geomorphic result has been entire river catchments or sub-catchments, for example, the upper Waipaoa River and its main tributary, the Mangatu, in Poverty Bay, in which the removal of forest some seventy years ago revealed gentle, even slopes bearing the surface forms of earthflow and slump and a few widely spaced normal stream

Fig. 11.44
Looking down Cave Stream towards Broken River and the Torlesse Range, Canterbury. The prominent terraces are underlain by gravels deposited when glacial melt-water torrents from the Waimakariri valley spilled through Craigieburn Saddle southwards into Castle Hill Basin and the Broken River catchment during the Blackwater Advances (I and II) of the Otira Glaciation. (Photo: D. J. Jones)

gullies. Forest clearance was quickly followed by renewed, vigorous slumping, earthflows and deep gullying, and the resultant overloading of the main rivers caused spectacular aggradation of their beds and severe flooding problems in the lower reaches. Thus 'Mangatu' became a name identified with the worst of New Zealand's soil erosion problems. After several successive attempts to check the discharge of debris from the disastrously eroding areas, a determined, vigorous policy of re-afforestation with quick-growing trees since 1960 seems at last to be re-establishing control.

A geomorphic puzzle remains. Why did a landscape of that type, modelled in the past by mass movements of waste, remain stable for periods of between a few hundred and a few thousand years prior to the dramatic upset caused by historic forest clearance? A mantle of volcanic ash may have brought about a hydrological regime under which the last generation of indigenous forest was able to establish itself and maintain stability, but much remains unexplained.

Similar phenomena on a much less extensive scale are to be seen in other areas of late Cretaceous-to-Eocene strata in eastern districts from Southern Hawkes Bay to Dunedin, creating many slope-stability problems for roading and railway engineers and for farmers.

Lakes

Every lake poses a problem for the geomorphologist because its very
existence challenges one of the important generalisations of his subject,
namely, that running water and gravitional mass movements are the
most widely effective agents in landscape modelling. Since the
continued operation of those agents depends upon maintaining
downward slopes, any enclosed hollow must appear as an anomalous
reversal of the normal direction of slope from land to sea. The most
familiar indication of such a reversal is a pond or a lake. On the whole,
lakes are temporary features in a landscape, destined sooner or later to
disappear as a result of sediment filling combined with the cutting
down of an outlet.

Explanations for the presence of lakes have been grouped in various
ways for different purposes, the number of types recognised by
different authors ranging from half a dozen to over seventy. Very few
lakes (in New Zealand, at any rate) fall simply into one single category.
The majority of them owe their origin to combinations of
circumstances and agencies. A small minority of lakes are due directly
to sagging or subsidence of the land from tectonic or volcanic causes,
and the remaining important exceptions are those in basins hollowed
out of rock by glacier ice (which does not respect the laws of hydraulic
gradients in the same way as running water) and volcanic craters and
calderas. Many conflicting explanations have been offered for some
lakes through the years, though in some cases the conflict was
unnecessary, more than one of the suggested origins being partly true.

Summarising the main, natural causes of obstruction of the drainage:
landslips and *rockfalls*; *lava flows*; *moraines* (and other glacial
accumulations); *alluvial fans*; *aggradation* by a stream in an adjoining
valley; and the blocking of coastal bays and estuaries by *sandspits*, *gravel
bars* and *dunes*.

Lake Taupo, our largest lake, though once the subject of some
debate is now accepted as occupying a sunken tract partly bounded by
faults in the Rotorua-Taupo Volcanic Zone. Subsidence is ascribed to
collapse of the crust following the ejection of many cubic kilometres of
material in a series of catastrophic events culminating in the great
pumice eruptions of 1850 years ago. Submerged explosion craters show
up in the lake-floor contours. Many of the smaller lakes in the
Rotorua-Taupo region, including Rotorua and Rotoaira are due also
to crustal sagging combined with recent faulting, blockage of drainage
by erupted material, and explosion craters. Rotoma and Rotoehu
occupy explosion craters in a downfaulted area. Well-known crater
lakes include the one usually present in the summit caldera of Ruapehu
and the Blue Lake on Tongariro. Away from the volcanic regions,
subsidence has resulted in sag-ponds alongside the Awatere Fault in
inland Marlborough, known by the names Sedgemere and Bowscale
Tarn.

Among landslide-dammed lakes, pride of place is given to
Waikaremoana, in the upper reaches of a branch of the Wairoa River,
Hawkes Bay. Temporary lakes have formed on the upstream side of
large landslides after some major earthquakes, as in the upper Buller

Fig. 11.45
Lake Coleridge occupies a rock basin hollowed out by an extension of the Wilberforce Valley glacier, which merged with the mighty Rakaia Glacier at several places. A tunnel under the saddle at left feeds the Coleridge power station on the main Rakaia valley floor. Gravel spits and bars reflect the strength of currents driven by prevailing north-westerly winds. (Photo: V. C. Browne)

Gorge in 1968. Generally, their existence is brief. Lake Minchin, in a branch of the Poulter River, Canterbury, is dammed partly by an old rock-slide that was added to during the 1929 Arthurs Pass earthquake. Blockage by a lava flow is neatly illustrated by Lake Omapere, in Northland, the drainage from which now flows westwards into Hokianga Harbour instead of eastwards down the Waitangi River to the Bay of Islands.

Few of the innumerable lakes of glacial origin in the South Island seem to be purely in ice-excavated rock basins or due entirely to damming by moraines. The larger lakes (except Ellesmere) occupy basins deepened to below the level of rock at the present outlet, where it can be determined. Old lake-shore features indicate in most cases that the outlets are now lower than when the ice finally retreated. Lake Tennyson near the head of the Clarence River is one that could be entirely due, in its present form, to blockage by moraine and outwash deposits. Boulder Lake in the Collingwood district of north-west Nelson may be cited as an example of a rock-basin lake. The term 'kettle' is applied rather loosely to small lakes or tarns which commonly occupy hollows in the uneven surface of a moraine, or between adjacent moraine ridges. Strictly, this term should be restricted to lakes that mark places where detached slabs of dead ice once lay buried in

fluvioglacial deposits. Kettle lakes usually have no regular outlet above ground. Marymere, one of the popular fishing lakes in the Waimakariri Valley not far from Cass, is a fine example. Lake Hayes near Arrowtown is very probably of similar origin.

Lakes impounded partly or entirely by alluvial fans are common throughout the South Island mountains. Such an origin is obvious for Lake Pearson alongside Highway 73, the Arthurs Pass route, and there is additional interest in that although this lake receives a substantial inflow from Craigieburn River, its overflow channel at the eastern end is frequently dry because most of the normal discharge passes beneath the Craigieburn fan to re-emerge in a series of springs which are the main source of Winding Creek.

The blocking of tributary valleys and valley-side run-off through aggradation by the main river has been the main origin of many shallow lakes in both islands. The margins of such lakes tend to be ill defined and marshy, as in the cases of Lakes Whangape and Waikare in South Auckland, both being blocked by alluvium of the Waikato River, and Lake Wairarapa, dammed by the Ruamahanga River fan. The small Rotoiti, near Kaikoura, impounded by gravels of the Kahautara River, is a clear-cut example. In South Otago aggradation in the lower Clutha valley has given us Lakes Kaitangata and Tuakitoto, while

Fig. 11.46
Lake Unknown occupies an ice-carved rocky hollow high above the Routeburn Valley, west Otago.
(Photo: N.Z. Geological Survey)

Fig. 11.47
Detached masses of stag-
nant ice buried under
gravels downstream from
the snout of the receding
Fox Glacier slowly melted,
causing collapse and the
appearance of 'kettle-holes'
on the river bed.

Fig. 11.48 (centre)
One of a chain of small
lakes dammed by alluvial
fans in a former glacier
channel near Lake
Coleridge.

alluvium of the Taieri River has obstructed its tributary, the Waipori, giving rise to Lakes Waihola and Waipori. River aggradation is involved also in retaining the water of some Westland lakes which lie against moraine ridges (e.g. Ianthe, Wahapo), but one would guess that lakes have occupied these positions ever since the final retreat of ice.

Ellesmere, one of the South Island's largest lakes, is really a lagoon separated from the sea merely by a long shingle spit of geologically recent origin. In other examples of blocked valley or fiord mouths, including Lake Oneke, southern Wairarapa, and Lake McKerrow, west Otago, the obstruction is partly the result of bar-building, with assistance from blown sands.

Much more could be written about practically all the lakes named above. Suffice it to end this section with a reminder that the origins of lakes are seldom as simple as they may first appear to be.

The Form of the Coastline
Some of the background for this section was provided in Chapter Two while tracing changes in the outline of New Zealand through later geologic time. There are points of similarity between the work of running water in sculpturing the landscape and the work of waves and currents in shaping the shore, both lengthwise and transversely. Changes of form are effected in both realms by wearing away rock here and depositing sediment there, and again the momentum of moving water masses combined with gravity provide the immediate supply of energy. Under conditions of excess sediment load the coast tends to be built out ('prograded'); in the reverse situation, excess energy is

directed to erosional attack along the shoreline which, if persistent, leads to retrogradation. Prolonged retrogradation may lead to the cutting of a rock-floored wave-cut platform at a level related to the mean level and range of tides, and a rising cliff at the rear.

Just as a stream tends to evolve a smooth, even profile along its channel, the coastal agencies in like manner tend to straighten ('rectify') an originally irregular coast through erosion of headlands, where wave energy is concentrated, together with building of bars and spits across the mouths of bays and inlets under the influence of longshore current drift, and infilling of the bays with sediment derived from the land at the rear. A change of sea level interrupts the process, which must begin anew.

Sandspits and Bars

Bars and spits across bays and at the mouths of rivers and estuaries at the sea coast and lake shores are seen by some to result from a process analogous with the deposition of sediment by a stream in places where it is needed to help it produce a smoothly graded profile adapted to its vigour and condition of sediment loading. The analogy should not be taken too literally but has some validity nevertheless where coastal cross-profiles are adjusting to changing conditions.

Sand bars form in the mouths of rivers and estuaries where there are changes in the directions and velocities of water movement in both stream outflow and sea currents, and where the contribution of sediment by the stream is being added to any that may be drifting along the coast. The type and strength of wave action helps to

Fig. 11.49
Lake Forsyth, on the south side of Banks Peninsula, ceased to be an open bay within the last thousand years or so, when it was cut off from the sea by the continually widening coastal plain of Birdlings Flat. Parallel lines of old beach-gravel ridges on the flat mark the successive stages of progradation. (P. 330.) (Photo: V. C. Browne)

Fig. 11.50
Development of coastal rock platforms and cliffs by wave erosion. (H. W. M.—high water mark.)

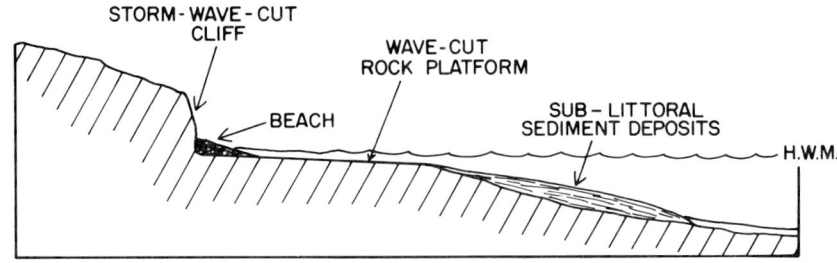

Fig. 11.51
Waves and currents armed with abrasive sediment carved this inter-tidal rock platform in Kaiata Mudstone at Rapahoe near Greymouth. Beach gravel and sand usually cover most of it. Similar deposits on the terrace above were laid down when sea level was higher during the Oturi Inter-glacial event. (Compare with Fig. 11.50)

Fig. 11.52
Progressive straightening ('rectification') of a deeply embayed coastline. Wave attack, concentrated around headlands, de-velops an intertidal rock platform bearing remnant islets and rocks ('stacks'). Strong coastal currents may build spits and bars across bay mouths, speed-ing up the growth of bay-head deltas and the in-filling of bays.

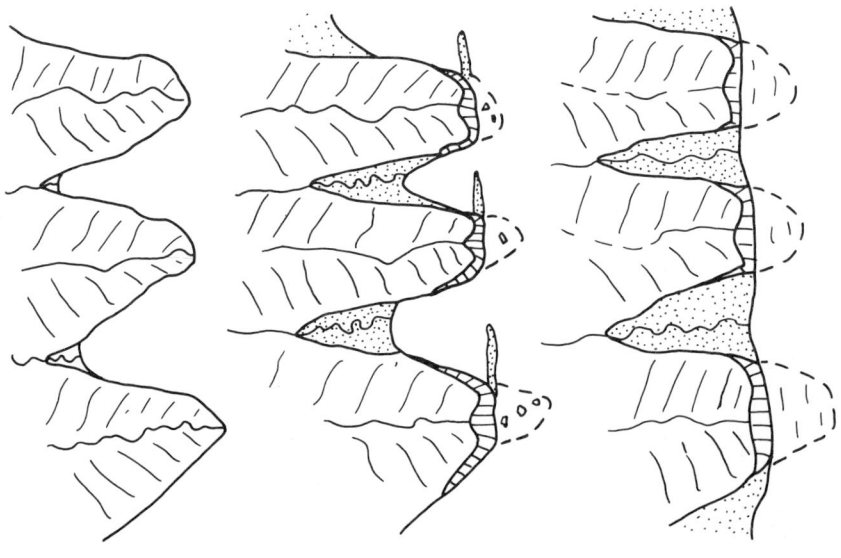

determine whether an accumulation of sand and gravel will build up, and indeed many continually varying factors are involved in determining when and where river-mouth sand bars will grow. They are notoriously subject to sudden changes of position and water depth.

Small streams which discharge at a coast where wave and current action is commonly strong and where a large amount of sediment is being moved along the coast in longshore drifting tend to be deflected parallel with the shore in the dominant direction of drift, sometimes for several kilometres. Typically, these deflected reaches are marked by sluggish flow and swamps, and along their entire length may be separated from the ocean merely by a storm-beach ridge of gravel. Many such streams indeed only succeed in opening a channel to the sea when in flood, their water at other times being discharged by seepage through the gravel. On each occasion of breaching, the point of outbreak may be at a different place. Okarito Lagoon is one of many examples of such stream deflection in Westland.

A powerful longshore current moving undeflected past a cape or a sharp coastal bulge and also bringing an abundance of sediment contributed by nearby rivers may build a spit of sand or gravel seaward

Fig. 11.53
The Boulder Bank enclosing Nelson Haven is a long gravel spit built southwards from Mackays Bluff into Tasman Bay under the influence of a drift from the north-east. The date and manner of its origin pose some questions. (See p. 334.) The present harbour entrance is artificial. (Photo: N.Z. Geological Survey)

for distances of many kilometres. Once established, the spit may be widened in the leeward direction by accumulating wind-blown sand, while a storm-beach ridge may be thrown up high enough above normal tide levels to provide a dry land connection with offshore islands and rocks (under calm conditions, at any rate). Examples of former islands thus tied to the land by sand bars or gravel spits are not uncommon, among the best known being the Mount Maunganui volcano at Tauranga; the Bluff Hill of Napier ('Scinde Island' was its earlier name); and Pepin Island at the eastern end of Tasman Bay, Nelson. Farewell Spit is 25 km long and continues still farther as a shoal; the gravel spit enclosing Lake Ellesmere, Canterbury, is 27 km long, and the Boulder Bank at Nelson 15 km altogether, including 5 km of boulder beach at the northern end behind which the lagoon is now filled with sediment. These longer spits have built since the end of the post-glacial sea level rise about 4700 years ago.

Boulder Bank is noteworthy for the unusually large size of the granitic boulders of which it is chiefly composed, and for the difficulty of explaining its length and direction under present conditions. The material appears to have come from the steeply cliffed coast between Wakapuaka and Cable Bay, yet today only the occasional cyclonic storm with gale-force north-easterly winds driving waves and currents into Tasman Bay reproduces the correct conditions. The prevailing westerlies are unfavourable, yet the Boulder Bank is stable and under no threat of destruction since removal of gravel from it has been banned. The main period of its building therefore must have been during an earlier climatic episode within the last 4700 years when strong north-easterly weather was more prevalent and the current circulation pattern in Tasman Bay was different from the present one.

Sandhill Landscapes

Vast, moisture-starved regions exist on all the major land masses of the world and among their best known characteristics are tracts of drifting sand, their surface forms usually thrown up into migratory dune ridges. These are regions in which running water is only rarely and locally effective as a geological agent of waste removal. The finer products of weathering move about, if at all, mainly through the agency of the wind, which may indeed export the finest dust far beyond the borders of the arid lands. A high proportion of the breakdown products from many rock types is reduced in size to the sand grade, whereafter the particles roll, slide, jump and are airborne over short distances at a time. In the process, a number of distinctive relief forms develop on various scales from surface ripples to sandhills and dunes tens of metres high. Dunes are not stable forms and characteristically migrate in directions determined by the prevailing winds.

Few areas of New Zealand could be described as chronically water-starved, the most notable examples being in situations shielded from rain-bringing winds by mountain ranges, as in the case of The Wilderness between Mossburn and Te Anau in western Southland and of the upper Awatere valley in Marlborough. These regions cannot be described as arid, for permanently flowing rivers are not far away and there is at worst an adaptive vegetation cover struggling to survive the ill effects of an earlier period of inappropriate pastoral management. Moreover they do not contain extensive sand dunes. As in other temperate climatic regions of the earth, the chief origin of our sand dune country is escape of sand from coastal belts. The sand content of sediments moving along the inshore littoral and beach zones can be carried inland out of reach of tides and waves where a persistent landwards airflow exists either in the form of steady and moderate, or of occasional and strong onshore winds. Dune belts of this origin up to several kilometres wide are more common behind the western coasts of the three main islands and occur to a more restricted extent along exposed tracts of the eastern coasts as well.

Belts of dunes migrating landwards have threatened and indeed overwhelmed farm lands in some areas, moving inland until a line is reached where the sand can retain enough moisture for stabilising vegetation to become securely established. Specially adapted plants, such as marram grass, have been introduced in these areas in order to achieve this result more quickly. The extent of our dunelands has varied in the past in line with our later climatic history. More than one area of coastal terrace land bears fixed dune sand deposits dating from late Pleistocene interglacial times of higher sea level. Probably our most extensive dune area in recent times, in which, between high dune ridges and out of hearing of the sea, one could imagine oneself in the Sahara, formerly covered most of the far northern isthmus between Awanui and the hilly country of North Cape, but much of this is now stabilised under planted exotic forests and pasture. It is in no sense an arid region, however, and at all times a permanent watertable existed at a little above sea level. Indeed, its history as a sandy waste was short. Forests grew on the northern isthmus when sea level stood lower during the Otira Glaciation, and relict trees have been found more than once under sand or swamp east of the Ninety Mile Beach.

An interesting kind of sand dune country, mostly stabilized under grasses and scrub since before European settlement, forms belts up to a few kilometres wide beyond the south-east banks of the lower reaches of major rivers in mid-Canterbury. Sand to build the dunes came directly from the river beds, and it is thought significant that these dune belts are on the leeward side of each river with respect to the westerly winds which prevail in the summer months today, but which during the cold climatic phases of the late Pleistocene were probably dominant throughout the year. Though derived from the same source, the blankets of finer loess silt were of course much more widely distributed.

Straight and Crooked Coasts

Those who enjoy sailing need no reminding that the winds and seas around New Zealand make it vitally important to know where one can run to for shelter, and that safe, all-weather shelter is scarce along great stretches of coast except for bar-bound river mouths that are treacherous to navigate at any time. Stretches of strongly indented

coastline are few and far apart. Let us now put together the main factors which determined where the present coast is straight and where crooked.*

In general, an indented coast with deep, branching bays and estuaries and numerous near-shore islands is a fair indication that the land has either remained stable since before the post-glacial rise of sea level or has been depressed, and that coastal processes have made little or no progress with straightening since the sea reached its present level. This may be because of the shortness of the time since that happened, or weakness of longshore drift currents, or inadequate supply of sediment, or a combination of these.

Straight or broadly curving coasts can mean a number of different things. In the relatively stable region of south-east Otago, for instance, parts of the coast are now fairly straight, but because of bay-filling and progradation rather than the cutting back of headlands. Vigorously prograded coastlines backed by broad coastal plains, alluvial fan plains such as those built out by large rivers draining glaciated regions, and coasts affected by abundant supplies of volcanic debris from inland are generally widely curving. Straight segments of rocky coast may reflect a regional structural grain expressed in alignments of stream valleys and ridges parallel with the coast, with major outlets far apart. Again, a straight coastal outline may be due to the region having been uplifted enough to bring up the offshore bottom, the contours of which are likely to be simpler than those onshore except near where there are submarine canyons. Finally, one would expect coastlines determined directly by active faults, or indirectly by older fault lines, to be straight, and the coasts of recently active volcanoes to be simple in plan.

Let us now apply these generalisations to the coasts of New Zealand:

(1) Indented coastlines resulting from the drowning of pre-existing valley systems or glaciated troughs occur on the east side of the Auckland peninsula, in northern Marlborough (actual subsidence of the land here as well), around Banks Peninsula, Otago Peninsula, Stewart Island and Fiordland.

(2) Previously indented but recently more or less straightened coastlines occur on the Auckland west coast, in eastern Otago and western Otago, from Jacksons Bay to Milford Sound.

(3) Coastlines which are straight chiefly because of their lying parallel with the structural grain of the country are to be found on the Auckland west coast south of Kawhia, in southern Hawkes Bay and eastern Wellington (in the latter two cases there have been crustal movements as well).

(4) Straight or openly curved prograded coastlines dominate in Bay of Plenty (thanks to the abundance there of volcanic debris from the Rotorua-Taupo volcanic zone), Manawatu-Horowhenua, central and southern Westland, mid-Canterbury, South Canterbury and Otago as far south as Oamaru; also much of the northern shore of Foveaux Strait.

(5) The coastal forms of Taranaki, northern Hawkes Bay, Poverty

Fig. 11.54 (opposite) Subsidence of parts of the Marlborough Sounds block combined with the post-glacial rise of sea level partially submerged an erosional landscape of ridges and valleys, producing an intricate, deeply indented coastline with many offshore islands and reefs. (Photo: N.Z. Geological Survey)

---

* Readers may wish to refer back to the early part of Chapter Two.

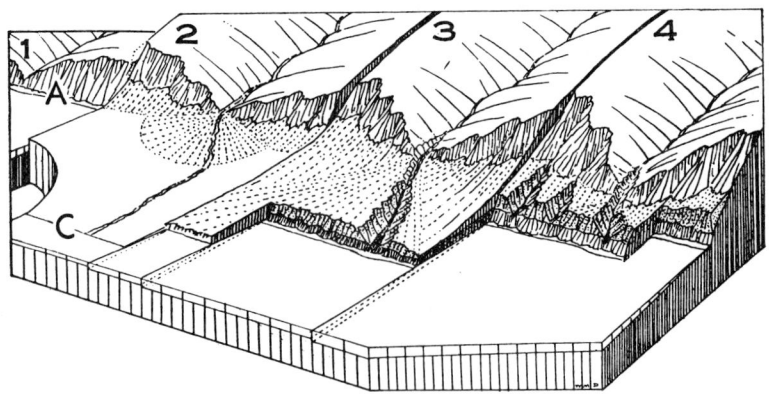

Fig. 11.55
One of C. A. Cotton's
instructive sequential
diagrams showing (1 to 4)
the development of coastal
terraces, such as those on
the northern shores of
Cook Strait, in response to
an uplift of the land which
initially causes the sea to
withdraw from A to C.
(Cotton, after Davis)

Bay, Kaikoura and the South Island West Coast north of Greymouth are much affected by crustal movements, predominantly of uplift or upwarping inland.

(6) Surprisingly,perhaps, little of the New Zealand coastline is directly determined by faults. The western shore of Port Nicholson (Wellington Harbour) is a well-known case of fault-determined coastline. The coast for some distance north of Westport is determined by the Kongahu Fault, and the general line of the northern Fiordland coast, though deeply indented in detail by fiords, reflects the offshore presence of the southwards continuation of the Alpine Fault.

(7) Simple, young volcanic coasts are exemplified around western Taranaki and Rangitoto Island.

Inevitably such generalisations must ignore minor, local exceptions and as with lakes so too with coastlines—the complete explanation is nowhere simple.

Uplifted Shorelines

Coastal terrace features in various parts of the world have been attributed to the activity of marine processes at times when sea level stood higher during Pleistocene inter-glacial stages. Rock platforms carved by waves, often with still-recognisable old sea-cliffs at the rear, beaches and near-shore types of sediment on them are found at altitudes of up to several hundred metres, and many examples have been observed around the New Zealand coasts. Various authors have described them, classified them according to height and otherwise generalised about them, but on not very consistent grounds. It was noted earlier, pp. 244–9, that the evidence for formerly higher stands of sea level provided part of the basis for the current scheme of classification of the Pleistocene Period in New Zealand, and that there are uncertainties because of crustal unrest even within post-glacial time. The history of coastal benches and terraces on the North Island east coast involves late Pleistocene and recent crustal movements as well as the effects of sea level fluctuations.

Examples of uplifted coastal platforms are very common around our coasts and readily recognisable by their form alone if not from deposits on them. Two areas deserve special mention. From the ferries between

Wellington and Picton one can obtain fine views of prominent coastal terraces on the northern side of Cook Strait between Cape Terawhiti and the mouth of Karori Stream. Similar features showing the effects of recent crustal warping are conspicuous between Pencarrow Head and Cape Turakirae. Highways near the South Island West Coast, especially between Westport and Greymouth, provide not only some magnificent coastal scenery but also frequent opportunities to see flights of coastal terraces and to examine roadside exposures of the sediments on them. Descriptions in some detail can be found in the booklets accompanying the *Geological Map of New Zealand* 1:63,360 Sheets S23 and S30 (Foulwind and Charleston) and 1:25,000 (Foulwind and Westport), both compiled by S. Nathan.

Volcanoes as Landforms

A chapter on New Zealand geomorphology would be incomplete without discussing the volcanic features which are not only prominent in the scenery of the northern half of the North Island but which stand prominently in parts of the South Island as well. To do so at length would mean repeating much that has already been written in Chapter Eight, but at the cost of minor reiteration, it is worthwhile to point out some important distinctions. The first of these separates features which resulted from constructional volcanic action from those which evolved as erosion attacked earlier volcanoes.

Taking these cases in turn, further distinctions may then be noted between (a) strong relief of recently built and currently growing volcanic edifices such as lava domes, lava cones and composite cones (e.g. Rangitoto Island, Ngongotaha, Ngauruhoe); and (b) constructional features with little or no relief, such as basaltic lava plains (e.g. west of Bay of Islands) and rhyolitic ignimbrite sheets (covering much of central North Island, but now with relief features due to erosion). Still further subdivision can be made on the basis of differences in the initial forms and structures characteristic of different petrographic types of lava.

Turning to features due to the erosion of volcanic accumulations, again there are differences in the course of erosion in different lava types, and differences also from increasing age. Original constructional forms naturally become modified increasingly with time, but for Pleistocene basaltic volcanoes, as in the Franklin district of South Auckland, the volcanic origin is usually still quite obvious. Through intermediate stages of increasing erosional modification (exemplified by Pirongia, Banks Peninsula, Otago Peninsula), the volcanic appearance fades to the volcanic skeleton stage when nothing is left but a plug of lava rock marking the site of the main conduit and outcrops of dykes representing subsidiary feeding channels.

Examples of deeply eroded volcanoes are scattered through eastern Otago and are also to be found in North Auckland. The lava plug of a vanished volcano frequently is enclosed within a conical hill, the result of differential erosion having removed weaker rock surrounding it, and such features should not be confused with constructional volcanic cones. The hill near Ruawai, northern Wairoa, known as Tokatoka and

'The Crater' and Mount Dasher on the flanks of the Kakanui Range, Otago, thought of as 'volcanoes' by some, will illustrate the point. Erosional landscapes developed over extensive tracts of older volcanic rocks—for example, the Waitakere Range west of Auckland City and the landscape eroded from the extensive Tangihua Volcanics of Cretaceous age in North Auckland strictly should not be regarded as volcanic landforms. Likewise, it is questionable whether a thick mantle of pumice or other tephra should be regarded as a volcanic landform in itself, or merely as a modification of what was already there, not necessarily of volcanic origin.

So far we have been considering raised features or surfaces of volcanic origin, but it would be as well to recall that 'negative' relief features also result from volcanic action, directly or indirectly. These would include explosion craters, pit craters due to the withdrawal of lava in the feeding conduit after an eruption combined usually with some surrounding gravitational collapse as well, and broader sunken areas attributed to subsidence following the discharge of large quantities of eruptive material. Some examples were mentioned when describing the lakes of the Rotorua-Taupo region in an earlier paragraph.

How 'Old' is the Landscape?

In writings about geomorphology, 'old age' applied to a landscape generally means that it is being compared with the stage of senility in the Davis evolutionary cycle. In another sense, it is sometimes remarked that New Zealand is a 'younger' land than, say, Australia, but it may not be clear whether the reference is to the land surface or to the rock underneath. In either case it would be a very broad generalisation. Leaving aside the special geomorphological meanings of 'youth' and 'old age', what exactly determines the age of a landscape?

To begin with, the age of any land surface obviously can be no greater than that of the youngest rocks beneath it, and no less than that of the oldest superficial deposits, soils or vegetation upon it. These limits may leave a long span of time within which the landforms have been generated. It is not always possible to tell whether most of the surface forms were shaped very soon after the underlying rock was formed, or much more recently, or whether by a long, slow, process of evolution covering most of the available span of time. The point may be illustrated by comparing the features of glacial erosion in the mountains of north-west Nelson, shaped by ice no earlier than the late Pleistocene but carved in Paleozoic rocks, with landscapes in parts of Africa and Australia underlain also by ancient rocks but having attained their present forms in the Tertiary Era or even before that. Although it is hard to conceive of surfaces remaining exposed to the atmosphere for long periods of geologic time without any changes whatever, there is no doubt that vast tracts of land of low relief in the continental regions must be regarded as being far older than any New Zealand landscape.

Where strong uplift has occurred in late geologic times and where geologically young mountain ranges are still being attacked vigorously by erosion, the geologic age of the eroding surfaces and of the surfaces of debris accumulating on adjoining lowlands is very young, virtually nil, in fact.

The New Zealand landscape as a whole has evolved in the course of erosional modifications of the crustal blocks that were pushed up during the Kaikoura Orogeny. The framework was determined by the initial arrangement of relatively elevated and depressed elements, and the relief details were worked up by erosion of the highlands and accumulations of erosional by-products. Local and regional complications were introduced by Pleistocene climatic changes, by volcanic events and by the continuation of crustal movements in some areas into recent times. Very little of the present land area lies far from the zone strongly affected by vertical displacements in late Tertiary and Pleistocene times, and very little of the landscape can date from earlier than that. A glance at the geological map shows how much of it is in fact underlain by deposits of Pleistocene and younger age. In more stable areas of North Auckland and east Otago, present surfaces conceivably have been evolving from Pliocene times, with only minor modifications since early Pleistocene, but the rest of our scenery has been produced much more recently, even in places where the underlying rocks are of Mesozoic or Paleozoic age. The case of virtually 'zero' age is well exemplified by the actively eroding western face of the Southern Alps, where relief has been maintained by continual elevation of the alpine axis into recent times.

Chapter Twelve

STIRRINGS OF THE CRUST

Earthquakes and Crustal Warping

Little by Little . . . ?

The comfortable illusion that the earth's crust beneath our feet—and our houses—is something stable and immovable receives a jolt, literally, from earthquakes and the knowledge of the displacements of the ground which sometimes accompany them. If further confirmation were needed that the crust is in a chronic condition of unrest, there is compelling evidence that slow distortion of the surface goes on persistently at measurable rates in some regions, that man-made structures have been depressed beneath the sea and have risen again, and that geological features of recent date such as stream channels and terraces have been visibly dislocated, raised and tilted over the last few thousand years. We are of course speaking of displacements from deep-seated causes, not those due merely to gravitational slumping. Ground subsidence and slumping do often accompany strong earthquakes, but these are secondary effects not counted with the true tectonic displacements.

The next question to arise is whether these short-term and minor displacements have anything to do with the longer-term, large-scale crustal disturbances we have dealt with in earlier chapters. Or are these a different phenomenon altogether? The case for unity is now stronger than ever. A convincing case can be made in support of the view that displacements of a few metres during recent earthquakes may be the continuation of crustal displacements going on through the later geological periods to produce major physiographic features like the high, eastern face of the Sierra Nevada of California and the straight western wall of the Southern Alps of New Zealand. Although no one doubts it now, this relationship had to be established as one expression of James Hutton's 'Uniformitarian' doctrine. Large displacements of the crust necessary to account for mountain ranges happened *not* as sudden, cataclysmic upheavals but as the total result of uncountable small increments of uplift through geological periods of time. Hutton, the 'father of geological philosophy', taught along these lines.

There are opportunities for testing the principle in New Zealand, by first measuring how much stream-terrace edges were offset (in both vertical and horizontal senses) by fault movements during the late Pleistocene Period, and then comparing the results with independent

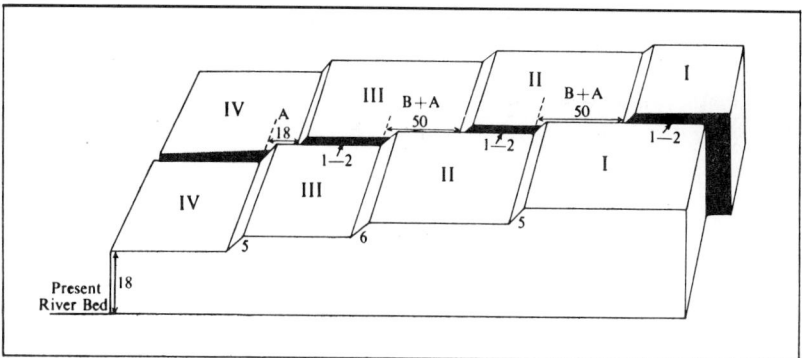

Fig. 12.1
A diagram (from Lensen, 1958) showing how intermittent strike-slip movement can be detected by the offsetting of a flight of river terraces. On the Wellington Fault at Emerald Hill, Hutt Valley, a total lateral displacement of 5.5 m (A = 18 ft) has occurred since the river began to cut below terrace IV, and an additional 9.7 m (32 ft; B + A = 50 ft) occurred after the downcutting which resulted in terrace III, but there is no evidence of lateral movement between the dates of formation of terraces III and II. Since downcutting by the Hutt River produced terrace IV, the block on the far (NW) side of the fault has risen about 0.5 m but there is no evidence of other vertical movements. (Mr Lensen informs me that more precise measurements have since been made, and will be published.)

geological evidence for greater amounts of displacement over longer geological periods. One of the first experiments of this kind was made by H. W. Wellman in 1952 in the Maruia valley close by the Lewis Pass Highway at a point where a distinct flight of river terraces is displaced systematically by varying amounts, vertically and horizontally, on the line of the Alpine Fault. Similar measurements were made also by G. L. Lensen and R. P. Suggate on fault-offset terraces in the Wairau and Awatere valleys of Marlborough and the Waiohine River of eastern Wellington. The geological ages of the various terraces could be estimated, and from these it is possible to calculate average rates of movement over specified time intervals, which worked out on the order of millimetres per year through the late Pleistocene.

Making use of radiocarbon ages of peat deposits on raised coastal platforms in Buller and in Westland, Suggate in 1965 computed average speeds of uplift in late Pleistocene times, which came out at rates of up to more than 1 m per century. Comparing these with the total amount of uplift as indicated from other lines of evidence for the whole of the upheaved mass from which the Southern Alps have been modelled, Suggate concluded that the rates of recent movement at different places are on the same order of magnitude as those which applied through the Kaikoura Orogeny and sustained the present mountain relief in the face of intense erosion.

It cannot always be determined whether the rates of deformation have been uniform, or intermittent, or faster and slower at different times, or even whether the directions of movement have been constant. At Pozzuoli in the Bay of Naples it is known that the ground supporting an early Roman temple was first depressed more than 6 m and later raised by about half that amount. In New Zealand most of the evidence so far points to consistent directions of movement, since the Miocene Period at any rate. It was once thought that small fault-scarps running along the trace of the Clarence Fault in Marlborough showed a reversal from the sense of movement responsible for raising the Kaikoura Range. Alternative explanations for this and other examples of apparent reversal include gravitational settlement of the lower mountain flanks with rotation and slight steepening toward the adjoining valley during earthquakes accompanying movements on other faults of the same system.

The case for 'little by little' progress towards great geological changes

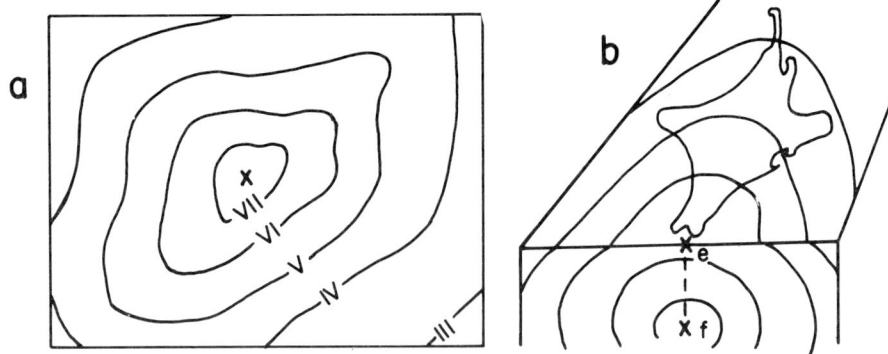

Fig. 12.2
a. *Isoseismal lines marking the increased severity of earthquake effects towards an epicentre.* b. *The epicentre (e) lies on the ground surface directly above the focus (f) of an earthquake shock. Distance e − f is the focal depth.* c. *To illustrate 'elastic rebound' (p. 347). Lateral distortion of the ground puts a double bend in the reference line (− − − −). A new reference line (−·−·−·) laid out prior to dislocation and earthquake is offset and doubly bent by the 'rebound', but the original line, now offset, is again straight.*

has been well established. Vast cataclysms, appealing though they have always been to the imagination, long ago ceased to be either necessary or credible explanations of the larger physiographic features.

Language of Seismology

There can be no need to define or describe an earthquake for the benefit of New Zealand readers. Few adults born in this country will not have experienced at least a few minor tremors, and many carry unforgettable memories of the great shocks of this century in Hawkes Bay, Wairarapa and Buller districts. News reports of major earthquakes sometimes use terms taken from the technical terminology of seismology, sometimes wrongly, and with the tacit assumption that everyone is familiar with the meanings of 'magnitude', 'epicentre', and so on, which is often not the case. As we accept earthquakes as a part of the New Zealand way of life (or worse), it will be as well to explain some of the more common terms and to outline some views as to the nature and causes of earthquake shocks.

Several different kinds of scales have been devised to classify earthquakes according to their severity. The ones now in widest use are of two distinct types, and their divisions have very different meanings. First, there are the scales of 'Felt Intensity', of which the one now mainly used is called the 'Modified Mercalli Scale'. The basic Mercalli Scale has twelve divisions, numbered by roman numerals I to XII, based upon compilations of eye-witness reports and the distribution of damage, if any. It is 'modified' to suit the circumstances of different regions of the world. Class I shocks are recorded by instruments only; Class II are felt only by people in favourable situations, as in high buildings or at rest in quiet places; Class XII signifies total havoc and destruction, loose objects being thrown into the air. After a strong earthquake seismologists gather all the information they can from press reports, direct inquiries and circulated questionnaires (not always welcome in devastated areas!) from which are compiled 'isoseismal' maps showing the distribution of areas of similar felt intensity, according to the classes of the Mercalli scale. Belts of similar felt intensity are roughly concentric about the place of origin of the shock. If the materials below ground were perfectly uniform in their properties in all directions the isoseismal zones would be arranged in

c

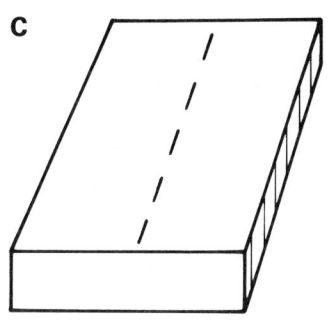 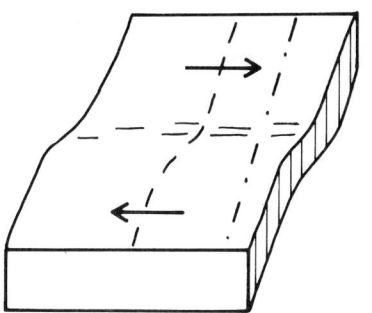 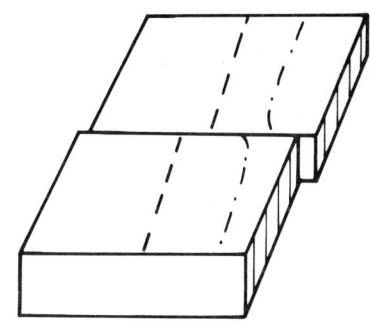

concentric circles about a point on the ground above the source of the shock. In fact, they are irregular and commonly elliptical in shape, elongated parallel with the structure of the region or with any prominent set of faults.

The great majority of instrumentally recorded earthquakes are not felt by people and only a minute percentage over a long period cause damage and casualties. The majority of seismographs are designed to record precisely the times of arrival at the station and the general direction of travel of vibrations originating elsewhere—they are usually put out of gear by strong, local shocks—and a rough measure of the strength of the motion of the ground at the station. Some types can determine the energy of vibrations, and from all these instrumental measurements as determined at a number of different stations, each shock is classified according to a scale of 'Magnitude' ('M-Scale'). Isoseismal values vary with the distance from the origin, but magnitude applies to the energy released and can have only one value for each shock. The maximum possible magnitude (limited by the amount of energy it is possible for rock to transmit without disintegration) is about 9. Shocks of magnitude 2.5 or less will probably not have been felt by people anywhere; those of magnitude 5 or more are likely to have caused some damage, and those above 7 will have had disastrous effects near the place of origin.

The place within the earth where a shock originates is the 'focus' (plural: 'foci'), the point on the ground surface directly above it is the 'epicentre' and the distance between these points is the 'focal depth'. All these values are calculated from the times of travel of three different types of earthquake vibration to seismological stations around the world. Generally the epicentre lies near the centre of the highest isoseismal zone, but this is not always the case with earthquakes of deep origin.

According to their depths of origin, earthquakes are classified as 'deep', 'shallow' or 'intermediate' shocks, but the definition of the boundaries between them varies somewhat according to the purposes of study, and to how the shocks tend to be distributed in depth in different regions. Conventionally, the zone of 'shallow' earthquakes extends down to 60 km, but the majority of 'shallow' shocks originate at depths less than 33 km, the average thickness of the crust; 'deep' shocks originate below 150 km. About two-thirds of all recorded earthquakes are 'shallow'. New Zealand and regions to the north are seismologically notable for the number of very deep shocks originating at depths of from 300 km to 600 km. The felt intensity of shallow shocks is likely to be relatively high over a restricted area whereas deep earthquakes are felt more moderately but over wider areas.

With regard again to felt intensity, the severity is much affected by the situation of the observer, and by the type of terrain. Thus, earthquakes tend to have more effect in areas of alluvium, soft rock and artificially built-up ground than in firmer situations. Damage is often more conspicuous on steeper country than on flatter ground because of landslides and other gravitational side-effects. Variations result also from the amount of water in the ground and from the direction in which a slope is facing relative to the travel paths of the shock vibrations. This was very clear after the 1942 Wairarapa earthquake. Damage to chimneys, walls, etc. varied considerably over the Wellington city area, and not only in terms of the age and condition of structures.

It is a common mistake to think that the shaking of the ground one feels during an earthquake and the visible ground-waves sometimes reported near the epicentre of a strong shake are the actual earthquake vibrations. In fact, the vibrations are much too rapid and their amplitudes too small to be visible directly. The 'felt' vibrations and visible waves are indirect effects set up in surface materials during the passage of the true earthquake vibrations through the underlying rock, travelling at between 3 and 9 kilometres per second and having amplitudes of vibration of at most a few millimetres.

The Causes of Earthquakes

Judging by the amount of energy that has to be released in geophysical exploration to produce an artifical 'earthquake' which can be picked up by sensitive detectors a few hundred metres from the source, the amount required to generate a strong earthquake can be compared only with that released in a major nuclear explosion or at the impact of a large meteorite. The question then arises as to where and how such enormous stores of energy can accumulate, awaiting sudden release to the accompaniment of shock waves that travel around the world.

From ancient times until the dawn of systematic scientific enquiry in the seventeenth century many fanciful theories were invoked to explain geological catastrophes such as volcanic eruptions and earthquakes— theories involving demons, the displeasure of deities and the escape of gases supposedly pent up in vast subterranean caverns. The terrible Lisbon earthquake of 1755 attracted the attention of enquiring minds throughout the world, but it was another century before the term 'seismology' had been coined, and experiments with artificial earthquakes begun. The notion of subterranean gases was evidently known to Shakespeare, and it would be a pity not to repeat here an apt quotation from *Henry IV, Part 1* which the New Zealand seismologist G. A. Eiby used to preface his useful little book *Earthquakes* published in 1957:

GLENDOWER: I say the earth did shake when I was born . . . .
          The heavens were all on fire, the earth did tremble . . . .
HOTSPUR: Diseased nature oftentimes breaks forth
          In strange eruptions; oft the teeming earth
          Is with a kind of colic pinched and vexed
          By the imprisoning of unruly wind
          Within her womb; which for enlargement striving,
          Shakes the old beldam earth, and topples down
          Steeples, and moss-grown towers. At your birth
          Our grandam earth, having this distemperature
          In passion shook.

Eiby notes that the idea of subterranean gases lingered on until early
colonial days in New Zealand, when Wellington settlers pondered over
the causes of the 1848 earthquake. Discharges of volcanic gases and
surgings of subterranean molten rock are not unreasonable
speculations in volcanic regions but not very helpful in accounting for
earthquakes originating thousands of kilometres distant from any
recent volcanic activity.

The only satisfactory general explanation involves the momentum of
large volumes of crustal material in sudden, temporary motion. Merely
to set in motion many cubic miles of rock from a state of rest requires
an enormous store of energy to overcome shearing strength, friction
and inertia. To stop such a mass again suddenly, even after a mere few
millimetres of motion inevitably means converting the momentum into
shock vibrations analogous with the hammering in a water pipe when
the flow is suddenly checked, or with the thump of a sledgehammer
striking the ground. The shock arises both from the reaction or 'recoil'
as movement begins and from the transfer of momentum when it
ceases.

Since early in this century seismological theory has favoured the idea
that stress is stored up by elastic distortion of the crust, and the energy
suddenly released in the rebound which follows the instant of failure. A
reasonable analogy is provided by the effect of bending a piece of
clock-spring until it snaps—and the fingers feel the shock. The 'elastic
rebound' theory was introduced by H. F. Reid in California when after
the 1906 San Francisco earthquake it had been shown, by
re-triangulation of old survey points, that the ground on either side of
the San Andreas Fault must have been distorted before the fault
suddenly moved as the earthquake occurred. From this arose a widely
accepted association of earthquakes with movements on faults. Doubts
have been expressed as to whether, at the depths where the deepest
earthquakes originate, the rock would have sufficient strength to allow
a great build-up of elastic distortion. Alternatively it was suggested that
very deep earthquakes might have some different cause, and that they
might trigger off movements on faults at higher crustal levels. The
doubts seem to have faded, and the elastic rebound theory remains
respectable. The seismic focus is now pictured as either a point but
more likely an area in a fault surface at depth where friction is
overcome and movement begins, or as a limited volume of rock within

the crust wherein the shear-strength is overcome and movement begins.

Modern diastrophic theories suggest various ways in which the stresses build up in the first place. Basically, they are supposed to derive from slow movements of thermal convection within the sub-crustal materials, transmitted to the upper mantle and the base of the crust by friction, and through the crust by compression. Up to a point, analysis of the stress-field within the crust is possible from the seismological records of earthquake motions and from earth-creep data, but only in a generalised way.

Precise surveys are repeated periodically across lines of active faults in California, Japan, New Zealand and Soviet countries. Although the build-up of distortional strain has been demonstrated in many cases, this has not yet led convincingly to the promise of being able to predict whether, where and when a sudden release (i.e. an earthquake) is imminent. Opinions indeed differ as to whether long continued creep of belts on either side of a known or suspected active fault trace is a 'good thing', meaning that stress is *not* being built up across it but being continually relieved, or whether a 'bad thing' in that *when* failure finally occurs there will be a strong earthquake, in proportion with the accumulated strain across the entire belt. The former seems more reasonable.

Earthquakes and Volcanoes—the True Relationship

It is a popular error to suppose that all earthquakes somehow are connected with volcanic action. Earthquakes certainly do originate in many places far from where volcanoes have erupted except in the distant geological past, but there is a distinct class of shallow, local shocks which precede and accompany eruptions. Sudden expulsion of gases from a crystallising magma, or sudden vapourising and condensing of water vapour might generate minor shocks, but the most likely general cause of volcanic earthquakes is again one of momentum—in this case connected with the sudden mobilising of the lava column beneath a volcano, or the sudden checking or diversion of flow of thousands of cubic metres of dense, liquefied rock by the blockage of a channel.

The more general relationship is one of association between zones of frequent earthquakes and chains of active volcanoes, both of which follow the earth's main belts of crustal mobility.

Earthquakes and Plate Tectonics

Earthquake distribution around the world is embodied in the newer diastrophic theories of sea-floor spreading and plate tectonics. The circum-Pacific and Mediterranean belts of high seismicity are associated with belts where the outer edges of expanding ocean floor segments are underriding slabs of continental crust, and where opposing convectional currents of sub-crustal material meet beneath continents in collision and descend into the mantle. Well-defined mid-oceanic lines of less intense and shallower shocks are related to the zones of magmatic upwelling where freshly solidified strips of new submarine lava are continually thrusting apart the ocean floors on either side.

The seismicity of the continental slabs is generally of a low order, and the Antarctic continent is reported to be seismically 'dead', but few regions of the world can safely be assumed to be entirely free from risk of earthquake damage, as the people of Australia learned from events at Meckering, West Australia, in 1968.

Earthquake Hazard in New Zealand

If a damaging earthquake can happen on the relatively stable platform of Australia, what are the chances that any part of New Zealand may be immune? The answer is: virtually none. Our tectonic situation encourages little optimism on this point because although it is true that the largely submerged crustal segment of which New Zealand is the largest emergent part has decidedly continental characteristics, the emergent part lies along a profoundly active crustal rift along which two of the major crustal plates are drifting in opposite directions.

The experience of one-and-a-half centuries of European colonisation shows that earthquake prevalence varies from area to area. Residents of the North Island east coast, Wellington and Marlborough are accustomed to frequent tremors, and are aware that disastrous earthquakes have struck those regions several times since settlement began and are sure to strike again. The people of the far north, where it is a rare event to feel an earthquake, were understandably complacent until 1963, when a series of sharp shocks in the Kaitaia area reminded us all that in no part of the country can we afford to relax the codes of building practice aimed at minimising the risk of injury or death, or to neglect earthquake disaster relief in our civil defence planning.

Looking now at the relative degree of risk in different areas, obviously it is not a matter simply of earthquake frequency, but rather of frequency of shocks above a given intensity. But there is no regularity in the spacing of shocks, even in areas of relatively high frequency, so the question becomes a purely statistical one—in any given seismic zone, what is the probability (chances in a hundred) that a shock of specified strength will occur within a decade, or a century? It could happen tomorrow, of course.

G. A. Eiby in *Earthquakes* goes to some length in disposing of the myth of an 'earthquake fault line'. Faults are present in large numbers in rocks of all ages throughout the country but only a very small proportion of them show signs of having moved in recent times, and very few of these can be connected in any direct way with historic earthquakes. Nor can many earthquakes be identified with movement on any particular fault exposed at the surface. The 1888 North Canterbury earthquake gained notice abroad because at that time it was one of the few well-authenticated cases of renewed displacement on a known fault coinciding with a serious earthquake. The Alpine Fault naturally has come in for consideration as the possible source of some South Island shocks, but after a thorough assessment, geophysicist F. F. Evison has discounted the possibility that activity on this fault is connected with any historic earthquake.

It is far from easy to find a satisfactory way of subdividing New Zealand into zones of varying degrees of earthquake hazard. It may seem simple to draw maps showing the epicentres of historic shocks, and reasonable to infer that areas including many epicentres will generate more shocks in the future than other areas, but this procedure is of no more than general help in predicting which locality will receive the next strong shock, when, and how severely. The situation is complicated by the fact that the areas containing the greatest numbers of epicentres are not the same for shocks originating at different depths. Generally, shallow shocks have originated over wide areas in the south-eastern half of the North Island and the northern quarter of the South Island and in Fiordland; shocks from deeper than 100-km are mainly within a well-defined belt from Nelson north-eastwards through the Rotorua-Taupo zone and out into Bay of Plenty; successively deeper classes of earthquakes originate mainly within similarly northeast-trending belts, lying progressively farther to the north-west with increasing depth. Since a significant number of shocks from each depth class are of magnitude 6 or greater, and therefore potentially damaging over their 'felt' areas, it is far from simple to designate, at the surface, zones of relative seismic risk taking into account shocks arising from all likely depths.

Of several schemes proposed over the past half century, the most credible so far is still based on analysis of seismic data rather than on geology. Accredited to Eiby, it recognises a 'Main Seismic Region' taking in the south-eastern half of the North Island and the adjoining ocean floor out as far as the Hikurangi Trench (southern branch of the Kermadec Trench), and the South Island north of a seismologically well-defined boundary line stretching from Westport to Waipara. The next most important is the 'Fiordland Seismic Region' and the offshore area to the south-west; then the intervening 'Central Seismic Region' taking in most of Canterbury, Westland, Central and eastern Otago. A narrow zone marginal to the Main Seismic Region on its north-west side includes Taranaki and South Auckland where there have been isolated occurrences of strong tremors. South-eastern Otago and North Auckland almost deserve the adjective 'aseismic', but the Kaitaia earthquake is a reminder that *absolute* aseismicity exists nowhere in New Zealand. The strongly seismic area of the Pacific Ocean north-eastwards from the Bay of Plenty to the Kermadec Islands is regarded as a separate seismic region.

The progressive north-westward shift of the main band of earthquake foci with increasing depth is quite consistent with a general pattern of depth-distribution shown by the numerous earthquakes that occur along certain boundaries between ocean and continent. It was noticed by seismologists in California that the deep earthquake foci around the Pacific and some other ocean margins lie within a thin zone, almost a plane surface, dipping at angles of 45° to 50° inwards under continental margins. This arrangement has been named after one of the discoverers as the 'Benioff Zone', and it is believed to mark a real surface of discontinuity which extends down into the mantle where spreading ocean floors are underriding continental margins. The

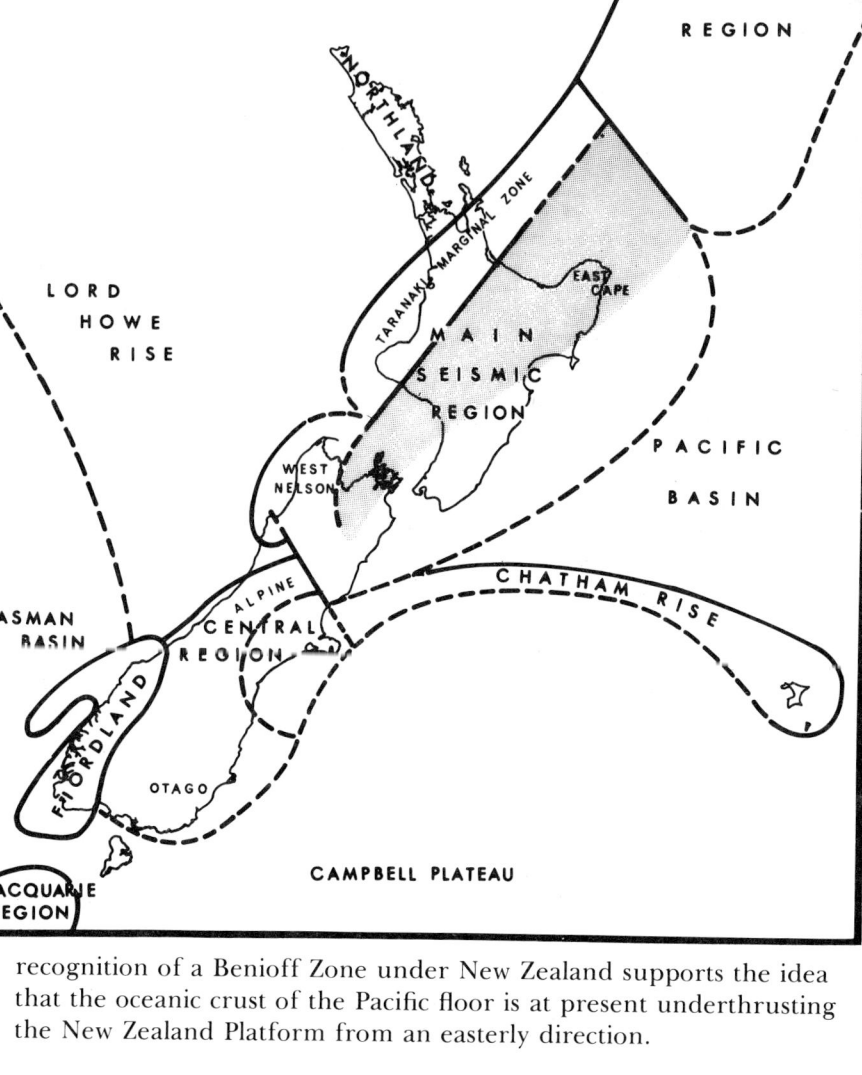

Fig. 12.3
The main seismic regions of the New Zealand area, according to G. A. Eiby. The shaded portion of the Main Seismic Region shows the extent of what Eiby calls a 'sub-crustal rift', a wedge-shaped structure which dips below the central North Island to an estimated depth of about 370 m beneath the Bay of Plenty, and within which almost all of our deep-focus shocks are believed to originate. (G. A. Eiby, 1971)

recognition of a Benioff Zone under New Zealand supports the idea that the oceanic crust of the Pacific floor is at present underthrusting the New Zealand Platform from an easterly direction.

Historic Earthquakes in New Zealand

We experience on the average about 120 earthquake shocks of magnitude greater than 4 in any year, while many more weaker ones, recorded instrumentally, originate in or near the New Zealand land area. Shocks severe enough to cause notable damage or casualties have occurred at an average frequency of about one in ten years, but they have been spaced out irregularly in time. For instance, there were five major earthquakes, each followed by long periods of after-shocks, between 1929 and 1934, but in the interval between the June 1942 Wairarapa and May 1968 Inangahua quakes, no shock attained magnitude 7 except one centred offshore from Milford Sound which caused only minor damage in Otago.

Fig. 12.4
The positions of the larger shallow earthquakes of the historic period in New Zealand. Major shocks of magnitude about 7 or more are indicated by large circles. (Numbers refer to date-key, top left.) Small circles show the positions of shocks between magnitude 6 and 7 that occurred between 1940 and 1968. The map shows also the major active faults and the broad belt of relatively high crustal mobility (shaded). (G. W. Grindley, 1974, after a map by the Seismology Branch, Geophysics Division, D.S.I.R., Wellington)

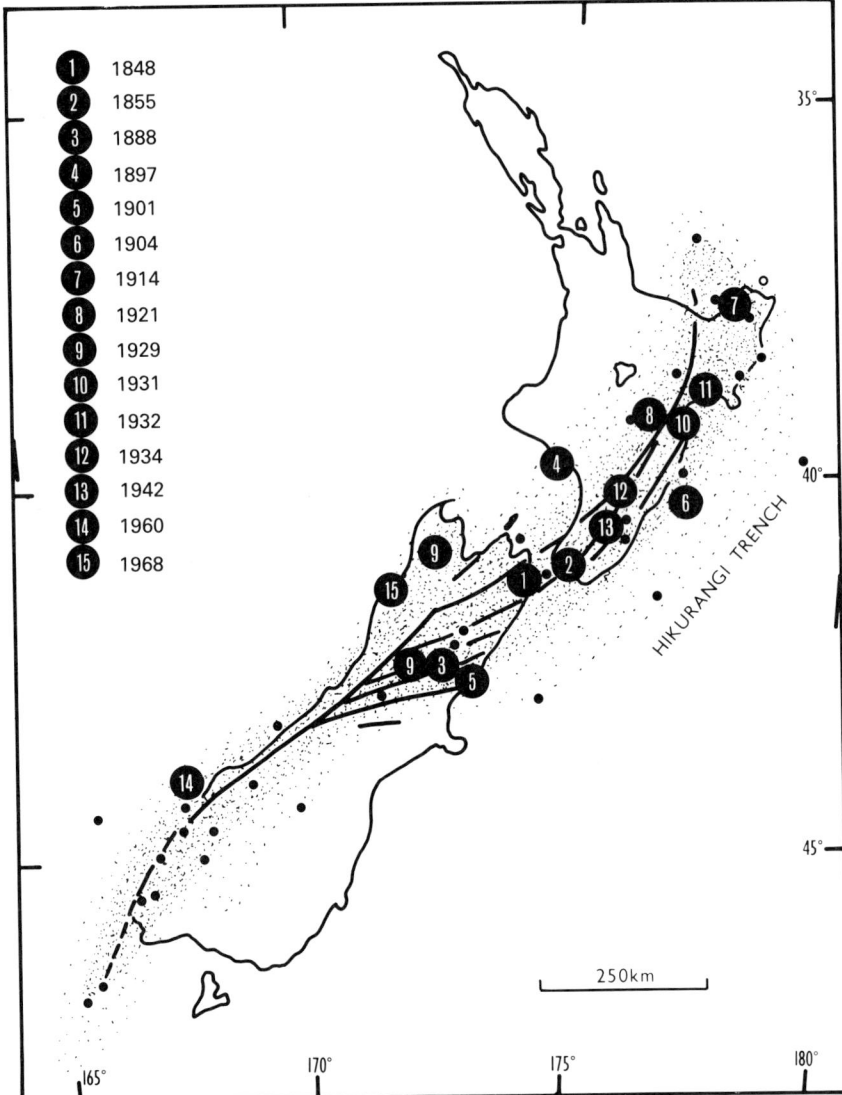

1	1848
2	1855
3	1888
4	1897
5	1901
6	1904
7	1914
8	1921
9	1929
10	1931
11	1932
12	1934
13	1942
14	1960
15	1968

Leaving aside for the moment Maori traditions and indirect indications of pre-pakeha earthquakes, there were several reports by sealing and whaling crews, notably from Fiordland. The earliest fully reliable description of a major earthquake or series of earthquakes with landslides, subsidences and local uplift of the land relates to the area between Dusky Sound and Milford Sound in 1826 and 1827.*

The first major earthquake, after systematic settlement by Europeans had begun, struck the Wanganui district in July 1843 and was followed by many aftershocks, some of them strong, during the ensuing year. Then in October 1848 Wellington was seriously affected by a shock originating in Marlborough, during which movement evidently

---

* A descriptive catalogue of New Zealand earthquakes prior to 1846 and another briefly annotated list of all shocks of magnitude 5 or more up to 1965, compiled by G. A. Eiby, are to be found in *N. Z. Journal of Geology and Geophysics* vol. 11, pp. 16-40 and 630-647, 1968. I have drawn from these and an earlier list published by R. C. Hayes.

occurred on the Awatere Fault. Three deaths were reported, being the
first recorded casualties from this cause in New Zealand, and brick
buildings were badly damaged. Less than seven years later Wellington
was stricken again. During what may be the strongest New Zealand
earthquake of the historic period, with an estimated magnitude of 8,
the entire Wellington Peninsula was raised by as much as 3 m on
January 23, 1855. The Te Aro Flats were drained and the rocky tidal
platform became exposed along considerable stretches of the south
shore of Cook Strait. The eastern limit of the uplifted block was
determined by the West Wairarapa Fault along the eastern base of the
Rimutaka Range, movement on which at that time has since been
traced from Palliser Bay most of the way north-eastwards for 140 km
into southern Hawkes Bay.

Since 1855 there have been 13 major earthquakes of magnitude 7 or
more in New Zealand, as listed below:

Sept. 1888 N. Canterbury	(fence displaced 3 m. in Hope Valley)
Nov. 1901 Cheviot	(one death; damage in Christchurch)
Aug. 1904 off Cape Turnagain	
Oct. 1914 Bay of Plenty	(one death; widely felt)
Jun. 1921 Inland Hawkes Bay	(widely felt)
Mar. 1929 Arthurs Pass	(widely felt)
June 1929 Buller-West Nelson	(17 deaths; displacements up to 4 m)
Feb. 1931 Hawkes Bay	(256 deaths; changes of level and horizontal movements up to 3 m)
Sep. 1932 Wairoa, H. B.	(severe local damage)
May 1934 Pahiatua	
June 1942 Wairarapa-Wellington	(6 deaths; long period of aftershocks)
May 1960 off Milford Sound	
May 1968 Inangahua	(3 fatalities; damage as far south as Greymouth)

Nearly all of these were followed by periods of aftershocks, some
approaching the strength of the initial shocks, completing the
demolition of damaged structures. After the Inangahua earthquake
there were more than 800 aftershocks, one nearly as strong as the main
shock and twelve of magnitude greater than 5.

Some seismic events are notable not so much for their severity but
rather because moderate shocks continue at intervals over periods of
many months. One such 'earthquake swarm' affected the Taupo district
from May until December 1922 with surface faulting and subsidence
along part of the lake shore. Another swarm affected Great Barrier
Island for two months in 1953.

Some change of level of the ground has been reported after most
earthquakes of magnitude 7 or more. Besides the uplift of Wellington
in 1855, another with far-reaching effects accompanied the 1931
Hawkes Bay earthquake, draining the Napier inner harbour area,
Bayview and Marewa lagoons, providing much of the land over which
Napier city has since spread. During the 1929 Buller earthquake
vertical displacement of over 4 m was measured on the White Creek
Fault where the Buller Gorge Road crosses it. Other vertical
movements are of the nature of gravitational slumps and subsidences,
secondary effects of the earthquake vibrations.

Fig. 12.5
A familiar spectacle in various parts of New Zealand where severe earthquakes have occurred in historic times. This large rockfall in the upper Poulter valley, Canterbury, is one of many in and near Arthurs Pass National Park produced by the earthquake of April 1929. (Photo: R. Speight)

Fig. 12.6
At the time of the June 1929 west Nelson earthquake the Buller Gorge road was dislocated vertically by about 5 m near Whites Creek. The displacement can still be seen on the terrace across the river, despite a half-century of vegetation growth. (Photographed in 1935)

Prehistoric Earthquakes

Maori tradition apparently has little to say that is recognisably about earthquakes. Eiby's catalogues include only two events inferred from that source of information. In the middle of the fifteenth century, according to the approximate, genealogical chronology, Miramar Peninsula was joined on to the mainland by an uplift of about 1.2 m which caused a channel on the site of the present Wellington Airport to silt up. The other event concerns subsidence at Lake Omapere, North Auckland, about the year 1600, but this may have been from volcanic rather than seismic causes.

Powerful indications of at least one pre-pakeha earthquake with strong effects in the Grey Valley, the Paparoa Range and an area to the

Fig. 12.7
The Porters Pass Fault may have had a long history, its belt of intensely sheared and crushed Torlesse rocks having originated as far back as the Rangitata Orogeny. It has moved laterally (i.e. by strike-slip) in geologically recent but prehistoric times, offsetting physiographic features so that ridges are now set opposite the axes of gullies. (Photo: R. Speight)

Fig. 12.8
'The Carriage Drive' was the old name for this surface trace of a geologically recent fault displacement near Lake Coleridge, Canterbury. Many similar fault traces are found in the inland parts of Canterbury and in other seismic regions. This one records an earthquake happening before European settlement but since the disappearance of glacier ice from these valleys, 13,000 years ago or earlier.

north of Reefton are provided by ancient landslides covered in a mature forest that appears to be uniform in age, and also by slumping of large tracts of mountain crest in places where the underlying rocks and structures are not particularly conducive to deep-seated gravitational slumping and subsidence. A. C. Beck drew attention to irregular trenches and low scarps which are prevalent along ridge crests in the Southern Alps. I have also seen them in the Victoria Range east of the Inangahua Valley. Though directly due to gravitational collapse of mountain flanks made very steep by glacier erosion, the movements are likely to have been triggered off by earthquakes. It is a reasonable speculation that these trench features, together with the large slumps and landslides, are the record of a strong earthquake a few hundred years ago.

Recent Crustal Movements

The comfortable illusion of terrestrial stability received another blow in this country in 1957 when M. T. Te Punga reported having established that the ground in parts of western Wellington was being heaved up into broad folds at so fast a rate that it should be measurable over periods of a few decades or even less. This was not the first time that slow, progressive distortion of the crust had been demonstrated in New Zealand, but the first time away from active faults and from the immediate scene of strong earthquakes. More about Te Punga's findings later.

Slow crustal deformation, not necessarily connected with visible surface faulting, has received a good deal of attention in recent years, partly because of the possibility that such studies might hold the key to the prediction of earthquakes, but also because the movements sometimes provide ways of testing theories about deep-seated movements as well. Another reason for the interest no doubt is the availability of new, very precise surveying methods which make it possible to detect very small continuing displacements over much smaller intervals of time than before. Studies of ground movements began in Japan after the 1923 Tokyo earthquake, and go on now in all the more tectonically active parts of the world, including New Zealand. In 1971 an international gathering of geologists, surveyors, seismologists and geophysicists representing twenty countries was organised in Wellington by the Royal Society of New Zealand to ensure the future co-ordination of work and interchange of information.

Studies of ground-surface deformation in New Zealand have been of three main kinds. Efforts to detect displacements not only across active faults but across regions subject to strong earthquakes began after the 1929 Buller earthquake and continued after the 1931 Hawkes Bay earthquake. The possibilities were limited in those times because of the low degree of accuracy of early trigonometrical surveys and the lack, until 1945, of a complete New Zealand-wide trigonometrical network. Then there have been repeated re-surveys of lines across active strike-slip faults, aimed at detecting whether slow, continual creep is going on along the faults and whether the country for some kilometres on either side is being distorted as well. Finally, active crustal deformation during recent geological time is being detected by its effect upon critical geomorphic reference lines, such as the profiles of stream beds and terraces of known geological age or dated by radiometric evidence.

Prompted initially perhaps by the success of Californian investigators in proving 'drift' along the San Andreas and other active faults, similar methods were applied on the Alpine Fault, the Awatere Fault and the Wellington Fault. No distortion along the Alpine Fault itself has been proved despite evidence that it must have moved within the last few thousand years, but measurements across the other ones have established drift at rates of as much as 2.6 cm per year on stations spread over belts up to 10 km in width. Changes of level have also been detected, differences of even millimetres being within the accuracy of modern instruments. It is indeed now so much quicker and cheaper to

obtain and repeat precise survey measurements that we can expect
interesting new information to appear about up-and-down and
horizontal relative movement in various parts of the country in future
years

Te Punga's discovery, referred to above, concerned what he called
'growing anticlines' in western Wellington from Levin north to
Pohangina and Marton. Broad warpings revealed by their distortion of
stream and terrace profiles transverse to the axes of warping are
growing, it appears, at rates of up to about 1 m per century at the fold
crest. This could be enough eventually to interfere with drainage in
certain directions. The folds are aligned approximately in a
north-south direction. Recently, K. B. Lewis published evidence that
the coastal region of Hawkes Bay is being distorted into a similar series
of folds with north-easterly trends and about 10 km from crest to crest.
These appear to have been growing at average rates of around 1.5 m
per thousand years during the last 120,000 years. The eastern coast of
Wellington has been under study by H. W. Wellman and others
working with him over the past ten years, showing foldings and
warpings that seem to have been consistent in direction of movement
and to have reached maximum rates of from 2.5 to 4 metres per
thousand years.

In an entirely different context, the floor of the main crater of White
Island, Bay of Plenty, has been shown by R. H. Clark to bulge upwards
at times apparently connected with movements of magma beneath. The
studies in fact were aimed largely at discovering whether continual
monitoring of crater floor levels would show it to be closely enough
linked with temperatures and the condition of the magma to enable
reliable predictions to be made of imminent eruptions.

Fig. 12.9
*Raised former beach-
gravel ridges behind the
present shoreline at Cape
Turakirae, southern
Wairarapa, have given
evidence as to the speed
and style of recent crustal
movements in the area. To
the left, beyond the mouth
of Orongorongo River, the
remains of formerly more
extensive coastal terraces
carved at higher levels can
be made out. Now much
reduced by later stream
and marine erosion, they
are also masked by debris
shed from higher ground.
(Photo: N.Z. Geological
Survey)*

Chapter Thirteen

GEOLOGY SERVES NEW ZEALAND

The Economic Incentive

The Purpose of Geology

It is hard to say just when 'economic geology' began, but certainly geological knowledge was being put to practical uses for a very long time before the phrase was coined. The word 'geology' is only about 200 years old. The science arose as a branch of learning out of a mixture of innate human curiosity about natural things together with a growing need to obtain useful substances from the earth. To begin with, the most sought-after materials were rock for building, clay for pottery, gemstones and the 'noble' metals gold and silver for ornamentation, and in due course man found out how to use the 'base' metals copper, lead, zinc and iron which occur naturally in the pure or native state or can be extracted fairly simply from certain 'stones', i.e. ores.

Metals were being mined systematically, one might say professionally, in Europe and Asia and perhaps in Africa too, long before the Christian era. In Europe in mediaeval times the mining and extraction of metallic ores developed into a skilled craft with its own lore concerning the nature and occurrence of mineral substances. Some sound ideas about the origin of mineral veins were expressed, and some fanciful ones as well. Attention was concentrated at first on the veins themselves, then on the nature of the rock containing the veins, and eventually it was realised that the origin of the rock was no less important. By the middle of the eighteenth century the fund of mining experience and observation was being organised on scientific lines, especially in southern Germany. The great leap forward in mechanical technology which we call the Industrial Revolution created unprecedented demands for iron ores, coal and clay from the beginning of last century onwards, and resulted in earth materials being sought and exploited on an ever increasing scale. During the same period, advances in the engineering arts were beginning to call for better understanding of rock and soil foundations, and indeed it was a critical moment in the history of geology when, in 1815, an engineer called William Smith, then engaged upon canal surveys, prepared for his own use a map showing the succession of strata across England and Wales.

Thus began the systematic and deliberate application of geology in the service of man, and economic geology had come into being. Since then, the practical applications of the science have diversified continually behind man's growing need to exploit—*and now to conserve*—the earth's non-renewable resources.

Early Geological Explorations in New Zealand

Before Europeans came to this country the Maori people had been gathering rock materials where they found them, fashioning them into weapons, ornaments and striking and cutting tools, trading with them, and also collecting iron oxides and clays to use as pigments and ornamentation. It has been suggested that the early European settlers turned their attention to the possibility of finding valuable mineral deposits chiefly because it was then either impracticable or uneconomic to export much agricultural produce to distant markets, but I suspect that man's age-old fascination with minerals, especially gold, would in any event have directed the eyes of the pioneers towards the ground.

In a foreword to C. A. Fleming's translation of Hochstetter's *Geologie von Neu Seeland* (1864), R. W. Willett noted the early interest of the respective provincial governments of the day in engaging Hochstetter to examine coal deposits south of Auckland and gold discoveries near Nelson in 1859. These are the earliest instances of public authorities in New Zealand seeking geological advice as to the potential of the country for mineral development. Hochstetter reported on other minerals as well, including ironsands on the North Island west coast. He was then about to return to Germany, and the Nelson Government, impressed by the favourable implications of the initial report, employed Hochstetter's companion in the field, Julius Haast, to explore the coal and other mineral potentialities of what was then the southern part of the province. This was done between January and August, 1860, and the report presented in 1861.

Other provincial governments rapidly followed suit. The Otago Provincial Council, spurred by gold discovery in 1858, decided in 1862 to establish a Geological Survey and appointed James Hector, who was later to become Director of the Colonial Geological Survey. Haast, having just completed his Nelson explorations, was hired in 1861 by the Canterbury Provincial Superintendent to survey the geology along the proposed alignment of the Lyttelton-Heathcote railway tunnel as a prelude to his appointment as Provincial Geologist. He held that post until 1868, prior to becoming founder and first Curator of Canterbury Museum. The tunnel survey is the earliest example of an engineering geology investigation in New Zealand.

The Westland region was then part of Canterbury Province and the gold rushes were about to begin. The appointment of a geologist was timely, though not without opposition. Besides preparing a number of special reports on coal, gold and other deposits, Haast while he was Provincial Geologist published several folio-sized descriptions of the geology of the region, illustrated by lithographs from his sketches, and also of the glaciers, in which he was particularly interested. Although not the original discoverer of the alpine pass west of Lake Wanaka which bears his name, Haast proved its usefulness as a potential route to the far west of Otago. A lasting outcome of his career as a geological explorer was a book-length account of the regional topography, geology and minerals published in Christchurch in 1879 under the title *Geology of Canterbury and Westland* —it is still worth reading.

The Province of Auckland did not establish a permanent geological survey, but it appointed F. W. Hutton to examine coal deposits in South Auckland and gold and copper at Thames and Great Barrier Island. In 1861 Wellington Province asked J. C. Crawford, settler and keen amateur geologist, to report on gold at Cape Terawhiti, on the geology of the Wairarapa and on the implications of the Wairarapa earthquake of 1855.

So far there had been little co-ordination of geological explorations initiated by the provinces. Following a Central Government decision to hold the first New Zealand Industrial Exhibition in Dunedin in 1865, Hector was put in charge of the task of compiling a geological map of the whole country. He was then called upon to establish an indefinitely continuing Geological Survey of the colony and a Colonial Museum, with a small permanent staff, including W. F. Skey, chemist, who made innumerable mineral assays and rock analyses before his retirement in 1900. Within a year, a geological map in colour had been printed on a scale of 1 : 2,000,000, and 1866 also saw the first of a long series of *Reports of Geological Explorations* which continued to come from the Survey under Hector's direction for the next twenty-six years.

Older and Newer Geological Surveys—a Brief Historical Outline
Report No 1 of the Colonial Geological Survey contained a descriptive account of coal deposits with analyses and results of steam-raising tests. It sign-posted the direction the Survey would be forced to follow. Although a great deal of purely exploratory geological study was published subsequently, with detailed descriptions of sequences of strata, classifications, and maps of local areas, the emphasis throughout the years of reconnaissance by the Colonial Museum and Geological Survey of New Zealand was upon the search for mineral deposits and problems of mining enterprises. It was especially marked during the waves of economic depression which swept the country in the latter part of the century, at which times it was seen as most urgent to locate new areas in which to prospect for gold and other metals, for the boom days of easily found and easily worked deposits had come to an end.

These pressures prevented Hector from getting on with what he originally saw as the main task, which was to produce geological maps in colour covering the whole country at a scale of one mile to an inch. He was forced instead to begin applying the reconnaissance findings immediately, before a sound background of regional geological knowledge had been built up. However, as Peggy Burton observes in *The New Zealand Geological Survey 1865-1965*, 'One essential role of a geological survey that was fulfilled under Hector was counselling to prevent minor mineral discoveries from using more labour and money than was warranted.'

The end of the Colonial survey came in 1892, when R. J. Seddon's government cut off the funds to support it, after which Hector's sole remaining assistant, Alexander McKay, became Mining Geologist in the Mines Department while Hector continued as Director of the Colonial Museum until he retired in 1903. These changes left something of a vacuum in the sphere of geological exploration. McKay, although now a respected authority, was no longer young. For the Mines Department he continued to produce geological reports, including some of the best accounts ever written about the auriferous deposits, but his great days of geological mapping for Hector were over. Mining in any case was in a decline and, as we shall see, when a need did arise for geological service to the mining community it could now be obtained from other sources.

When Hector went to Wellington in 1865 to set up a Geological Survey his former post as Otago's Provincial Geologist remained unfilled, but eight years later F. W. Hutton who had been on Hector's staff since 1871 resigned to become Curator of Otago Museum and Professor of Natural Sciences at the University, from where he published many papers and popular articles on geology and minerals. Otago University also established a separate School of Mining and brought G. F. Ulrich over from New South Wales to be the first Professor. Hutton and Ulrich worked together, and in 1875 jointly published a book, printed in Dunedin, describing the geology and gold-mining of Otago. It was a notable achievement, and the first comprehensive account of regional geology to be produced in the colony. James Park, another of Hector's men, succeeded Ulrich as Professor of Mining in 1891, and besides building up the school to a high level of international repute, he did a great deal of what we now call consultant work in geology, mining technology and metallurgy in various parts of the country after the collapse of the Colonial Survey.

Greatly improved economic conditions and a marked return of mining prosperity in the first years of this century led to the Geological Survey being revived within the Mines Department in 1905 under the energetic direction of a young Canadian, James Mackintosh Bell. With an initial staff twice as large as Hector ever had, it embarked again upon a plan to map the whole country systematically on the mile-to-an-inch scale, with accompanying *Bulletins* describing the mapped areas geographically as well as geologically. Seventy-three years later that plan remains uncompleted, and a new scheme of mapping on a metric scale is about to commence, but the new Survey was vastly more successful than Hector's in that respect. From the very beginning, however, Bell's conviction that the mining industry must be served first was obvious. All of the regional geological *Bulletins* that came out before Bell resigned in 1911 described either active mining areas or areas thought to have mineral or petroleum potentialities. Most of his field staff had backgrounds or interests relevant to mining geology. Background research in other fields was neglected by the Survey in Bell's time, and pursued mainly within the university colleges until after the First World War. Under Bell's successor as Director, Percy Gates Morgan, paleontology was, however, greatly promoted by the appointment of J. Allan Thomson and continued under Thomson's inspiration even after he become Director of the Dominion Museum in

1914, until 1920, when the Survey acquired J. Marwick as its first permanent full-time paleontologist. Thomson established the important principle that stratigraphic work in New Zealand would have to be based upon our own fossil succession, and an independent geological time scale built up here before trying to link up with the Periods defined overseas.

In 1926 the Geological Survey was transferred from the Mines Department to the new Department of Scientific and Industrial Research, and thenceforth its basic task of regional mapping came more into focus. Under the vigorous policies of the departmental head, Ernest Marsden, there was expansion into other special fields including soil science, geophysical exploration, micropaleontology and petrology during the inter-war years, the first two of these fields separating eventually from the Geological Survey to become independent divisions of the D.S.I.R. The Survey now has a large staff and excellent facilities for its diversely specialised functions, and is able to provide most of the geological requirements of coal-mining, civil engineering, the ceramic industries, water supply, gas and oil exploration, etc., while still carrying on its background task to complete and up-date the geological mapping on suitable scales.

University geologists continue to do as much consultant work as their research interests and normal academic duties will allow, and a small number of self-employed geological consultants are now offering their services. Petroleum and mineral exploration companies have tended to employ their own geological teams, but the Geological Survey continues to be a source of geological advice at all levels, from the scale of major hydroelectricity dam-site investigations downwards, and on all manner of questions such as where a county engineer might turn for road-metal, where to sink a water well on a farm (without using a divining rod!), whether enough limestone is in sight to justify a proposed new limeworks, and possible geological reasons for bumps heard in the night.

In the following sketch of geological applications in New Zealand, I have drawn freely from G. J. Williams's *Economic Geology of New Zealand* (an extremely valuable compilation which earned Professor Williams the McKay Hammer Award of the Geological Society of New Zealand in 1976), the Survey *Bulletins*, and sundry other sources.

The Search for Gold

In terms of real and lasting worth to the community, it may be debated whether the discovery of gold, or that of coal in economic quantities was the more fortunate. Gold finds were an immediate source of capital and employment and an enticement for overseas investors, all badly needed at times when the country's pastoral and agricultural prosperity was flagging and before there were any manufacturing industries to speak of. Yet, at the price of gold in those early days, a modern cost-benefit analysis of New Zealand gold-mining operations as a whole, taking in public expenditure on roading and other services, may well have turned out unfavourably. Coal on the other hand, though increasingly a practical necessity, always lacked the glamour and get-rich-quick illusions attaching to gold. The presence of traces of gold certainly did not escape the notice of early explorers and settlers in Coromandel, north-west Nelson and Otago, though I suspect that some sightings were really of 'false gold', most likely flakes of mica with a golden glitter in the streams.

Earliest attempts to mine alluvial gold at Coromandel in the early 1850s were not encouraging, and successful mining began really with the discovery of rich quartz reefs in 1861. In part the delay was due to the fact that many of the prospectors hailed from Victoria and expected to find gold by the same signs and methods as they had been using in Australia. Alluvial mining in the Collingwood district of north-west Nelson was well under way by 1857 and continued after the early rushes, on a steady if unspectacular scale for many decades. In Otago, though reports of gold in the Upper Clutha region and elsewhere had reached Dunedin much earlier, Gabriel Reid's discovery near Lawrence in 1861 marks the real beginning of gold-mining in that province. Similarly, gold had been known to be present in the inaccessible and inhospitable West Coast region for several years prior to the decline of mining fortunes in Otago and Marlborough in 1864 which induced prospectors to rush to the western El Dorado in large numbers.

How Gold Occurs

Metallic gold is present in veins of quartz penetrating Haast Schist, the ancient sedimentary rocks of the Aorere, Greenland and Preservation groups, and to a limited extent in the Torlesse greywackes and slates, though rarely in economic amounts and still more rarely visible to the naked eye even after the quartz has been crushed and the gold concentrated by washing in a prospector's pan. In the Hauraki mining belt from near Matamata to Great Barrier Island it occurs alloyed with silver and in compounds of tellurium, along with copper, lead, zinc and other sulphides, in veins and more irregular masses of quartz set in hydrothermally-altered andesite lava rock of Miocene age. Debate about how the veins in the older rocks were formed has been outlined in Chapter Six. Those in the Hauraki andesite rocks are attributed to solutions emanating from magma intruded below, the same solutions as were responsible for metasomatic alteration of the 'country rock' in which the veins of ore minerals were deposited in fault-fissures and fractured zones. Some of the gold is contained in crystals of pyrite and other sulphide minerals, requiring special methods of extraction. The Hauraki gold deposits are unusual, arousing much interest in the heyday of the field and causing metallurgical problems which led to new extraction methods being first developed in New Zealand.

From its ultimate sources in veins or disseminated through the country rock, gold particles found their way into younger deposits of detritus from weathering and erosion. Being both of high density and very durable, they became concentrated by phase after phase of re-erosion and re-deposition to a degree of richness which caught the early prospectors' eyes in alluvial terrace gravel and beach sands. A. McKay was one of the first to point out that Pleistocene glacial deposits,

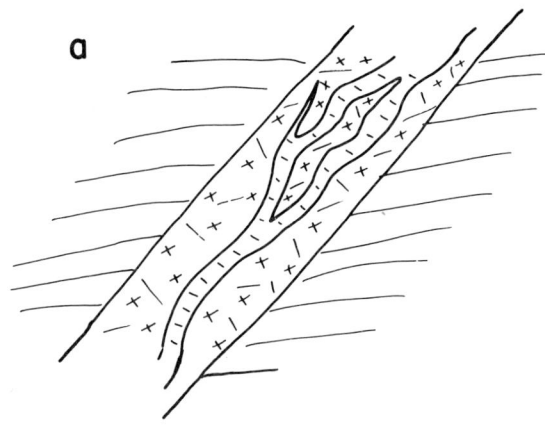

Fig. 13.1
An underground photograph of the auriferous quartz vein which was exploited for nearly half a century by the Blackwater Mine, at the southern end of the Reefton goldfield, until the mine closed in 1951. The vein here was from 30 to 40 centimetres wide, occupying a mineralised crush-zone in greywacke rocks of the Greenland Group. It was worked from the surface down to a depth of more than 750 m, maintaining an unusually even gold content throughout.

outwash gravel rather than till or moraine, were the immediate source of gold particles won from stream beds and banks—a theme reiterated with even stronger conviction in recent books about New Zealand goldmining by the late W. F. Heinz.

Essentially similar in origin, though older and deposited in a rather different setting, the so-called 'auriferous cements' which occur at many places at the base of the younger sedimentary succession, represent a concentration of the more durable products of weathering while older rocks containing quartz veins were being eroded at the time of formation of the Late Cretaceous Peneplain. These compacted gravels made up largely of quartz debris, our old friend the Quartzose Coal Measures in another role, were the immediate source of the gold mined at Gabriels Gully, and indirectly of the gold in some younger alluvial deposits in Otago and Southland. In the West Coast region they were mined at Lankeys Creek near Reefton, the 'cements' being crushed in the same kind of stamper equipment as is used to crush vein quartz. Experienced prospectors could usually tell whether the gold in the pan came from the 'cements' from younger gravel deposits or directly from vein-quartz outcrops in the catchment of the stream they were fossicking.

Geological Assistance in Gold-mining

Original gold discoveries were mostly made by explorers and prospectors, and the geologists cannot claim credit for any major new finds. Their assistance, however, was sought on many occasions by the operators of vein-quartz mines when deciding how best to prospect by drilling or other means in search of the continuation of a vein cut off by a fault, or of another vein to take the place of one nearly worked out. Gold-miners are notoriously independent, as well as optimistic, but there is no doubt that such advice has helped to prevent some waste of money and human energy upon hopeless enterprises. An understanding of how the glaciers of the Pleistocene Ice Age eroded and deposited has suggested where to look for deeper gold-bearing

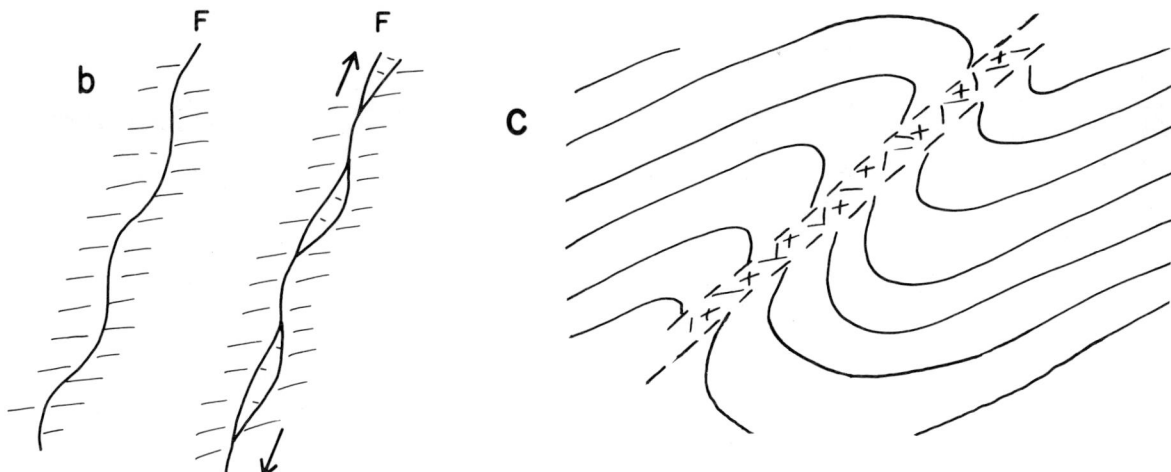

gravels in old, infilled gutters and channels, former meltwater sluiceways that were abandoned when the ice was retreating.

Suitable places for state-subsidised gold prospecting ventures in the mid-1930s were chosen with the help of special geological surveys and geophysical exploration of areas where the potentially ore-bearing rocks were hidden under gravel or other deep overburden (see later). The Reefton quartz-mining field, then in its final decline, was the scene of such investigations, aimed at discovering whether there were any special relationships between rock-folds in the Greenland Group and the location of rich lodes. They did in fact show that the ore bodies worked by the more successful mining ventures of the past and some of the few remaining ones had been aligned along the axial planes of folds which could be traced by surface mapping (and with geophysical help) over distances of several kilometres. The discovery would have been very useful had it been made twenty or thirty years earlier. As it was, the circumstances of the Second World War and a complete loss of momentum in the quartz-mining industry prevented the more promising indications from being followed up. Even with the present high price of gold, the cost of prospecting, which would now involve deep drilling, tunnelling and shaft-sinking to open new mines or re-open old ones, has come to be regarded as too high in view of the financial risk. In any case the twenty years that have elapsed since the last producing gold-quartz mine closed down are long enough for the essential skills of the hard-rock mining craft to have vanished from the region.

The Search for Coal

Unless we exclude shallow pits and short tunnels driven into outcropping coal beds by settlers to obtain coal for their own use it would be hard to say exactly when coal-mining began in New Zealand. The presence of coal was certainly known about long before any was regularly mined. In Otago, for example, whalers are believed to have broken coal from coastal outcrops at Shag Point in 1830 or thereabouts

Fig. 13.2
Types of fissure vein.
a. The branching body of ore has replaced part of the crushed and ground-up rock (gouge) between the walls of a fault. b. Fissure-veins (F) in which ore material has filled up spaces produced by renewed movement along a slightly uneven earlier fault surface. c. Zones of crushing and shearing in the overturned or over-steepened limbs of asymmetrical folds provided favourable sites for the deposition of gold-bearing quartz veins in the Reefton goldfield.

Fig. 13.3
One of the three largest (and also the last) gold dredges to be built in New Zealand, on its original claim at Kanieri, near Hokitika, in 1938. It worked glacial outwash gravels profitably here for many years before being transferred to the Kumara district.

to heat their try-pots. Coal was noticed in 1844 during the reconnaissance preceding the Otago settlement, but the first mine at Saddle Hill near Dunedin did not open until 1849, only a year after the main shiploads of settlers arrived. The Westland explorers Brunner and Heaphy in 1846 discovered what was to become a very important bituminous coal area in the Grey Valley but the Brunner Mine did not open until 1866. According to the Nelson historian J. N. Newport coal from West Haven (known as West Wanganui in the early days) was shipped to Wellington in 1840, and there may have been even earlier shipments than that. Small mines were opened on the fringes of the infant town of Nelson about 1841, and in the Takaka and Collingwood districts in the early 1850s. It is surprising that although Hochstetter noted that coal outcrops had been opened up near Drury in 1858, the first commercial mining in this area, so near to the growing town of Auckland, did not begin until 1862, but it has to be remembered that Maori-pakeha relations were not very cordial at that time.

With ample supplies of good firewood—especially manuka—available in most districts for domestic uses, firing sawmill boilers and the like, the incentive to mine coal was not strong before steamships became common around our coasts and on our few navigable rivers. From 1870 onwards, the rapid expansion of the railway system, town gas supplies and later dairy industries caused the demand for coal to grow at a faster rate than the population until about 1920.

Coal being a commodity of relatively low value in relation to its bulk (low unit-value), deposits had to be found near settled areas, and where bulk transport could be provided to get it to the market. Outcrops were found in wooded areas by tracing the source of coal debris in streams, and most of the mining was by small concerns which simply 'followed the coal' in from the outcrop, and were easily discouraged if the seam thinned out or was interrupted by a fault. Much unworked coal was subsequently lost through spontaneous fires. Some of the earliest mines on a larger scale were at Kawakawa, Bay of Islands, and Huntly, on the Denniston plateau, at Brunner and at Kaitangata. Exploration by shaft-sinking and boring was carried out from long before the end of last century to test the extent of the coal beds underground.

Fig. 13.4
An aerial view of the extensive Maramarua open-cast coal workings, a major source of sub-bituminous coal for electric power generation in the Waikato coalfield. (Photo: N.Z. Geological Survey)

Geological surveys have been of more direct assistance in prospecting and mining for coal than for gold. It was already known in Hochstetter's time which formations were the main producers of coal, and it has been noted that the attentions of Hector's Survey were directed to coal-mining problems from the outset. All the important coalfields were included in the regional surveys during Morgan's directorship, and the departmental Annual Reports record many special investigations of faults and other mining problems. The coal-mining industry, both state and privately owned, has made much use of geological advice through the years when planning drilling programmes and developing new mines. The geology of our coalfields, with few exceptions, is not simple and it has not been possible to introduce mechanised methods of winning and transporting coal to the same extent as overseas, but predictions can generally be made on geological grounds with more confidence than in gold-quartz mining. The development of at least one successful new mining field, at Garvey Creek, Reefton, arose out of favourable inferences from geological work in adjacent areas. Re-surveys of the main fields, on a scale of precision unprecedented in New Zealand and accompanied by sampling and analysis programmes, began in 1937 as part of a national stocktaking of our natural resources, and were virtually completed by 1960. With coal returning to a position of relative importance as an energy source, these surveys have been amply justified. Suggestions can also now be made as to where entirely new coalfields may be sought.

Grades of Coal

Different grades of coal suitable for such different uses as steam-raising, gas and coke making, slow-combustion stoves, etc. are identified by a variety of technical terms, according to which of several schemes of classification is appropriate. Most people are familiar with a range of coal grades known as lignite, brown or sub-bituminous, bituminous or coking, and anthracite, and many are also aware that these form part of a series of varieties increasingly altered by pressure and heat from peaty vegetable accumulations which were the raw materials. This kind of variation is referred to as the 'rank' of the coal.

An independent series of differences related to the kind of parent plant material is recognised as coal 'type'. Besides rank and type, classifications of coals to determine their best uses also take account of the potential heat energy ('calorific value') measured by standardised assay methods, and the amount of sulphur.*

New Zealand coal deposits encompass an unusually wide range of both rank and type, while the way in which our coals from within a single bed or formation may vary in rank from place to place, rapidly but regularly, is exceptional too. In most other parts of the world the bituminous coals are more or less confined to strata of late Paleozoic to Triassic age. In New Zealand, neglecting some thin, impure coal in Jurassic beds, the productive coal measures are all Cretaceous or younger in age, and the rank is no guide at all as to age. This was not only a puzzle to the early geological explorers, but a source of past mistakes in the mapping of some areas on the West Coast.

This independence of age and rank was discovered in the course of re-surveys of West Coast coal areas during the Second World War years. It was found that the variations in rank of coal in any one bed at different places could be accounted for simply in terms of the maximum thickness of younger strata that had been deposited on top afterwards, as noted in Chapter 7 (p. 183). By using this principle one can sometimes predict that if coal is found by drilling in a new area, it should be within particular limits of rank.

Geological Settings of New Zealand Coals

It will be useful here to put together and summarise information already included in Chapter Seven. The workable coal deposits of New Zealand fall into four groups distinguished by the circumstances in which the parent peat was deposited, thus:

(1) Coals formed from peats that accumulated inland in fresh-water lakes and swamps at the close of the Rangitata Orogeny in mid-Cretaceous times. These are low-sulphur, low-phosphorus coals, ideal for making metallurgical coke and town gas. They are confined to the Paparoa Group and have been mined only in the Greymouth Coalfield. Reserves are not large.

---

* In order to judge the suitability of coals for various uses and for purposes of scientific study, routine methods for sampling and analysing them have been developed and standardised to ensure as far as possible that the results obtained by different samplers and laboratories and in different countries can be compared. Complete assays of coals expressed as percentages of the main component elements (carbon, hydrogen, oxygen, nitrogen, sulphur, etc.) are called 'ultimate analyses'. They are needed in coal research, but are time-consuming and expensive when, as is usually the case, very large numbers of samples have to be put through. For practical purposes, as a guide to the best use for a coal, the composition is determined and expressed in terms of the percentages of arbitrary 'components' as estimated by a standardised procedure:
  (i) 'Moisture' (loss of weight of a sample heated to 105° C under specified conditions).
  (ii) 'Volatile Matter' (loss of weight when dried coal is heated to 900° C under specified conditions).
  (iii) 'Ash' (percentage of incombustible residue).
  (iv) 'Fixed Carbon' (arbitrary term for the non-volatile constituents not driven off by (ii)).
  This is called 'proximate analysis'. Coals are assayed also for coking properties, yield of gas, etc. Classifications of coal into type and rank are mainly based on proximate analyses, but some require ultimate analyses and others bring in the heat yield (calorific value) as well.

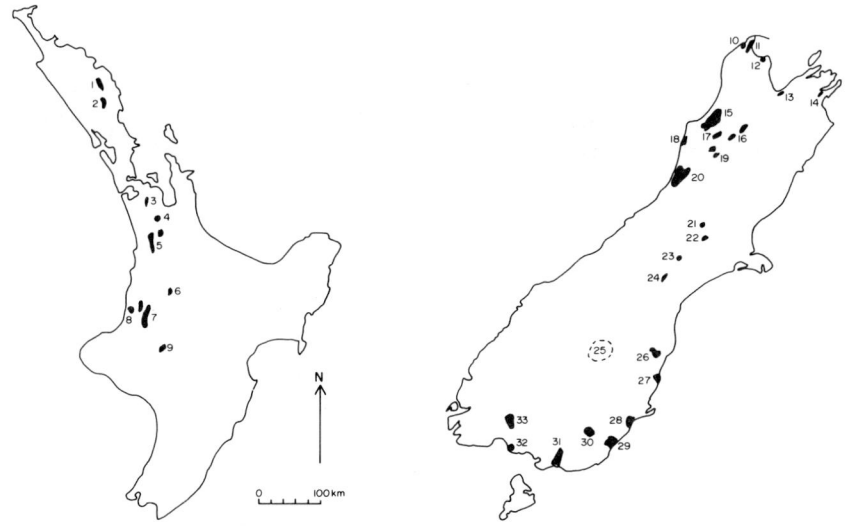

Fig. 13.5
Locations of important coalfields, present and past. 1, Kawakawa; 2, Kamo-Whangarei; 3, Drury; 4, Maramarua; 5, Huntly-Pukemiro; 6, Mangapehi-Benneydale; 7, Ohura; 8, Mokau; 9, Retaruke; 10, Manga-rakau (West Haven); 11, Puponga; 12, Takaka-Motupipi; 13, Nelson; 14, Picton; 15, Denniston-Millerton-Seddonville; 16, Murchison and Owen River; 17, Buller Gorge-Inangahua; 18, Charleston; 19, Reefton-Garvey Creek; 20, Greymouth; 21, Broken River; 22, Malvern Hills; 23, Mount Somers; 24, South Canterbury; 25, many small areas in Central Otago; 26, North Otago; 27, Shag Point; 28, Green Island; 29, Kaitangata; 30, Pomahaka; 31, Mataura; 32, Orepuki; 33, Ohai-Nightcaps.

(2) The coals of the Quartzose Coal Measures (see Chapter Seven) formed from peat that accumulated in fringing swamps in advance of transgressing seas between late Cretaceous and Oligocene times. In rank they range from lignite (as at Mataura or Charleston) to high-rank bituminous coal (as at Brunner, Garvey Creek and Stockton) and they tend to contain a good deal of sulphur. This group has produced the great bulk of the coal already mined in New Zealand, and includes most of the potential reserves.

(3) Coals formed in the opposite circumstances to (2), that is, while seas were retreating irregularly from emerging land during the onset of the Kaikoura Orogeny in the Miocene Period. Coal beds come in above marine formations and are succeeded upwards by non-marine conglomerate and sandstone. Sulphur content is high and rank mostly low except at Owen River and Matakitaki Valley near Murchison, where the upper of two sets of coal beds formerly mined there (Longford Coal Measures) are of bituminous rank. Apart from Murchison, these coals have been mined only in South Canterbury, and on a very small scale. Potential reserves are limited, perhaps negligible.

(4) Coal beds interstratified with the Miocene marine rocks in inland Taranaki mark a temporary, local shoaling of Tertiary seas. The sulphur content is high, and the coals have been worked only in the Ohura district, where, however, there are significant potential reserves.

Anthracite, the highest rank of coal, does not fit into the above scheme. In New Zealand it has been found only where coal of much lower rank has been 'baked' locally by heat from igneous intrusions. The only important anthracite deposits are in inland Canterbury, notably in Malvern Hills and in the Acheron River coal area near Lake Coleridge, where the heat was provided by thick sills of basalt or dolerite injected along the roof of the coal bed, and by dykes cutting across it.

Case (2) alone provides good prospects for the discovery of new, hidden coalfields remote from present or past mining areas. Much drilling would be required to discover and prove them worthy of

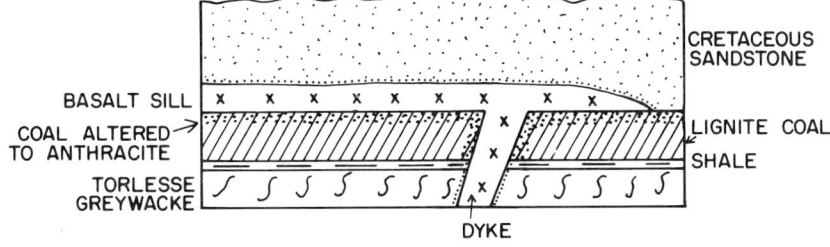

Fig. 13.6
Bituminous coal outcropping at the entrance to an abandoned mine on the Stockton plateau, north of Westport. In places the main coal bed in this coalfield, contained in the Brunner Formation (the Quartzose Coal Measures in this region), exceeded 20 m in thickness. Nearly all the coal produced in the area now comes from open-cast workings, favoured by very thin overburden, as in this view.

Fig. 13.7
Anthracite deposits in Canterbury, for example at the Acheron Mine (as illustrated diagrammatically here), occur where fairly low-rank coal has been altered by the heat of intruded basaltic dykes and sills.

mining. It is unlikely that any coal thus found would be of bituminous rank, although there are several areas that could yield substantial quantities of sub-bituminous and lignite coal. High quality bituminous coals of group (1) are already known to occur in the heart of the Paparoa Range, at high levels and rather inaccessibly placed.

It is also unlikely that much of the coal in these potential areas could be worked by open-cast methods, for the thickness of covering strata would be considerable. This would be a disadvantage in view of the relatively high cost per tonne of winning coal in underground mines.

The Search for Oil and Gas

Petroleum (Lat. = 'rock oil') has been used by man since prehistoric times, but until 1859 when E. L. Drake drilled the first successful oil well in Pennsylvania—in fact, it was the first well drilled anywhere specifically to find oil—it had been obtained from dug pits and natural seepages. In New Zealand the Maori people certainly knew about oil seepages and gas vents long before the arrival of the pakeha, but the earliest mention of petroleum indications in this country, near New Plymouth, came from the German explorer E. Dieffenbach in 1839. The outlook must have seemed bright indeed when, in 1866, the first of a series of productive wells was sunk at New Plymouth. How ironical it is that despite hundreds of millions of dollars spent on investigations in other districts where the initial oil showings were at least as good, the small Moturoa field at New Plymouth remained the only continuously productive one until the Kapuni discovery in 1959. The flow from Moturoa is small indeed by comparison with major oil fields of the world—at most a million or so litres per year—but enough to warrant the small refinery that was built in 1930. Even more important, perhaps, was the encouragement it gave to persist with oil exploration in New Zealand for nearly a century until a commercial gas and light liquid petroleum deposit was at last discovered in the same district.

Gas vents and small oil seepages occur in scattered places throughout the country but are most numerous in East Coast districts of the North Island, Taranaki and the Murchison area. Some of the largest seepages are in inland Poverty Bay, so it is not surprising that the second locality to be tested by drilling was Waitangi Hill near Gisborne, where a pit and well to a total depth of 100 m was sunk in 1874, without success. From that time onwards company after company, New Zealand or overseas owned, separately and in partnership, drilled at least fifty deep wells, some to depths of between 1000 and 2000 metres, in the East Coast region but found no more than scant traces of oil and gas. Responding to the strong interest in that region, the Geological Survey produced three descriptive regional *Bulletins* with one-mile-to-an-inch maps and a *Paleontological Bulletin* dealing with it between 1910 and 1931, and its staff was involved from time to time in joint investigations with oil company geologists, examining critical areas again and again, and improving the precision of mapping with the help of advances in paleontology. A great impetus was given, in fact, to the development of micropaleontology, especially of the foraminifera. Detailed mapping was followed later by geophysical explorations and then by more drilling, until only recently the last of the major East Coast operators, the Shell-BP-Todd consortium, relinquished its concessions. This probably means the end of on-land explorations in that region.

The third area to attract serious attention was Westland. An oil seepage was found near Kotuku, in the valley of the Arnold River, in 1897. In 1899 and again on several later occasions shallow pits and borings to depths of a few hundred metres were put down, yet despite the impressive amounts of surface seepage only slight flows of oil and gas were encountered in any well. Apart from a drilling in the Mangles Valley near Murchison in 1926 the western South Island region was neglected from about 1914 until the Second World War. During the war period different companies drilled a few deep test wells and many more subsidiary ones to obtain stratigraphic information in the area between Hokitika and Nelson Creek. No flowing oil was found, but traces in the cores from the Quartzose Coal Measures (Brunner Beds) at the bottom of one hole aroused some interest because a waxy petroleum residue was found about the same time by the Mines Department while drilling for coal at Dobson, and there had been a similar occurrence in the same area in 1902. Geophysical surveys were carried out near Kotuku in 1955 in the hope of discovering some clue as to the origin of the seepages there. Wells have since been drilled in

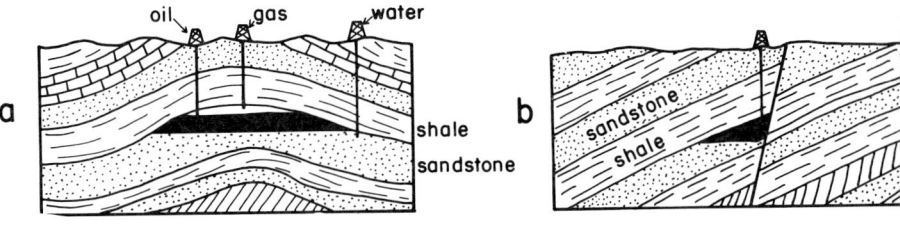

the Arahura, Taramakau and Arnold Valleys and one on the foreshore
at the Waiho River mouth in South Westland, but no oilfield has yet
been found in that region.

Largely because of the continuing small production at the Moturoa
field, but also because the known geological conditions (as distinct from
the number of natural seepages) have always seemed most favourable
there, Taranaki has been the scene of the greatest amount of deep
drilling in New Zealand, and also of all the important finds so far.
During the Second World War period, five major wells were sunk to
depths of up to 3000 m, all of them dry. Then in 1959 came the first
notable success, achieved by Shell-BP-Todd at the Kapuni-1 site on the
south side of Mount Egmont. This was not only our first economic
natural gas well (discounting a few minor ones from which methane
was obtained), but it yielded some of the lighter liquid hydrocarbons as
well ('condensate'). Six wells altogether were drilled in that area in the
next few years, down to a maximum depth of 5000 m, and only one
was reported 'dry'.

Numerous dry wells have been put down in various other parts of
the country from Waimamaku in the Hokianga district to Tuatapere in
western Southland, a few on sound grounds but a good many others
that were hopeless as well. No fewer than 54 wells were drilled
altogether between 1955 and 1970, and the same period saw the
introduction of new 'remote sensing' methods of investigating
sub-surface conditions on land and beneath the continental shelves.
Meanwhile, the Kapuni field was being developed to supply the whole
of the North Island with natural gas for a long period, but we still
lacked flowing oil except at New Plymouth.

Offshore drilling began in 1968 in the North Taranaki Bight. The
great event so long awaited happened in the following year when
Maui-1, drilled to a depth of over 3000 m below the open ocean floor
25 km offshore from Cape Egmont, struck gas, condensate and oil in
commercial quantities. Within the succeeding eighteen months Shell
had followed up their initial success with three more deep, productive
offshore wells in the same area. The Maui group of wells has
established the presence of a major gas field with substantial amounts
of oil as well. The mammoth task of getting the gas and oil ashore is
now well under way. Further offshore wells have since been drilled for
other companies to the south of Maui in the western mouth of Cook
Strait, but without further success.

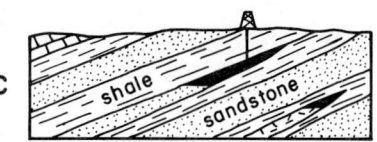

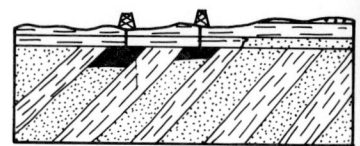

On-land drilling has begun again recently in Taranaki and Westland, and further offshore operations in the stormy waters south of Fiordland where features of the sea floor and geophysical investigations suggest the presence of favourable rock and structures.

The essential geological requirements for an oilfield are these:

(1) *Source rocks*—these are generally believed to be marine oozes and shales which originally contained abundant remains of animal and plant micro-organisms, the alteration of whose body substances after burial produces hydrocarbons; there are reasons for thinking, though, that the original source of some New Zealand occurrences was in non-marine sediments.

(2) *Permeable reservoir rock*—sandstone, conglomerate or fractured rock with the necessary amount of pore-space to hold oil in quantity and yield it to wells.

(3) *Cap rock* a cover of compact shale or other impermeable rock.

(4) *Structure*—one or more of several kinds of rock structure, including domes and anticlines, which make it possible for oil in the reservoir rocks to be retained under the cap rock until tapped by the drill. (Fig. 13.8)

Within a typical reservoir, gas saturates the highest parts, then oil, with water (often brine) below. An exhausted well usually yields water after the oil has ceased to flow.

The late Cretaceous and Tertiary strata in many parts of New Zealand abound in mudstone beds that should be good oil-source rocks, and not infrequently these will emit a kerosene smell when freshly broken from the outcrop. A good place to try this is at outcrops of Kaiata Mudstone, the dark grey formation along the coastal highway near Rapahoe, north of Greymouth. Despite the general belief expressed above that oil source rocks must be of marine origin, there is a growing conviction that the heavy oil struck by wells at Dobson and seeping into certain Greymouth coal mines has actually been generated in the coal measures, and did not migrate downwards from marine strata above.

Suitable reservoir rocks are present in almost all areas of thick Cretaceous and Tertiary beds and a considerable number of theoretically favourable structures have been detected by surface mapping, exploration drilling and geophysical investigations. The East Coast region contains enormous thicknesses of Cretaceous and Tertiary strata and many gas vents and oil seepages. However, it also has a most

*Fig. 13.8
Some of the simpler oil-retaining structures:
a. Anticline or dome. In the case of an anticline, there must be some obstacle to longitudinal movement of the petroleum up the 'plunge' of the fold; this might, for example, be a transverse fault.) b. Typical fault-trap. c. 'Pinch-out' trap; up-dip thinning of the reservoir formation. d. Permeability-trap; the reservoir bed changes from permeable sandstone, for example, to impermeable shale in the up-dip direction, preventing migration of oil towards the surface. e. Unconformity-trap. To the right, where the younger capping strata change to permeable sandstone, oil in the underlying reservoir beds has not been retained.*

complex structure (which has baffled more than one generation of geologists) and innumerable faults, so the chances that large 'pools' or oil-saturated rock still remain below perhaps were never very good. It is more difficult to understand why Moturoa and Kapuni remain the only commercially successful onshore finds so far in Taranaki, where structural conditions are quite as favourable as those in major oil-producing regions like California; and again, why has there been no success in Westland?

Some petroleum geologists feel that the explanation may lie in the combination of the following circumstances: (1) the potentially oil-bearing rocks of New Zealand are steeply dipping in many places; (2) their outcrops extend up to appreciable altitudes above sea level; and (3) most of the country receives an abundant rainfall. In short, they suspect that the oil which may once have been in our Cretaceous and Tertiary rocks has largely been flushed out. The chief exception to the above generalisation, in two important respects, is Taranaki, where the only successes have been achieved.

This does not mean that we must rule out the possibility of success onshore in any area where there are substantial thicknesses of the appropriate sedimentary formations. Many possible areas have been drilled, but a few still remain, always with some faint promise of some real return for the vast sums already spent, and it is hardly necessary to stress the advantages to New Zealand of an important oil find *on land*. It must be faced that in the long term the prospects for more offshore successes are better, even if the waters are deep and the weather often stormy. But whether the offshore production could be marketed competitively may be another question.

Iron Ores

The needs of the earliest settlers for iron were modest compared with those of later generations, but it was nevertheless a major task in sailing-ship days to transport enough of such a heavy commodity half way round the world. Naturally, there was interest in the possibility of finding ores to smelt in New Zealand, and the western blacksand beaches could hardly fail to be noticed. Attempts were made from as early as 1849 to produce iron from blacksands at New Plymouth. Twenty years later, blast furnaces were built at Onehunga and further experiments at making iron and steel from west coast ironsands were tried from time to time until the First World War. Although pig-iron was successfully produced, every attempt failed ultimately because titanium oxide was intimately blended with the iron oxide, and its presence made the slag so refractory that after a time it would always clog the furnace. The metallurgical problems of smelting ironsands were solved only in recent times.

Meanwhile the brown limonite ore deposits of north-west Nelson also had been receiving attention. Limonite from Onekaka and Parapara near Collingwood had been used by the Maoris as a pigment, and it was the basis of a paint industry started in Nelson in the early 1870s. Successive attempts from 1873 onwards to establish an iron industry at Parapara failed despite the attractive quality of the ore, partly because

Fig. 13.9
The results from under-
ground prospecting for
limonitic iron ore at
Onekaka, north-west Nel-
son, in the late 1930s were
disappointing. 'Mountains
of iron' proved to have
merely a thin mantle of ore
rubble at the surface and
discontinuous, irregular
masses of ore underneath.

the extent of the deposits was uncertain, which discouraged investors, but also partly because of the remoteness of the area from the main markets.

Eventually a blast furnace which had been used for blacksand trials at New Plymouth after the war was moved to Onekaka, and from 1924 onwards pig-iron and cast iron pipes were manufactured regularly until the enterprise fell victim to the economic depression in 1931. In any case, doubts expressed as far back as 1908 by geologist J. M. Bell as to the true thickness of the ore which mantled Onekaka hills began to seem justified. Marble showed up in the floor of the main ore quarry. Aiming for an industrial expansion that would not be completely dependent upon overseas supplies of iron and steel, the Government in 1937 began a major programme of tunnelling, drilling and assaying of the ore deposits between Onekaka and Parapara, and called in experts from abroad to advise on establishing a steel industry. The outcome was disappointing, only 9 million tons of iron ore averaging 40 per cent iron content was proved, and although it was shown that up to 26 per cent of blacksand could be added to the Onekaka limonite ore without making the slag too refractory, the scheme was finally abandoned after World War II. An increasingly adverse economic factor was the high cost in New Zealand of coal suitable for making metallurgical coke, then indispensable in the extraction process.

Things were going better with the blacksands. Continual experiments in New Zealand and overseas laboratories finally came up with a solution to the metallurgical problems of using titanium-bearing sands, and since the industrialisation of this country now assured an economic scale of operation, the present steel works were built near Waiuku. It did not take very long to prove vast reserves of ironsands in Pleistocene beach, terrace and dune deposits on the Auckland west coast and in Taranaki.

The ultimate source of the main ore mineral, titanomagnetite, is andesite debris from the south Auckland west coast and Taranaki Pleistocene volcanoes, enriched in the worked deposits through selective removal of the lighter quartz, mica and other weathering products by wind and waves. The deposits compose a set of raised coastal terraces up to 70 m above present sea level dating from recent back to interglacial times. Blacksand beaches north of Waikato Heads containing ilmenite rather than titanomagnetite are derived from the weathering of Miocene volcanic rocks north of the Manukau Harbour, while those on the South Island West Coast and Southland beaches, containing magnetite and ilmenite, have been concentrated similarly from weathering products of mafic and ultramafic igneous rocks and Haast Schists.

The source and origin of the north-west Nelson limonite ores were less immediately obvious. The basement rocks of the region are not exceptionally rich in iron. Occasional veins of pyrite and pyritous quartz could explain the scattered surface accumulations of limonite representing the oxidised 'gossan' capping of the veins, but could not account for the three main belts of massive ore between Onekaka and Parapara. Eventually it was realised that the main limonite masses originated while the Late Cretaceous Peneplain was developing, here evidently under moist, warm climatic conditions which favoured the fixation of iron weathering products in laterite soils and superficial deposits (in this area, these were the Quartzose Coal Measures), as in tropical regions today. The large limonite accumulations now surviving escaped later erosion because they were protected in fault-angle depressions of post-Cretaceous date, and, where the underlying basement rock consists of marble, in giant sinkholes and smaller solution cavities into which the limonitised materials had slumped.

Other Base Metals

The term 'base metal' originally meant a metal 'inferior' to the 'noble metals', gold and silver, because it was less durable under the rough corrosive treatment given to it by alchemists in their experiments. The term is still used, but as a convenient grouping of all the industrial metals other than iron and aluminium, especially copper, lead and zinc, but often including antimony, molybdenum and other metals frequently occurring with them.

The search for base metals in New Zealand makes a strange and rather depressing story, for although they occur in great diversity, all attempts to mine them have been short-lived. We have an ironical situation, recalling the early success of oil drilling in Taranaki, in that although copper, chromium and manganese were all found and mined in a small way in the earliest days of the colony, they never contributed

Fig. 13.10
An aerial view of the Waipipi ironsands mining operation in an area of dunes underlain by former beach deposits west of Wanganui. Floating pipelines carry the sand raised by the two dredges to the concentrating plant, also floating on a pontoon. (Photo: N.Z. Geological Survey)

significantly to the national wealth. Manganese ore was mined briefly on Kawau Island in the 1840s and later at Taieri, Otago. Chromite was produced for a while in the early days of Nelson from ultramafic rocks in the Dun Mountain area. Copper ores were worked quite extensively for a time on Kawau Island. Visitors to the island are likely to have seen the old flooded shaft and pumphouse at Miners Point, and some may have noticed the blocks of smelter slag on the foreshore at Mansion House Bay. It was mined also on Great Barrier Island (at Miners Head) and from serpentine rock at Nelson in the middle of last century, but there has been little mining activity at any of these localities since 1900.

The variety of ways in which copper minerals occur in New Zealand is interesting, even if the economic possibilities are not. Here is a summary: (1) as veins of solid chalcopyrite, or of quartz containing this and other copper sulphide minerals, in most of the older sedimentary rock groups, in the granites, and in the gneisses, the latter especially in Fiordland; (2) as sulphide and other mineral impregnations in fault-crushed greywackes and iron-rich chert or jaspillite in rocks of the New Zealand Geosyncline (often detected in these rocks by the exaggerated indications of green surface staining); (3) as sulphides and native copper in serpentine and other altered ultramafic rocks in the Nelson 'Mineral Belt'; (4) in the zones of contact metamorphism surrounding some granitic intrusions, as at Mount Radiant near Karamea; (5), as (1) but in the Haast Schists; (6) together with sulphide and other compounds of the other base metals in the altered andesite country of the Hauraki gold-silver mining fields. It is from this last source, and essentially as a by-product, that copper has most continuously been produced in New Zealand, though only in minor amounts.

Fig. 13.11
A vein of galena and other
sulphide ore minerals with
quartz in the Tui Mine,
Te Aroha. The walls of the
vein are composed of
hydrothermally-altered
andesite rock. (Photo:
N.Z. Geological Survey)

Lead, zinc, antimony and the other base metals occurring mainly as sulphide minerals in quartz veins are not rare in the older rocks of many areas, but from the point of view of possible economic development the promising area has always been the hydrothermally altered andesite rocks of the Hauraki district. Galena, sphalerite, stibnite, along with chalcopyrite and the other copper minerals are common amidst the 'complex sulphide ores' there. Quartz is the usual gangue mineral, but calcite and others occur too. In the other quartz mining areas, base metal possibilities have rarely been the incentive to mine, and in fact the presence of base-metal sulphides, especially stibnite, makes it more difficult to extract the gold and silver. The rarity and small size of rich sulphide veins and the lack of facilities for smelting such ores has made the search for them unattractive to investors. For the most part, concentrated sulphide residues from gold extraction were sent to Australia for final processing.

High market prices for metals in the post-World War II period promoted a mineral exploration boom in New Zealand, where for the first time since the early days, the base metal ores rather than gold became prime objectives of search. Also for the first time in New Zealand mining history there was interest in finding large quantities of low-tenor ore rather than smaller, rich bonanzas, but no major discovery had been made before the slump in metal mining in the early 1970s brought an end to large-scale mineral exploration in this country. The only mines to work sulphide ores in recent years have been those seeking mercury in North Auckland and the Tui Mine at Te Aroha. The latter, re-opened in 1963 after long idleness, followed two veins of complex sulphide ore containing traces of many other metals besides lead, copper and zinc. It is not operating at the present time.

Long known to exist in the ultramafic igneous rocks and serpentine belts of Nelson, Westland and western Otago, compounds of nickel and cobalt and the unique native alloy of nickel and iron called awaruite (first described in 1886) received attention during the boom years, along with a search for new deposits of asbestos (chrysotile). Some of the areas then prospected had been rather inaccessible before the advent of the helicopter. Knowledge of the regional geology and of the distribution of minerals in hitherto remote areas was advanced, but no mining has resulted.

There are other minor occurrences of base metal ores, but it is impossible to deal with them all here. Readers are again reminded of the valuable compendium of information about New Zealand mineral occurrence in Professor G. J. Williams's *Economic Geology of New Zealand* (1965; 2nd ed. 1976).

Radioactive Minerals

Under a mantle of security, beach and stream sands in various parts of New Zealand were examined during the Second World War for traces of radioactive minerals, and the search for possible sources of uranium was stepped up in the decade following. Although low-level radioactivity and some occurrences of uranium-thorium minerals were picked up, the first significant find was made by two prospectors, Jacobsen and Cassin, using a geiger counter along the Buller Gorge road. They found that breccia of the Hawks Crag Formation was radioactive close to a dyke. This was in 1955, and for the next two years all areas of Hawks Crag and other breccias of similar type and age were vigorously prospected. Indications were found in many places in the northern Paparoa Range area, but no workable deposit turned up.

Uranium minerals were found in dykes of porphyry and lamprophyre cutting through the Hawks Crag Breccia, and G. J. Williams believes that they were deposited in a time of mineralisation due to solutions which accompanied the intrusion of the dykes, and probably not long after the formation of the breccia in early Cretaceous times. It seems the most satisfactory explanation.

Non-metallic Earth Products

All earth substances obtained for purposes other than to extract metals from them are referred to as 'non-metallic', which is not very logical because nearly all of them are chemical compounds containing one or more of the metal elements. Besides the fossil fuels (coal and oil, which strictly are not minerals either) the materials under this heading fall into the following groups: industrial substances like clays for brickmaking, talc for cosmetics, silica sand for glassmaking, sulphur for fertilisers, etc.; agricultural supplies, mainly limestone; engineering construction materials such as gravel and sand for concrete aggregate,

Fig. 13.12
A potential source of quartz sand in great abundance for use in a variety of industrial operations is offered by the Quartzose Coal Measures in some areas, as exemplified here in an abandoned railway cutting near Ngapara, North Otago. Quartz debris released by deep weathering and erosion from the basement rocks (here, from the Haast Schists) was concentrated in superficial deposits on the Late Cretaceous Peneplain (p. 171).

crushed stone for road surfacing and fill, etc.

The gross value of such products is surprisingly high even in New Zealand, although the unit-value is far lower than for most metallic ores. In the United States, still a large producer of metals, the non-metallics outstrip the metallic ores by two to one, and in New Zealand they amount to a multi-million dollar business.

Limestone. Though not top of the list in aggregate value of annual production, limestone is vitally important to our farming and pastoral industries. Fortunately, limestone deposits are widely distributed in both main islands. The greatest amount comes from Oligocene limestone formations, widespread thanks to the mid-Tertiary marine transgression (Chapter Seven), but in the far south the source is in Miocene rocks and in Hawkes Bay in the Pliocene. Pre-Tertiary limestones are used only in a few places. Cement manufacture uses the largest amounts of limestone, followed by agriculture and manufacturing industries using lime, such as glass-making and tanning.

Limestone rarely consists of pure calcium carbonate, and for various uses there are maximum tolerances of 'impurities' such as clay, sand, and pyrite or other iron compounds. I use inverted commas around 'impurities' because the substances are natural constituents of the rock, and impure only in the sense of some specific application in which they have adverse effects. It is hard to judge by eye the amount of carbonate in a limestone—it can vary greatly both laterally and stratigraphically—so routine analyses are always necessary. Recrystallised limestone (marble), as from the Paleozoic rocks of Nelson, is no longer in great demand as a building or ornamental stone but is still used in crushed and size-graded fragments to improve the appearance of concrete building blocks.

Clays. To begin with, it should be noted that the word 'clay' can mean rock substance composed either of particles of the finest size grade, that is, a few thousandths of a millimetre, regardless of composition, or of minerals having the composition of the 'clay' group, which are hydrated alumino-silicate produced by decay or alteration of various types of rock. The largest amounts of clay go into the manufacture of bricks, tiles, drainpipes, pottery and other ceramic products. Other important uses are as fillers or absorbents in industrial processes, including paper and rubber manufacture, and special types are used to raise the density of water used in the drilling of deep wells in rock.

Clays suitable for brickmaking are commonly found as residual deposits of the products of weathering of sedimentary, igneous or metamorphic rock in place, or as transported clay that has been washed and redeposited elsewhere. Some brick-works have used unweathered Tertiary siltstone and claystone, as at Karoro near Greymouth. Christchurch and Dunedin brickworks have used loess, which is really a very uniform silt, not a true clay (Chapter 9). Wellington bricks have been made from clays produced by weathering of the local greywacke rock. Not only the nature of the parent rock but also the length of time and the climatic conditions in which the weathering occurred help to determine the character of the clays.

Porcelain and china clays should contain mainly the mineral kaolinite, with preferably only quartz as an impurity. The best kinds come from the decomposition of igneous rocks rich in potassium feldspars, e.g. rhyolite, pegmatite. Clays formed by the weathering of such rocks in 'fossil' soils beneath the Late Cretaceous Peneplain have given us the best 'whiteware' clay in New Zealand, at Mount Somers in Canterbury. Refractory clays with a very high fusion temperature have come from fossil soils and decayed basement rock under coal seams, but in order to make a good firebrick the clay should be free from iron oxides, which act as fluxes. The Huntly fireclay, a once-famous refractory clay from beneath coal at the former Brunner Mine near Greymouth, and the Kakahu clay of South Canterbury are notable examples in New Zealand.

The name 'bentonite' does not signify a single clay mineral, but a special group of clays, including notably the mineral montmorillonite, some of which have the properties of expanding as they absorb water and of remaining dispersed in water suspensions for long periods without settling out. The swelling bentonites, which are the ones rich in sodium rather than calcium, have many special uses as fillers and absorbents in chemical industries and particularly as drilling muds. In New Zealand bentonite has been produced for many years from marine mudstones of Eocene age at Porangahau, Hawkes Bay, and more recently from Pliocene or early Pleistocene non-marine lake clays near Coalgate, Canterbury. The latter are believed to be derived from fine volcanic ash, erupted at the same time as the lavas forming the Harper Hills. Bentonites occur widely in beds of late Cretaceous and

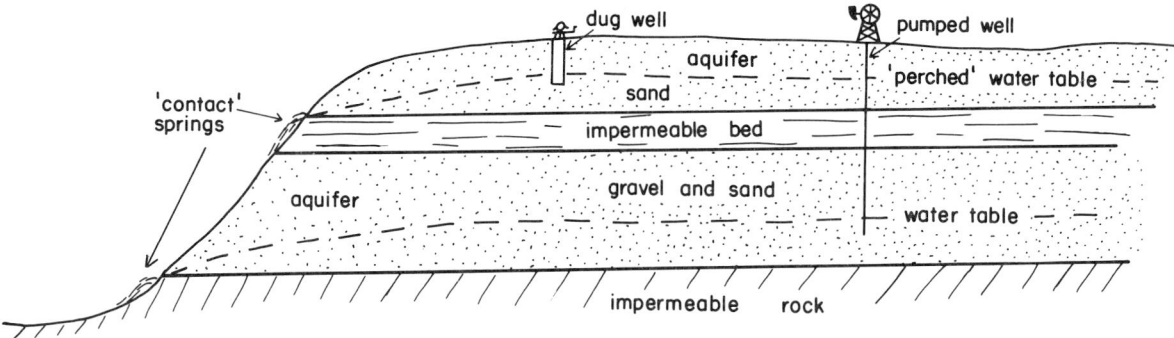

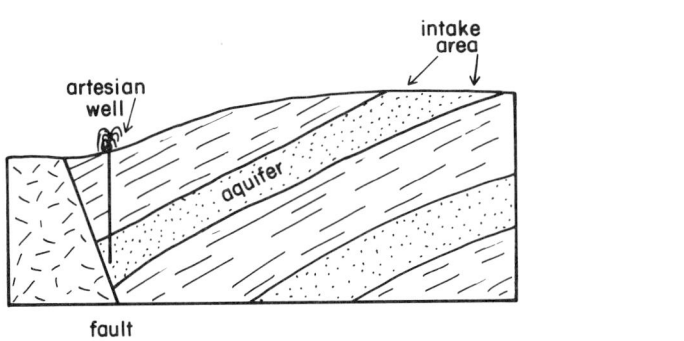

Fig. 13.13
Types of underground water supply, greatly simplified. The essence of an 'artesian' situation is that the aquifers are charged with water at a higher level than the draw-off point, so that the water can rise to (or above) the ground surface without being pumped. In the lower diagram, natural escape of water up the plane of the fault would be an 'artesian spring'.

early Tertiary age, and because of their expanding properties and slippery nature when wet they cause slumping of the countryside and serious problems for engineers and farmers.

Sand and gravel for road material and aggregate is produced in large quantities and tops the list of non-metallic earth products. The days are long past when any kind of stream or beach gravel was acceptable for those uses. Rigid specifications now apply as regards kinds of rock, amount of clay, particle shape and size grading. Sand and gravel for both purposes are still obtained from recent river bed alluvium, river terrace deposits and beaches, but are now more often screened into different size grades and re-combined in the specified proportions.

Some types of cement cannot be used with aggregate that contains certain igneous rock varieties because of the risk of chemical interactions which will cause the concrete to swell and rupture. Specifications for road-metal are also strict nowadays, requiring durability and uniformity to stand up to modern traffic conditions. Most parts of New Zealand are adequately provided with suitable stream gravels or hard rock suitable for crushing, which is fortunate since haulage costs are high. Northern and western parts of the North Island are less well favoured, and some country roads in Taranaki in earlier days were paved with locally-fired bricks because of the scarcity of suitable road-metal.

The Search for Water
The water demands of our style of living and modern industries have raised the importance of public water supplies to an unprecedented level, and indeed the availability of water can be a major limiting factor

in urban expansion. Most parts of New Zealand are favoured with adequate, reliable rainfall and in some there are still relatively unpolluted streams in suitable country for building reservoirs. Many communities have developed underground water sources by means of borings rather than from the dug wells or natural springs which served the needs of earlier times.

Many people are still unaware that underground water only exceptionally occurs in 'pools' or 'underground streams' in open cavities. Generally, water is drawn from gravel, sand or porous rock beds below the upper limit of water-saturation which we call the 'water-table'. The depth distance below ground to the water-table varies greatly according to the climate, kinds of rock, ground slopes, and other factors, and changes also from season to season and from year to year. It may lie at anything from a few centimetres to hundreds of metres below the surface.

Essential conditions for underground supplies suitable for human use are summed up below:

(1) A stratum or other body of rock with open space between its component particles ('porosity'), the spaces being interconnected to give it 'permeability' to allow water to move through it; such rock material may be an 'aquifer', a water-bearer.

(2) For human use, to prevent pollution of the water by direct downward percolation, the aquifer should be sealed above by a sufficient thickness of impermeable silt or clay.

(3) Water from rainfall or other surface source must be able to get into the aquifer in some area where there is no risk of pollution, or far enough away for impurities to be filtered out while it is percolating through.

(4) Natural outflow from the aquifer should be restricted or blocked off, for example by a lateral change to impermeable material or by a fault.

An 'artesian' situation exists where an aquifer capped by an impermeable layer takes in water at levels higher than the point of discharge. The water may then return to the surface as a natural artesian spring, by way of a fault fissure, for example, or it may rise above ground level in a well-pipe without being pumped.

Stream gravel and sand provide the most common aquifers, but any kind of porous rock material may serve as such. In thick gravel deposits, like those that underlie the Canterbury Plain, several separate aquifers at different depths may be separated by impermeable layers. Limestone beds riddled with caverns and smaller solution cavities also may provide aquifers but the water in such cases will be hard. Moderately compacted but still porous sandstones and conglomerates, and even fractured greywacke have provided small local supplies for farms and dairy factories, but the water from such sources is likely to carry iron and other minerals in solution and the flow is usually limited.

Public water supplies from below ground in New Zealand come almost exclusively from river gravels and sands of Pleistocene and younger age. Thanks to the post-glacial rise of sea level, the lower courses of most rivers have built up sufficiently thick deposits to provide deep aquifers protected from surface pollution, affording adequate storage, and with sufficient permeability to allow water to percolate rapidly enough towards wells. Some town supplies come from shallow wells alongside rivers, the recent stream deposits serving as a natural filter against sediment pollution but not always against biological impurity, so that chlorination is necessary.

Artesian sources have been found in many areas, notably in Hawkes Bay, Horowhenua and Southland. Best known of all artesian systems is that which supplies the Christchurch area. For many decades before a reticulated, high-pressure service was provided, houses, factories, schools, etc. obtained water from innumerable individual wells, which were artesian over a wide area. The water comes from several distinct aquifers below the surface of the plains down to depths of more than 100 m, and it was long debated as to whether these were charged directly by rainfall in inland areas, or by soakage from the channels of the Waimakariri and other rivers; the answer is 'probably from both'. Natural discharge seawards is restricted both by the flanks of Lyttelton volcano and by changes in the texture of the Canterbury Plain gravels into finer silts and clays.

On the whole, the chances of finding underground supplies are best where there are thick gravel deposits, and poorest in compact sandstone, mudstone and hard-rock terrain. As you can see by looking at any river bank alluvium exposure, the texture of stream sediments can vary over short distances, which means that there is no certainty of obtaining an adequate flow at similar depths near an existing, successful well. A second well seldom doubles the available supply, and excessive pumping can lower the water table and make nearby shallower wells run dry. A rich fund of experience in groundwater hydrology is now available to ensure wise management of underground sources, and to guide the search for new supplies.

Geology and Civil Engineering

Geological factors affect the selection of sites for engineering works, the design of foundations, the supply of construction materials and the prediction of problems likely to arise during construction. New Zealand is notable for the number and magnitude of its engineering undertakings in relation to the size of the population, and also for the diversity of problems our engineers have had to overcome. Our largest hydroelectric schemes are also among the world's largest. The earlier schemes on the Waikato River encountered unprecedented problems arising from peculiar structural features in the ignimbrite rocks. The ignimbrite sheets, in the course of cooling, developed a pronounced, vertical joint system which destroyed the coherence of the whole mass. At the Arapuni station, the weight of water-filled headworks caused movement of the adjacent rock, leading to structural failures, soon after the work was completed in the early 1930s. At other sites, the designers had to cope with wide variations in the mechanical strength of rock in contact with dam abutments, and therefore with the extent of yielding to be expected when the dams were filled. Large schemes in the Upper Waitaki catchment involved moving glacial and glacifluvial deposits in enormous amounts, with many problems from the enormous size of buried, glacially-transported blocks that were encountered in the excavations, not to mention hidden, gravel-filled gutters in the bedrock. Much help has been given by geophysical methods for remote-sensing both the depth to bedrock at places between testing drillholes, and the physical characteristics of hidden materials.

Major tunnelling undertakings in New Zealand, including the Otira, Tawa, Rimutaka and Kaimai railway tunnels and the Manapouri-Doubtful Sound water tunnel have contended with belts of severely fault-crushed rock, saturated with water under pressure and difficult for the tunnellers to support ahead of the linings. Because of the depth of cover and the mountainous nature of the country above, it was not always possible to give accurate geological predictions as to what would be encountered in the tunnels.

Geological advice and assistance for engineers in this country has been provided mainly by the Geological Survey, which for some years has maintained a section of its geological staff to specialise in engineering problems. Local authorities and the large overseas contractors have also employed other geologists. To some extent, our problems are unusual if not unique, so that the New Zealand trained and experienced geologist enjoys some advantages. By virtue of its diverse geology, New Zealand has proved to be a first-class training ground for the engineering geologist, while also attracting a strong interest in geology among our civil engineering fraternity.

Perhaps the severest of all responsibilities placed upon civil engineers by geological factors in New Zealand is the need to design and build safely in regions subject to a high chance of strong earthquakes. The locations of the recently active faults are known to most engineers, who appreciate also the risks of damage not only from the shocks themselves but also from secondary effects, especially landslides. Perhaps not so well appreciated is the difficulty of 'zoning' the country rationally in terms of degree of earthquake risk (Chapter Twelve).

Mention of engineering hazards connected with earthquakes brings to mind displacements of the ground directly due to movement on active faults. However, in most cases the direct cause of ground movements causing engineering problems is simply gravity aided by a tendency towards instability in the soil or the rocks beneath the structure concerned. It is reasonable to guess that engineering geologists nowadays spend more of their time in checking the geological factors concerned with stability at the site and its surroundings which might affect the project or be affected by it afterwards than in any other single line of investigation. To take just one typical example, the building of a hydroelectric dam; not only will the finished dam and impounded lake affect loadings and other stresses in the surrounding rocks, but the raised water level (especially seasonal fluctuations of level) will also have consequences affecting the stability of land slopes above the lake shoreline as well as those below it. Slope-stability has become big business for engineering geology in New Zealand. (See also 'Geology and the Environment', p. 387.)

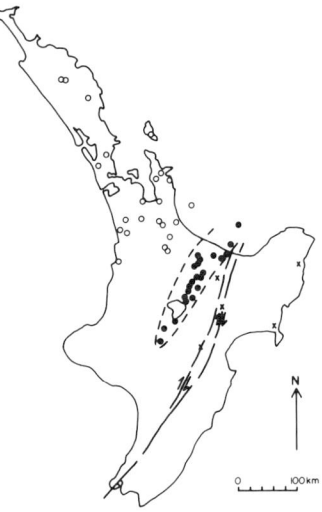

Fig. 13.14
Locations of North Island thermal springs and steam vents in relation to major active faults. Open circles: relatively low-temperature waters, believed to result from the presence of igneous magmas at moderate depths in the crust, but not related to recent volcanic action. Closed circles: volcanic springs, geysers, fumaroles, in the Rotorua-Taupo volcanic zone. Crosses: Non-volcanic warm springs, due to penetration by surface waters to such a depth as to encounter warmth due to the normal geothermal gradient. (After Healy, Grindley and Williams)

Energy from the Interior

The subject of geothermal heat was introduced earlier (Chapter Five page 119). We will now consider it again, as an energy resource for human needs.

The temperature below ground rises at an average rate of 30°C per kilometre of depth (roughly 1°F per 110 ft) due to the normal outflow of the earth's internal heat by conduction. The ratio of temperature-rise to increasing depth is the 'geothermal gradient', and it varies from place to place around the world. In regions where the crust is mobile and has recently been depressed to accommodate thick, young sediments the gradient is lower than the average, which for practical purposes means that it is necessary to penetrate more deeply to gain a given amount of temperature rise. Conversely, where mountain chains have lately been elevated the geothermal gradient will be steeper.

So far, this is independent of recent igneous action. In regions where magma has been injected into or erupted through rocks in the upper levels of the crust at temperatures of 1000°C or higher the gradient is steeper still, and in areas of present-day volcanic activity or where hot springs and fumaroles are numerous it can be many times higher than the average. Geothermal energy is thus brought within reach of human exploitation.

The heat outflow in the Rotorua-Taupo Volcanic Zone has been found to be as much as 500 times the average. It is inferred that a live body of magma exists not many kilometres below, but whether it is the same one as that which supplied the 1886 Tarawera eruption and recent minor outbursts from the Tongariro volcanoes is not known. Unusually large amounts of heat are reaching the surface, not only by thermal conduction through rock, but also by the agency of ascending water and volatile products emanating from a cooling and crystallising magma below, and rising along fault fissures or through zones of fractured rock. The very high temperature of gases in some fumaroles is consistent with this view.

Hot springs and geysers result from the transfer of magmatic heat to ground water that has percolated down from the surface. Surface water that has been able to penetrate deeply into the crust will be under high enough pressure to remain liquid well above normal boiling-point. Because of the high 'specific heat' of water (of all known substances, water requires the greatest input of heat energy to raise its temperature by a given amount) these deeply circulating water systems represent a great quantity of stored heat. Upon returning to the surface and lower confining pressures, the water boils and generates geysers, boiling springs and steam fumaroles which discharge large amounts of heat into surface waters and the atmosphere.

By drilling into the ground, superheated waters above boiling point at atmospheric pressure can be tapped and brought under control to the surface where the superheat can be transferred through heat-exchange equipment into boilers to raise steam for electricity generation or industrial power. This is the basis upon which the Wairakei geothermal power plant works. Experiments have shown that as the draw-off has increased, so has the depth at which superheated waters turn to steam. As a side effect, silica, calcite and other minerals are released from solution, clogging the wells and pipes and reducing the permeability of the hot-water aquifers. Obviously, the whole system needs to be managed carefully if it is to last for many years.

Besides at Wairakei, geothermal energy is harnessed also at Kawerau, where steam bores supplement the power from conventional boilers to operate the paper mill. Another geothermal steam field has been proved to exist beneath Waiotapu, and it is reasonably certain that further potential areas may be found in the Rotorua-Taupo Volcanic Zone. Regional geological mapping, assisted by geophysics and exploratory drilling have shown that the important aquifers at Wairakei and Waiotapu are pumice breccia beds and sandstones made up of volcanic debris under an impermeable capping of lake silts, all of late Pleistocene age, and in places include fractured andesite below.

From observations during twenty-five years of operation at Wairakei, and from measurements at other fields as well, it seems probable that an output of energy equivalent to at least a few hundred megawatts can be maintained indefinitely provided all fields are managed so as to avoid the 'quenching' of superheated steam by cold, surface waters which would occur if the rate of draw-off were too high.

In terms of return for capital investment, if one counts in all the developmental costs of a long, almost unique experiment (the Italian, Icelandic and Californian geothermal steam fields differ from ours in some respects) the whole operation may not compare so well with conventional forms of energy generation on the same scale. But the ultimate energy source is cost free, and future developments will not have to pay again for past research. It would be well to remember, though, that our geothermal energy fields are subject to an ever-present possibility of damage by volcanic explosions which, even if far less severe than Tarawera 1886, could be sufficiently destructive to put the installations out of action, temporarily or permanently.

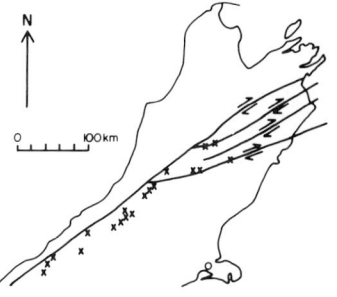

Fig. 13.15
Locations of South Island thermal springs in relation to major active faults and the Pliocene volcanoes of Banks Peninsula. (Refer to Fig. 13.14 for meanings of symbols.) (After Grindley and Williams)

Geology and the Environment

'Environment' means different things to different people. Lately the term has been used so loosely and with so much emotion, by dedicated groups, as almost to have lost its original scientific applications. Basically, the word means 'surrounding circumstances', with no implications concerning quality or value. As once used almost exclusively in a biological context it summed up all the chemical and physical factors that bear upon the organic world. Later it was extended into other realms, for instance by paleontologists concerned with the ecology of ancient life, and by petrologists to embrace the physical and chemical conditions affecting the composition, the texture and the structure of all kinds of rock at the time of their formation and afterwards. Astronomers, atomic physicists, geographers, sociologists, economists, politicians—for their own purposes all have adapted the term, in its general sense of a summation of surrounding circumstances. The emotionalism that now invades all discussions of the effects of man's activities upon the biological environment is generally

unhelpful to the worthy cause of preventing the unnecessary or avoidable environmental damage which may result from neglect, short-sightedness or commercial greed.

Long before 'environmental geology' became a recognised branch of the subject, geologists were already involved in many questions arising from human exploitation of the earth, including those from historic mining operations and from ancient forest clearance. Only since the first quarter of this century have the operations been on so large a scale as to induce possibly irreversible changes which ultimately could be

Fig. 13.16
Geothermal steam power development at Wairakei. The complex system of pipelines serves to collect the steam from many bores and to feed it to the heat exchangers and electricity generating plant. (Photo: N.Z. Geological Survey)

global in their effects. A well-known American 'environmentalist', Senator Edmund Muskie, dates the beginning of urgent public awareness of humanly caused environmental deterioration from the time of the first major catastrophe due to an offshore oil well blow-out. This happened at Santa Barbara, California, in 1969 and, after a bitter political and legal fight, it resulted in some safeguards against further similar disasters. Many countries besides the United States now have laws to protect or compensate citizens where there are hazards from mining, oil-drilling, engineering works, deforestation or any other kind of earth exploitation.

Environmental impacts in which geology is involved include water pollution (above or below ground) from dumping of industrial wastes, sewage and city refuse; subsidences due directly to withdrawal of underground water (as at Mexico City), or of oil and gas (as at Long Beach, California) or of coal (as in most coalmining regions, notably South Staffordshire, England, and in Southland, New Zealand); earthquake swarms triggered by injecting chemical waste fluids under pressure into deep wells (as at Denver, Colorado); coastal erosion due to interference by man-made structures with longshore currents (as at Timaru, New Zealand); landslides and slumping due to overloading of unstable slopes with dumped mine wastes or other rock debris (as at Aberfan, Wales, and some New Zealand coalfields); landslides, slumps and gully erosion due to engineering and urban developments, especially from poor disposal of rainfall run-off, from leaky water mains, sewers, swimming pools, etc. (as in San Francisco, Anchorage, Alaska, and the Port Hills, Christchurch); and accelerated erosion from unfortunate agricultural and pastoral practices (a world-wide problem).

Environmental geology now seems to cover also the advice that sometimes can be given to authorities and individuals regarding hazards beyond human control, such as eruptions and earthquakes. Accurate predictions of when an earthquake is going to happen, or a volcanic outburst for that matter, are not yet possible, and may never be, but there is much interest in the kind of general advance warnings that might be obtained by very precisely monitoring displacements of the ground in seismically active regions, and around active volcanoes, in the latter case along with routine observation of ground and spring water temperatures.

Geological aspects are nowadays nearly always considered in the preparation of the environmental impact reports which must, under law, precede the granting of permits by public authorities for any kind of proposed activity that might affect the environment. In many cases the complete answer cannot be given without unacceptably long investigations, but even a short-term appraisal is better than none, if well supported by theoretical considerations and past experience in comparable situations.

Geophysical Exploration

This book would be incomplete without some reference to the important part played in many geological investigations by persons whose basic training is in physics but whose interest and profession is in applying the principles and techniques of physical science to the exploration of the earth's crust and interior. These are the geophysicists.

Geophysics, like most other branches of science, has so-called 'pure' and 'applied' aspects, and as usual the distinction is vague because there is a great deal of feed-back in both directions. The main object in both cases is to discover, by means of physical measurements made from or near the surface, information about sub-surface conditions. In exploration geophysics, with which we are chiefly concerned here, the results are put to some immediate practical use.

Geophysical exploration is sometimes described as 'remote sensing' of earth environments. Most methods depend either upon the way in which contrasting physical properties (e.g. density, electrical conductivity, magnetic effects) in adjacent rock masses below ground affect 'signals' sent down in the form of mechanical vibrations (sound waves or artificial mini-earthquake shocks), electric currents, etc., or upon the way they affect the earth's gravitational or magnetic fields locally. Other methods detect spontaneous 'signals' from radioactive components of the rocks.

The practical applications are many. For more than half a century geophysics has played a vital role in the search for mineral deposits, oil and gas fields and underground water sources. For nearly as long, it has been used in engineering investigations for dams and other large works to extend observations beyond what can be seen at the surface or tested by drilling. In petroleum search, geophysical methods are employed to measure the distance from the surface down to a particular stratum at a number of points along a surveyed line or 'traverse', from which information, in combination with surface geological mapping and exploratory drilling, it is possible to map structural features (anticlines, faults, etc.) in the search for favourable deep-drilling sites. Continual close co-operation is needed between geophysicists, geologists and drillers.

Geophysical exploration was introduced to New Zealand in 1927 by private companies concerned in oil exploration. The Department of Scientific and Industrial Research was then headed by Dr E. Marsden (later, Sir Ernest), a physicist who was quick to see opportunities for geophysical work to direct the efforts of state-subsidised gold prospectors in the depression years. He appointed N. Modriniak and a staff of young enthusiasts to set up a Geophysical Survey in the early 1930s. All of the main methods were tried and some new techniques are said to have been pioneered here.

The Geophysical Survey became involved in the search for gold by indicating the depths of auriferous gravels down to the 'bottom' where payable quantities of gold might be found, especially in old, hidden gravel-filled gutters in the schist terrains of Otago. By measuring variations in the electrical conductivity of the ground, mineralised fault-zones were traced across areas where rock did not outcrop in some former quartz-mining regions of Central Otago, Reefton and the Hauraki district. Oil exploration was assisted by tracing rock-folds in Cretaceous and Tertiary strata in the eastern North Island and in Westland. Probably the earliest application to engineering questions in New Zealand was in Otago, to determine the depth of weak, crushed rock above the solid schist through which a tunnel for the Waipori power scheme was to be driven. Later, the Survey was heavily involved in the investigations of large dam sites throughout the country.

In 1951 the Geophysical Survey became an independent division of the Department of Scientific and Industrial Research. Private companies engaged in mineral and petroleum search have employed their own geophysical teams as well.

Some Basic Geologic Structures

The simplest of all geologic structures is the layering ('bedding' or 'stratification') developed in most sedimentary and some igneous rocks at the time of their formation. Since in the majority of cases we can take it for granted that layering was originally horizontal or almost so, any marked departure from horizontality suggests that the rocks subsequently may have been tilted, folded or in other ways deformed, and gives some indication of the kind and the amount of deformation. From such information we can build up a picture of past tectonic events that have affected the region. It is important, though, to be aware of the situations in which the stratification is likely to have been inclined from the horizontal initially, e.g. the sloping layers of sediment deposited on the submerged, advancing front of a stream delta. Layered metamorphic rocks such as foliated schist and gneiss and the cleavage surfaces of slate also provide reference surfaces which will show up any further deformation of a later date than the metamorphism. Besides tilting and folding, the nature and amount of dislocation of rocks by faulting can often be partly or completely worked out where stratification is present, but rarely where it is not.

The diagrams and accompanying text which follow illustrate (in a grossly simplified way) the more common kinds of deformed structures in layered rocks. It is always necessary to remember that what is seen in a roadside bank or quarry face is only a two-dimensional picture, whereas rock structures are three-dimensional. One has to learn to visualise them 'in the solid', to see into the rock as if with X-ray eyesight.

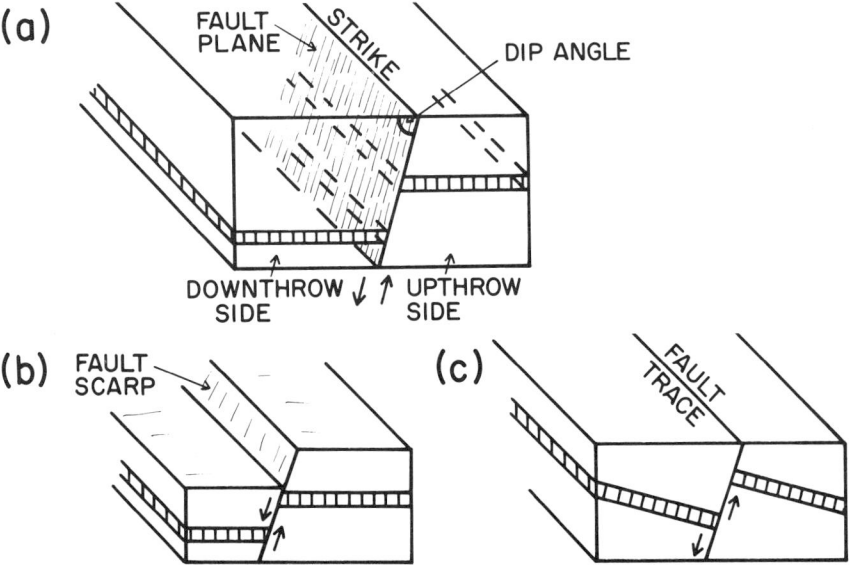

Cartoons of Simple Fault Description

Fault geometry

(a) The attitude of a fault plane is described in terms of 'dip' and 'strike', as with stratification. 'Upthrow' and 'downthrow' are relative terms; it is seldom certain whether either side remained stationary.

(b) A 'fault scarp' means that the ground surface has been displaced. (Refer to Chapter Eleven, p. 309.)

(c) 'Fault trace'; strictly, the intersection of a fault plane with the ground surface; also applied to line along which ground surface has been offset by recent fault movement.

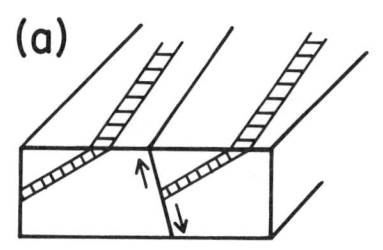 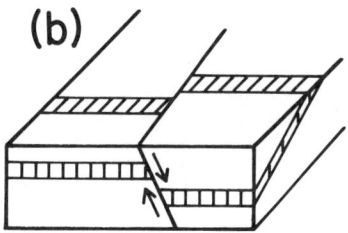 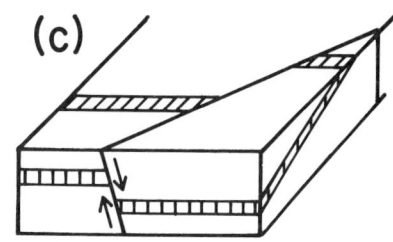

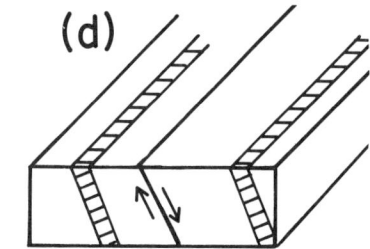

Strike of fault relative to attitude of strata
(a) 'Strike-fault'; parallel with strike of strata.
(b) 'Dip-fault'; parallel with direction of dip of strata.
(c) 'Oblique-fault'; parallel with neither strike nor dip direction of strata.
(d) 'Bedding-fault'; fault plane lies parallel with bedding of strata.

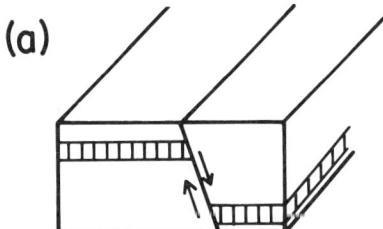 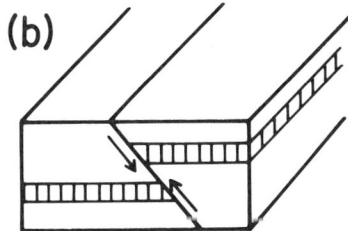

Dip of fault relative to downthrown side
(a) 'Normal fault'; fault plane dips towards relatively downthrown side.
(b) 'Reverse fault'; fault plane dips towards relatively upthrown side.

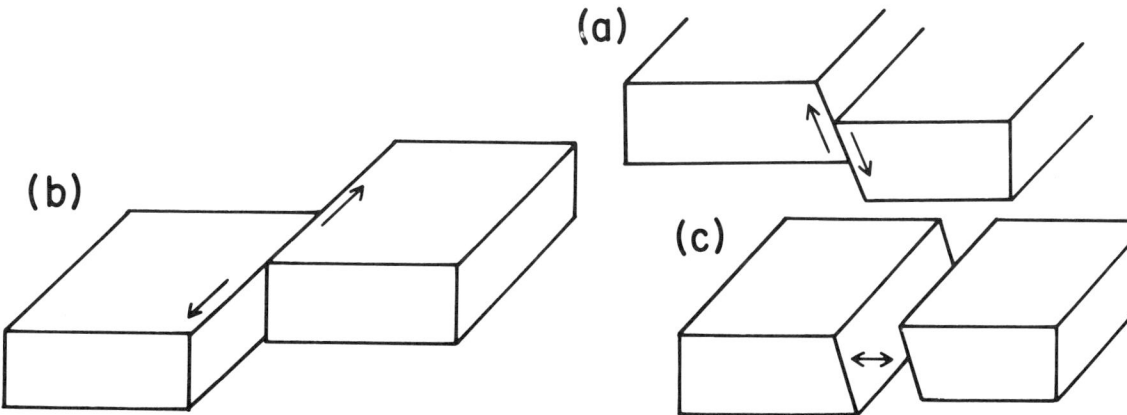

Apparent direction of movement ('slip') relative to attitude of fault plane
(a) 'Dip-slip'; movement parallel with dip of fault plane.
(b) 'Strike-slip'; movement parallel with strike of fault plane.
(c) 'Oblique-slip'; movement oblique to both dip and strike directions on fault plane.
N.B. The true direction of relative movement is often not determinable.

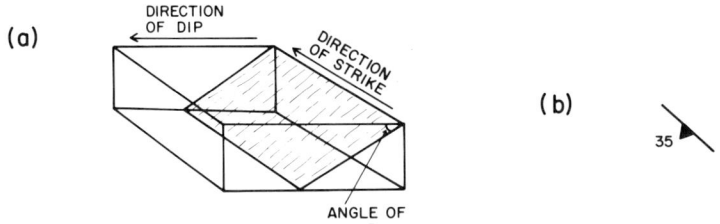

Dip and Strike

(a) The 'strike' of stratification, or of any other planar surface in rock, is the direction of its line of intersection with the horizontal. The 'dip' direction is that in which the surface slopes downwards; the 'angle' of dip is measured from the horizontal.

(b) One of the simplest symbols (there are others) for showing on a geological map the attitude of layered rock, as observed at the place indicated. The line shows the strike direction; the apex of the triangle the dip direction, and the angle of dip is given in degrees from the horizontal. More complicated symbols will be found in map legends.

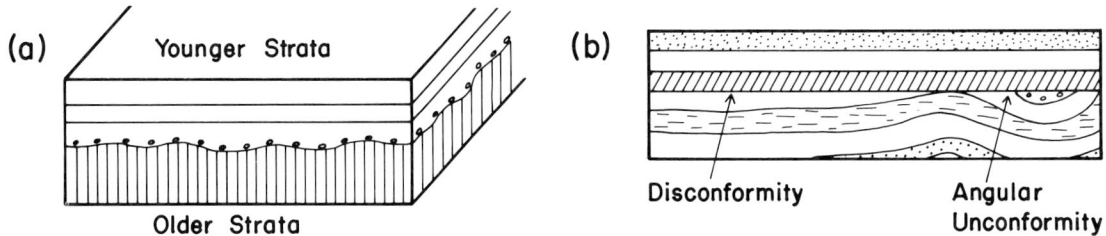

Unconformity and Disconformity

(a) In the wider sense, 'unconformity' means that an interval of time, unrepresented by deposits at that place, intervened between the formation of two successive sets of beds.

(b) In the structural sense, it signifies that an older set of beds has been tilted or otherwise deformed, eroded, and younger strata deposits then laid down, truncating the older stratification ('angular unconformity'). 'Disconformity' signifies that a time-gap can be demonstrated (e.g. by difference in the ages of fossils above and below it) although the strata show no difference of attitude.

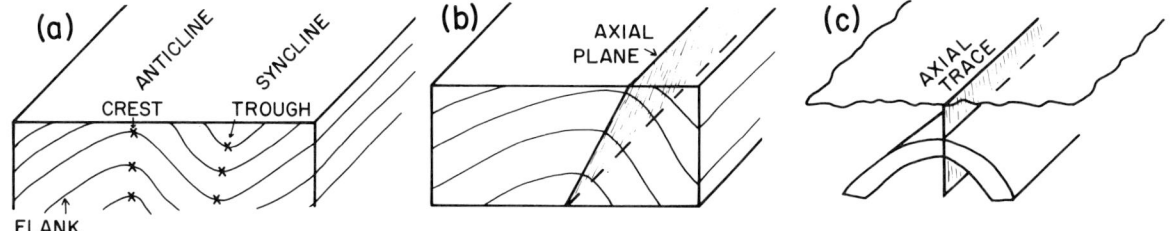

Geometry of Folds

(a) 'Anticline' (upfold); 'syncline' (downfold). Terms 'crest' and 'trough' apply respectively to highest and lowest points on each stratification plane; between them, the 'flanks' or 'limbs'.

(b) The 'axial plane' is an imaginary surface dividing the fold symmetrically.

(c) The intersection of the axial plane with the ground surface is the 'axial trace'. If the axial plane is approximately vertical the axial trace coincides with the 'crest' (or 'trough') 'line', often loosely called the 'axis', but strictly, this means something else.

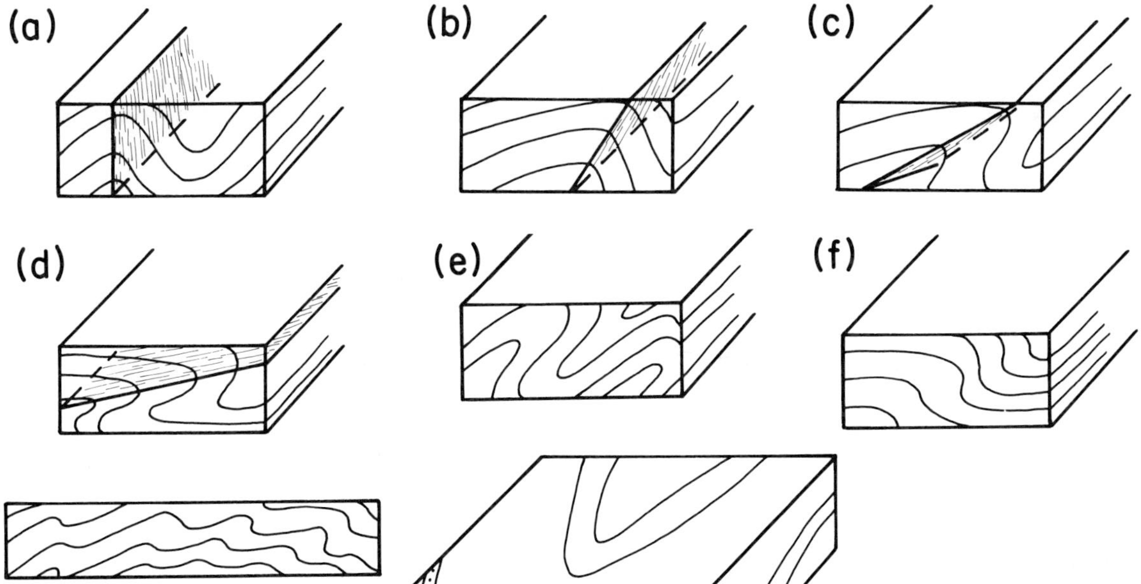

(a)

(b)

(c)

(d)

(e)

(f)

(g)

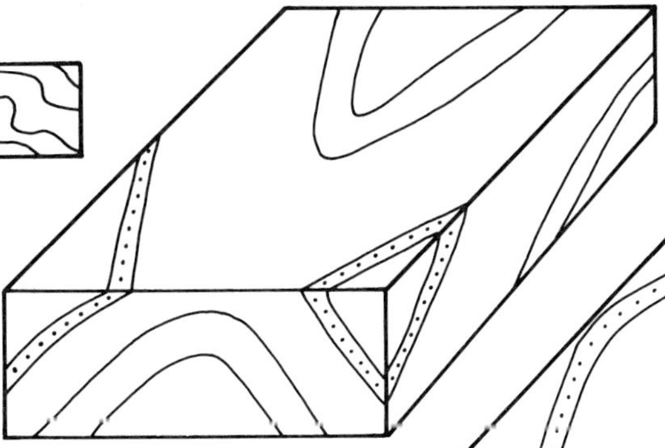

(h)

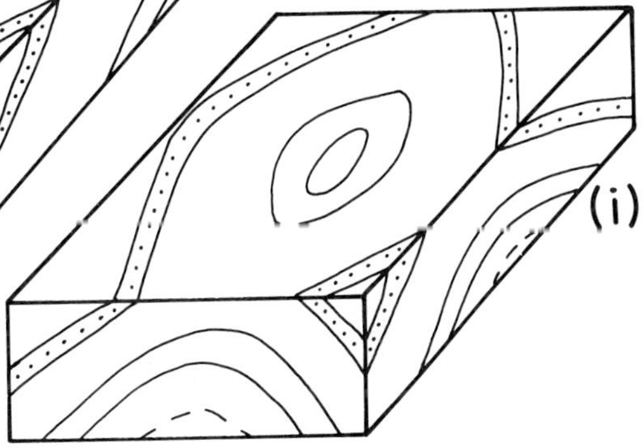

(i)

Types of Folds

(a) 'Symmetrical' or 'erect' folds (axial plane vertical).

(b) 'Asymmetrical' folds.

(c) 'Overturned' folds (one limb is inverted).

(d) 'Recumbent' folds (limbs are approximately horizontal; literally, folds 'lying down').

(e) 'Isoclinal' folds (both limbs dip at approximately the same angle; literally, 'equal slope').

(f) 'Monoclinal' fold (a downwards flexure in generally horizontal beds; literally, 'one slope').

(g) 'Anticlinorium'; a composite anticline; smaller folds superposed upon a larger one.

(h) 'Plunging fold'; direction of plunge, in this case towards the front of the diagram.

(i) 'Dome' structure; the dip of the strata everywhere is outwards, away from the crest of the dome. In 'basin' structure the dip should be inwards towards the bottom of the structure.

Fracture Cleavage

One of several kinds of rock cleavage, usually most obvious where highly compacted strata of varying rigidity and strength have been folded. The cleavage develops noticeably in the weaker (usually the finer-grained) layers, and tends to be parallel with the axial plane of the fold.

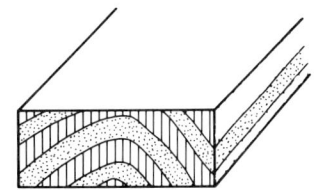

Some Characteristic New Zealand Fossils

As explained in the Foreword, this Appendix is not intended as a definitive outline of New Zealand paleontology, but has instead the limited objective of giving the reader some idea of what the more important groups of fossils characteristic of our rocks of various ages actually look like. But let me at once say that the accompanying sketches depict specimens that have been 'prepared' from collected material. Specialised techniques and skills have been used to extract the original organic material safely from the rock, or in many cases to make a cast or mould from impressions left in rock by a fossil the substance of which is no longer present. Some sketches give an impression of the supposed appearance of extinct creatures, of which only fragmental remains have survived as fossils.

A large number of man-hours of work is usually needed to prepare each specimen for study and comparison, and for ultimate storage in reference collections. Only the hard parts of most plants and animals are usually found fossilised, for example bone, limey or pearly shell, chitin, lignin, etc. Moreover, the original substance often has been changed chemically, having been replaced by another mineral substance that was more stable in the physico-chemical environment of the containing rock during its history after the fossil became embedded in the original sediment. Thus the aragonite of which many bivalve shells are made has usually been converted to the more stable mineral form of calcium carbonate, i.e. calcite. Fossil substance may also have been replaced by silica, pyrite, glauconite, limonite, etc. while still retaining more or less perfectly the original shape of the fossil hard parts.

Fossils embedded in finer-grained, compressible sediment are usually to some extent recognisably distorted as a result of compaction. Likewise, fossil shapes become distorted in rocks that have been deformed in the initial stages of metamorphism, and indeed enough is sometimes known about the true original form of fossils from undeformed rock to enable us to use the change of shape as a guide to the amount and kind of mass distortion the rock has suffered.

Very often the fragility of a fossil is too great to permit of its being freed completely from the matrix, even after the application of special cementing treatments. In such cases a number of specimens, partially exposed by removal of as much of the enclosing rock as can be achieved with safety, have to be considered together to give a complete picture of the fossil from all aspects.

The diagram depicts mainly examples of the larger fossil forms (macrofossils), from the larger vertebrates down to those easily seen by the unaided eye or with the help of a hand lens. The microfossils, which are so small as to require microscope study, are represented here only by two examples of the marine protozoan foraminifera, by some minute crustacea and by one of the distinctively formed needles ('spicules', usually of silica in fossils) which make up the skeletal framework of sponges. Incidentally, though shown here only as Cambrian fossils, sponge spicules occur in strata of all ages from Cambrian onwards. By no means all the useful groups of fossil micro-organisms could be represented in a diagram of this kind; perhaps the most noteworthy omission is that of the plant microfossils—mainly pollen grains and the spores of the lower plant families—which have become extremely important in New Zealand stratigraphy, especially of the Pleistocene and Tertiary strata and in connection with ancient climates.

Practically every major group of fossils representing sea life has been found in New Zealand strata of various ages, but, as might be expected, much of the dry-land and fresh-water life which evolved from Cretaceous times onwards, i.e. after the New Zealand region had become isolated from the other continental regions, fails to appear as fossils in our rocks. On the other hand, we have evolved some of our own endemic forms in isolation; our flightless or nearly flightless birds for example—whereas we were separated too early for any land mammals (other than the bat) to reach us before man brought them here.

CENOZOIC / **TERTIARY**	**RECENT** 10,000 years	Lower river terraces and lower raised beaches	All of the species shown are now extinct, except *Notornis*
	PLEISTOCENE 1,500,000 years		1 Moa (*Dinornis*) 5 Goose (*Cnemiornis*) 2 Moa (*Pachyornis*) 6 *Notornis* 3 Eagle (*Harpagornis*) 7 Seal Jaw 4 N.Z. Swan
	PLIOCENE 5,000,000 years		1 Moa (*Dinornis*) 5 Foraminifer (*Bolivinita*) 2 Fan shell (*Pecten*) 6 Sand dollar 3 Oyster (*O. ingens*) 7 Leaf (*Knightia*) 4 Sea snail (*Struthiolaria*)
	MIOCENE 22,500,000 years		1 Bivalve (*Cucullaea*) 5 Foraminifer (*Orbulina*) 2 Sea snail (*Struthiolaria*) 6 Ostracod 3 Lamp shell (*Pachymagas*) 7 Shark's tooth 4 *Dentalium* 8 Coconut (*Cocos*) (*Carcharodon*)
	OLIGOCENE 36,000,000 years		1 Giant Penguin 5 Fan shell (*Janupecten*) 2 Whale tooth (*Kekenodon*) 6 Sea snail 3 Crab 7 Sea egg (*Bathytoma*) 4 Foraminifer (*Rotaliatina*) (*Hemipatagus*)
	EOCENE – PALEOCENE 65,000,000 years		1 *Monalaria* 5 Foraminifer (*Hantkenina*) 2 *Dicroloma* } Sea snails 6 Solitary coral 3 *Speightia* 4 Nautiloid (*Aturia*)
MESOZOIC	**CRETACEOUS** 143,000,000 years		1 Reptile (*Mauisaurus*) 5 *Inoceramus pacificus* 2 Belemnite (*Dimitobelus*) 5a Prismatic shell } Bivalves 3 Sea snail (*Conchothyra*) 6 Trigonia of 5 4 Ammonite (*Tainuia*) (*Iotrigonia*)
	JURASSIC 212,000,000 years		1 Ammonite (*Phylloceras*) 5 Brachiopod 2 Belemnite (*Belemnopsis*) 6 *Cladophlebis* } Leaves (*Kutchithyris*) 3 *Inoceramus* } 7 *Taeniopteris* } of 4 *Buchia* } Bivalves Extinct Plants
	TRIASSIC 247,000,000 years		1 Reptile (*Mixosaurus*) 5 Brachiopod (*Spiriferina*) 2 *Monotis* 6 Leaf (*Lingulfolium*) 3 *Halobia* } Bivalves 4 *Manticula*
PALEOZOIC	**PERMIAN** 289,000,000 years		1 *Atomodesma* 4 Productid 1a Prismatic shell of 1 5 *Neospirifer* 2 Sea Lily (Crinoid) 6 Solitary Coral 3 Pleurotomariid
	CARBONIFEROUS 367,000,000 years		1 *Ligonodina* 2 *Cavusgnathus* } Conodonts 3 *Gondolella* 4 *Hindeodella*
	DEVONIAN 415,000,000 years		1 Crustacean 4 Coral (*Cyathophyllum*) (Trilobite) 5 Crustacean 2 *Hipparionyx* } Brachiopods 3 Spiriferid
	SILURIAN 440,000,000 years		1 *Conchidium* 2 Acrospiriferinae } Brachiopods 3 Rhynchonellid 4 Rugose coral 5 Crinoid ossicles
	ORDOVICIAN 500,000,000 years		1 } Trilobites 4 Conodont (*Cordylodus*) 2 3 Crustacean 5 — 7 Graphtolites { Colonial animals, each point is an individual
	CAMBRIAN 570,000,000 years		1 Polymerid Trilobite 2 Agnostid Trilobite 6 Monoplacophoran 3 Sponge spicules 7 Conodont 4 — 5 Inarticulate Brachiopods
PRECAMBRIAN 4,000,000,000 years			Organic — Walled microfossils (Achritarchs)

Affinities between sea-dwelling fossil creatures and their near relatives or identical forms in rocks of other lands have been used to build up a picture of former seaways along which they could have migrated as adult or larval forms to and from the site of New Zealand. Barriers to migration were also inferred to explain absences of certain groups from New Zealand rocks; these barriers could be either land masses or climatic belts. In fact this kind of information played an important part in earlier theories of land and sea distribution in the geological past, of former arrangements of the climate zones, and hence of inferred past positions of the poles. Owing to the long isolation of New Zealand and our distinctive present-day flora and fauna, our paleontological relationships with other parts of the world have always been of considerable interest. The great advance in global paleogeography which in recent years followed the paleomagnetic studies of rocks on land and beneath the sea floor has made it necessary to revise some of the older conceptions of land and sea connections with the New Zealand region.

Some readers are sure to have recognised the accompanying diagram as very similar to one they have seen before. It was developed from a pictorial chart of the New Zealand fossil succession drawn some forty years ago by the late Dr J. Marwick for a publication by the radio broadcasting authority of the times, and it re-appeared with revision in widely circulated reprints of an article by C. A. Fleming in the Victoria University of Wellington scientific journal *Tuatara* in 1949 and again in 1962. The present version is the result of a further revision.*

It should be clear enough to readers that the fossils are depicted on a wide range of scales; compare, for example, the moa reconstructions with the fossil leaves. A few less obvious cases deserve special comment. Thus, the foraminifera, ostracods, some of the crustaceans (in the Silurian and Devonian, for example), and the sponge spicules are microscopic objects. Others, including the graptolites, are reproduced a little smaller than actual size. It will be noted, too, that the existence of strata of Carboniferous age in New Zealand has been demonstrated so far only on the evidence of the enigmatic microfossils known as conodonts. Believed possibly to have had some kind of dental function, they have not been identified as belonging to any known group of fossil animals. Perhaps they were the only hard parts of some worm-like creature in Paleozoic and early Mesozoic oceans, but their composition (calcium phosphate) suggests an affinity with the vertebrates. Many people will already be familiar with the belemnite 'guards' (Jurassic, Cretaceous) as being analogous with the 'bones' of cuttlefish and squids. *Dentalium*, a common fossil in Tertiary marine strata in New Zealand, is a relative of the molluscs which lived a secluded life in its open-ended curved tubular shell in the sea-bottom muds.

* *Acknowledgement.* I am particularly indebted to Mr Ian Keyes of the N.Z. Geological Survey, Lower Hutt, for providing this updated version of the diagram, which takes into account such important advances as the recognition of Silurian and Carboniferous fossils in New Zealand rocks. It has been redrawn by the Paleontological Section of the Geological Survey and is reproduced with permission.

Notes on the Geology of the Outlying Islands

It was emphasised in Chapter Two that the North Island, the South Island and Stewart Island are merely the largest areas of the New Zealand subcontinental platform which at present stand above sea level. Next in order of size are the Chatham Islands, followed by Auckland and Campbell islands, the remainder being mainly collections of relatively inaccessible small islands and barren rocks. No doubt because its administrative ties are with Australia, Norfolk Island is never considered part of the New Zealand archipelago, yet its geological affinities are with this side of the Tasman Basin. Again because of administrative responsibilities, we tend to think of the Kermadec Group as included in our outlying islands though they are not much nearer than Macquarrie Island in which we have had little interest.

The *Chatham Islands* lie about 850 km east of Banks Peninsula. Besides Chatham Island (originally Rekohu; Wharekauri after the Maori invasion of 1835) and Pitt Island which are permanently inhabited, there are numerous smaller islands, rocks and reefs. Geological observations on the Chathams date from a visit by the New Zealand Company's naturalist E. Dieffenbach in 1842. Most of the geological information has to come from coastal cliff exposures and the rest from isolated hills protruding through the widespread turf and peat blanket. Chatham Island is about 56 km across at the northern end and about 50 km from north to south but much of the intervening area is taken up by lagoons, broad bays and swamps.

Metamorphic rocks similar to the Haast Schists form the basement and are exposed in the northern part of the island. They are generally regarded as a continuation of the schist belt of eastern Otago, having originated also in the New Zealand Geosyncline. Presumably after an uplift corresponding to the Rangitata Orogeny in New Zealand, the metamorphic rocks were worn down to a surface of very gentle relief, upon which non-marine conglomerate, sands and lignite were deposited during the Cretaceous Period. The land was then invaded by the sea in a transgression comparable with that affecting the mainland in the same interval of time. Coarse sandstone, limestone and greensand beds of early Eocene age overlying the lignite beds contain small amounts of volcanic material, but there is no suggestion that eruptions were then occurring near at hand. Local submarine eruptions were soon to follow, however, and large areas of southern Chatham Island and Pitt Island are covered by basaltic tuff beds of middle to late Eocene age, interfingering with limestone beds that must have formed as shallow-water reefs. The early Tertiary strata are very little affected by later crustal movements, tilting by more than 5° being rare, and yet there is a remarkable break in the succession of strata. In contrast with the mainland of New Zealand where sedimentary and volcanic deposits of mid-Tertiary age are widespread and in places very thick, there are no Oligocene or Miocene rocks in the Chathams. Perhaps they were deposited and then stripped away essentially by wave erosion before the final episode of late Tertiary and Pleistocene deposition began.

The early Pliocene to early Pleistocene sequence of beds begins with shallow-water limestones succeeded by thick volcanic tuffs of basaltic and other mafic compositions. The latter part of the Pleistocene, after the eruptions had ended, is marked by dune-sand, dune-sand deposits, fossiliferous near-shore marine sands, and peat. Some of the sand deposits take the form of coastal terrace and raised beach deposits; these are supposed to have accumulated while sea level was relatively high during one of the later interglacial periods.

Moorland peat deposits blanket much of the gently rolling upland areas of the Chathams and they are a common feature of all the larger southern outlying islands. Chatham peats have been considered more than once as a possible source of hydrocarbons alternative to petroleum.

The geology of the Chatham Islands has been examined on several occasions since the time of Dieffenbach. A good geological map on a scale of 1 : 100,000 together with detailed descriptions of the deposits are presented in *N. Z.*

Geological Survey Bulletin 83, 'Geology of the Chatham Islands' by R. F. Hay, A. R. Mutch and W. A. Watters (1970).

Auckland Islands. The main Auckland Island is about 43 km long, north and south, and 26 km wide at the southern end. The group lies about 300 km south of Stewart Island, and after the small Snares group it forms the next land area of appreciable size near New Zealand. It includes Enderby Island at the northern end, Disappointment Island off the west coast and several other islets and rocks, the whole assemblage representing the eroded remains of a Pliocene volcano, a double basaltic lava dome in some ways resembling Banks Peninsula.

Carnley Harbour at the southern end of Auckland Island is a drowned valley system eroded from the heart of the southern volcano, while the northern one (Ross Volcano) centred on Disappointment Island has been very largely destroyed by erosion. A basement of plutonic rocks including granite and gabbro and intruded by trachyte dykes and sills is now exposed beneath the centre of Carnley Volcano, recalling the situation on Onawe Peninsula at the head of Akaroa Harbour. Trachyte lava and tuff beds overlie the granitic rocks, succeeded in turn by a conglomerate possibly of Cretaceous age comprising pebbles of all the underlying rock types, and by marine, fossil-bearing basaltic tuffs of Oligocene age. On top of these rises the basaltic dome of Carnley Volcano, built up of lava flows in which an older and a younger set are recognised. Despite erosion it still reaches a height of approximately 610 m. To the north, the narrow part of Auckland Island is made up of many layers of basaltic lava, tuff and volcanic breccia (agglomerate), being all that is left of the eastern flank of Ross Volcano. During the Pleistocene Period Auckland Island supported glaciers which moulded the valleys and laid down sequences of till and other glacial deposits which are seen particularly well on Enderby Island to the north.

Campbell Island. Lying about 160 km south-east of Auckland Island, Campbell Island has a more complete geological record. Though only about 50 km in circumference it has a basement of gabbro rocks (and some evidence that schist is not far away, though nowhere exposed) planed by an ancient erosion surface that is suspected of being part of the Late Cretaceous Peneplain of New Zealand. It carries a veneer of sediments analogous with the Quartzose Coal Measures of the mainland, over which the sea transgressed in the Paleocene Period. Marine oozes continued to accumulate until the Miocene, but there could be a gap in the sedimentary record prior to the beginning of basaltic volcanic eruptions late in that period. Volcanic breccias from under-sea eruptions then raised the area up to sea level, and during Pliocene times repeated lava eruptions built up a lofty volcano, the remains of which still rise to heights of more than 600 m. The present form of Perseverance Harbour and Northeast Harbour, deep bays on the east coast, are attributed to Pleistocene glaciation, though glacial deposits are not as prominent as in the Auckland Islands.

Antipodes Island lies almost at the extreme eastern edge of the New Zealand Platform, the slopes of which here descend directly to the floor of the South Pacific Basin. It is the remains of another basaltic lava and scoria volcano or group of volcanoes rising now to heights of about 400 m above sea level. As with all the sub-antarctic islands exposed to the weather and storm waves of the 'Roaring Forties' (and 'Fifties'), vertical cliffs as much as 200 m in height almost encircle the island. The basalt lava and scoria have been described as having a 'recent' appearance. R. Speight remarked that they are unlikely to be able to withstand erosion in this exposed situation for very long, and it seems likely that the volcano dates from late in the Pleistocene Period. Supporting this, peat deposits in one locality have been 'coked' by contact with hot lava, giving rise to false rumours of coal deposits.

The Snares, so called because they lie hazardously across the sailing-ship course from Australia to Cape Horn, at about 100 km south from Stewart Island, are composed of granite. The largest island is about 6 km long and rises to heights of 120-170 m above the sea. Above the encircling cliffs, an undulating upper surface is blanketed with peat to depths of up to 3 m.

The *Bounty Islands*, a group of barren, storm-swept rocky islets composed of granite, also lie close to the extreme eastern edge of the New Zealand Platform.

The *Kermadec Group* is a line of islands on the inner edge of the Kermadec Trench about 1000 km NE of the Aucklands. They are of volcanic origin and Raoul Island (or Sunday Island, to give its other name), the largest of the group, is an active volcano which has erupted more than once in this century. Raoul has been described as a volcanic dome about 8 km across with a large crater or caldera open to the sea on one side. It consists of tuff and volcanic breccia beds having the composition of andesite, which shows that it belongs to the western or continental side of the Pacific boundary. Earthquakes are frequently felt, some of them severe.

Norfolk Island consists of a pile of basalt lava and ash rocks, but there is some hint of a hidden basement of granite or other plutonic rock from the presence, in fine tuffaceous rock on neighbouring Phillip Island, of quartz, orthoclase and plagioclase feldspar and hornblende grains. Norfolk Island rises to the sea surface from one of the submarine ridges continuing the north-westward trend of the north Auckland peninsula.

Sources. Apart from the Chatham Islands, which have been dealt with in recent years in a Geological Survey *Bulletin*, geological information about the outlying islands is scattered through many papers and reports. Although published as long ago as 1909 (by the Philosophical Institute of Canterbury), the massive, two-volume report of the 'Hinemoa' expedition of 1907, *The Subantarctic Islands of New Zealand*, remains the only convenient, comprehensive account of them. It is readily available in the larger libraries, and the geological descriptions by R. Speight, P. Marshall and A.M. Finlayson are reliable as far as they go.

Sources of Information about New Zealand Geology

A number of books and scientific journals have been mentioned in the foregoing text. These and other sources are brought together in this list of generally useful and reasonably accessible publications, many of which can be purchased; others may be consulted in the larger libraries.

Geological Maps

By far the most important source is the Geological Survey Division of the Department of Scientific and Industrial Research, Lower Hutt. The following list omits some earlier issues of maps which are now obsolete.

Geological Map of New Zealand, 1:2,000,000; one sheet; 1958

Geological Map of New Zealand, 1:1,000,000 (approx. 10 km to 1 cm); 2 sheets; 1972.

Quaternary Geology of New Zealand–North Island }
Quaternary Geology of New Zealand–South Island }   1:1,000,000; 1973

(Mainly shows the glacial, volcanic ash and other superficial deposits laid down since the beginning of the Pleistocene Period; specially prepared for the Ninth Congress of the International Union for Quaternary Research, Christchurch, 1973).

Geological Map of New Zealand, 1:250,000 (approx. 2.5 km to 1 cm) 27 sheets; 1959-1973.

Geological Map of New Zealand, 1:63,360 (approx. 640 m to 1 cm). These are mostly issued with the *Bulletin* series. A few accompanied the now extinct *D.S.I.R. Geological Memoir* series. Some have been published separately, with explanatory pamphlets. It is often hard to get hold of the older *Bulletin* maps now out of print because most of them were supplied loose in pockets at the back, and have long since been 'borrowed' permanently from library copies. Other geological maps on various scales have been issued by the Geological Survey from time to time, such as the 1:250,000 map of the Tucker Glacier area of Ross Dependency, Antarctica, in 1963, the memorial *W. N. Benson Map of Dunedin* (1:50,000), 1965, maps on scales of 1:25,000 and 1:15,840 with *Bulletins* describing the coalfields, and recently, separate 1:25,000 maps covering Westport and Cape Foulwind areas. Virtually all of the maps mentioned above represent by means of colour the distribution of formations and rocks of different geological ages, and most are accompanied by geological cross-sections. A large number of special maps, depicting special aspects such as structure (by means of structural contour lines), faults, estimation of coal reserves, etc., are included in Geological Survey publications. They are not usually available separately. Others accompany the publications of the New Zealand Soil Bureau and the New Zealand Oceanographic Institute. Finally, many geological maps in colour and in black-and-white and on a wide range of scales have appeared with geological papers in the scientific journals (see below).

Unfortunately, there is no general index to available geological maps, and it is often hard to find whether a map covering a particular area exists unless one has an extensive and up-to-date knowledge of the scientific literature.

Books

Apart from the the Geological Survey *Bulletins* described in a later paragraph, few books have been written exclusively on geological topics in New Zealand. Those written last century, such as Hochstetter's account of the geology in 1864 (translated into English by Fleming in 1959), Hutton and Ulrich's *Geology of Otago* (1875) and Haast's *Geology of Canterbury and Westland* (1879), though remarkable achievements for their times, and of considerable historical interest, have little relevance to modern New Zealand geology. The same may be said about the rival geologies of New Zealand by Park in 1910 and Marshall in 1912, the latter an excellent book in its day. Short summaries have appeared from time to time in booklet form, usually to mark some international scientific gathering in New Zealand or Australia, and regional summaries are included in books like *The Natural History of Canterbury* (ed. G. A. Knox; Reed, Wellington, 1969) and the National Parks handbooks. One of the best general summaries is in *An Encyclopedia of New Zealand*, Volume One, pp. 769-804 (Government Printer, Wellington, 1966). Another, issued as Geological Survey *Bulletin 66* in 1959, really a greatly extended legend for the 1:2,000,000 *Geological Map of New Zealand*, is now rather out of date as regards the names and ages of stratigraphic divisions. Also out of date, but to be replaced by a new edition being prepared in New Zealand, is the New Zealand volume of the *International Stratigraphic Lexicon* published in France in 1959. It is still an invaluable guide for those who need to know how, when and by whom stratigraphic units were defined.

The following are a few general books which are either still in print or readily available in libraries:

New Zealand Geomorphology (a collection of reprints of C. A. Cotton's early, classic papers), N.Z. Univ. Press, Wellington, 1955.

Earthquakes, G. A. Eiby; Muller, London, 1957; (a lot about New Zealand seismology).

City of Volcanoes, E. J. Searle; Paul, Hamilton, 1964; (Auckland volcanology and geology).

Economic Geology of New Zealand, G. J. Williams; Australian Institute of Mining and Metallurgy, Melbourne, 1965; 2nd edn. 1976.

Marwick's Illustrations of New Zealand Shells compiled by C. A. Fleming; Department of Scientific and Industrial Research, Wellington, 1966.

Geological Evolution of Australia and New Zealand, D. A. Brown, K. S. W. Campbell and K. A. W. Crook; Pergamon, Oxford, 1968.

Handbook of New Zealand Microfossils, N. de B. Hornibrook; D.S.I.R. Information Series No. 42, Wellington 1968.

The Geological Structure of New Zealand, J. T. Kingma; Wiley, London, 1974 (a remarkable book, probably too specialised for most readers, and very much one man's view).

Rugged Landscape, G. R. Stevens; Reed, Wellington, 1974 (geology of Wellington and Cook Strait region, mainly).

The Geological History of New Zealand and its Life, C. A. Fleming; Auckland University Press and Oxford University Press, 1979. (An expanded version of his articles in *Tuatara* referred to herein; a very useful companion volume to this book.)

Geology of New Zealand (2 vols), ed. R. P. Suggate; Geological Survey, 1978. A definitive treatise, keenly awaited.

Geological Survey Bulletins, etc.

The old Geological Survey under Hector produced between 1866 and 1892 a series of twenty-two volumes of *Reports of Geological Explorations*. They contain much sound descriptive writing and are historically important, but hardly count as a modern source of information. The present *Bulletin* series, started in 1906, has mainly dealt with regional descriptions but from time to time has included special topics such as national and regional limestone resources, geology of particular mining areas, petrology of serpentine rocks, greywacke and schist, geothermal power development, volcanic ash showers and recent fault movements, and with descriptive geology of parts of Ross Dependency and other territories of New Zealand interest in the South Pacific. With a few unfilled gaps, the serial numbering had reached 93 by 1977. Some may be puzzled by the leters 'n.s.' preceding the *Bulletin* numbers; librarians have insisted upon this to avoid the negligible risk of confusion with an earlier *Bulletin* series of Hector's Survey which ceased after No. 1, the present *Bulletins* being regarded, forever it seems, as a 'new series'!

The *Paleontological Bulletins* are strictly for the research scientist, dealing with taxonomic description, classification and naming of various groups of fossil organisms. The *Geological Memoirs* of the Department of Scientific and Industrial Research, of which nine were issued between 1928 and 1954, were smaller in scope than the Survey *Bulletins* and some dealt with specialised topics.

In addition to these regular serials, Geological Survey staff publish through a confusingly large number of other outlets. Among these are the *Bulletin* series of the D.S.I.R. (entirely separate from the Survey series and of different format), dealing with such varied topics as reports on the major earthquakes and the geology of New Zealand coal; and the *Information Series* of the D.S.I.R. which presents semi-popular descriptions like D. R. Gregg's very readable *Volcanoes of Tongariro National Park* (1960), Hornibrook's *Handbook of New Zealand Microfossils* (1968; listed above), and W. A. Sara's *Glaciers of Westland National Park* (1970). The several contributions by the Geological Survey to the National Resources survey of the 1960s and early 1970s include useful, regional accounts of the geology, aimed at non-geological readers. In the bibliographical shadow-land of 'semi-publication', there is a mimeographed (and bound) series of *N.Z.G.S. Reports*, mainly for the professional geologist.

Important recent additions to the above sources are *New Zealand Geological Survey Urban Series*, begun with *Map 1* and accompanying 'Geology of Nelson Urban Area' by M. R. Johnston (1979); and the first three of a series of inexpensive, paperback, local *Guidebooks*.

Other miscellaneous Geological Survey publications cannot be enumerated here.

Many of the earlier serial publications of the Geological Survey are out of print, but the only sure way to find out is to write to the Director, N.Z. Geological Survey, P.O. Box 30-368, Lower Hutt, asking for the latest list of available items. These may be purchased or ordered at the branch shops and bookseller agencies of the Government Printer in the larger towns.

Scientific Journals

The great bulk of New Zealand geological papers appears in either the *New Zealand Journal of Geology and Geophysics*, one of the successors (in 1958) to the former *New Zealand Journal of Science and Technology*, or in the former *Transactions* and present *Journal* series of the Royal Society of New Zealand. Geological papers have appeared in other New Zealand scientific serials, including the short-lived *Earth Science Journal* and the *New Zealand Science Review*. A substantial number also are published overseas. Non-subscribers must seek these publications in the scientific libraries of the universities and of the larger regional branches of the Royal Society of New Zealand.

Bibliographies

New Zealand geologists are prolific writers. The two modern editions of *Bibliography of New Zealand Geology* issued by the Geological Survey as *Bulletin 65* (1967) and *Bulletin 93* (1977) contain over 10,000 entries covering publications up to 1969. Alas, both of these list publications by authors' names only, and apart from subject indexes printed annually for certain journals (or upon completion of a volume) there is no easy way to trace the literature on a particular topic. Overseas abstracting services catch up with most New Zealand material eventually, but nevertheless the situation is quite a trial for the professional geologists, and hopeless for most amateurs. A recent addition to this kind of bibliographic source is an 'Author Index' to the *Transactions* and *Proceedings* of the Royal Society of New Zealand (formerly New Zealand Institute) from 1869 to 1971, published in 1978. An eventual subject index is promised.

Geological Survey Regional Offices

The Survey at present maintains district offices at Auckland (Otara), Rotorua, Nelson, Christchurch and Dunedin in addition to the Head Office in Lower Hutt.

Glossary of Rock and Mineral Terms

The purpose of this glossary is to summarise the meanings of rock and mineral names and a few other terms that are mentioned but not explained adequately (or at all) in the preceding pages. The subject index should also be consulted. It must be stressed that most of the definitions given here are very condensed and simplified, leaving out properties and distinctions that can be observed only with the aid of the microscope and other laboratory facilities. The glossary is not aimed at providing a key to the identification of rock and mineral specimens.

ACTINOLITE. One of the *amphibole* group (q.v.) of complex hydrous silicates of lime, magnesium and iron, usually occurring as pale green or white long needle-like crystals, often arranged in radiating clusters. Found in metamorphic rocks.

AGATE. A form of quartz, cryptocrystalline (q.v.), usually filling cavities in igneous rocks. Elegant colour bandings can make it a lapidarist's joy.

AGGLOMERATE. A coarse-textured aggregate of angular fragments of igneous rock embedded in a finer matrix of similar material, resulting from explosive eruptive activity.

ALBITE. The soda-rich end member of the *plagioclase* series (q.v.) of feldspar minerals, found mainly in pegmatite and in metamorphic gneisses and schists.

AMETHYST. One of the gem varieties of crystalline quartz, tinted mauve or purple by a trace of manganese impurity. Found in cavities in igneous rocks, especially andesites.

AMPHIBOLE. A group of complex, calcium, magnesium and iron hydrous silicates, usually occurring in long, glassy prismatic or needle-shaped crystals. The most familiar variety is the dark green or black *hornblende* found in mafic igneous rocks and in schists formed by metamorphism of basaltic material.

ANDALUSITE. An aluminium silicate mineral occurring usually as hard, squarish prismatic crystals of variable colour where shale has been altered by contact metamorphism.

ANDESITE. A volcanic rock of the mafic-intermediate family, composed of plagioclase feldspar and usually augite, hornblende or biotite; dark grey or black, dense or almost glassy matrix with phenocrysts usually of the dark minerals.

ANTHRACITE. A hard, greyish, metallic-looking coal of the highest rank (i.e. most altered from the original peat). In Canterbury anthracites, lignitic coal was altered by the heat of intruded basaltic sills and dykes.

ARAGONITE. One of the mineral forms of calcium carbonate, differing from the more usual *calcite* in crystal form (roughly rectangular cross-sections) and less easily cleaved. The pearly lining of most sea shells is aragonite, but in fossilisation it has usually been altered to calcite.

ARGILLITE. A hardened, fine-grained sedimentary rock. Many of the finer layers in the Torlesse Supergroup strata that have been called 'argillite' are more correctly siltstones.

AUGITE. The most common of the *pyroxene* group (q.v.) of complex, calcium, iron, magnesium and aluminium silicates; usually dark green or black, glassy prismatic crystals (also occurs in more stumpy shapes). Because crystals tend to cleave in directions about 90° apart, cleavage fragments are usually square in cross-section (unlike the otherwise similar hornblende). Found in most mafic igneous rocks.

AWARUITE. An unusual and rare metallic mineral, a natural alloy of nickel and iron (2:1) found in sands of coastal west Otago, derived from peridotite and serpentine rocks.

BASALT. The most voluminous of all lava rocks; the volcanic member of the mafic family. Fine-grained, dark grey or black matrix composed of plagioclase feldspar and pyroxene (commonly augite), often with magnetite and olivine. Visible phenocrysts usually augite or olivine.

BENTONITE. A light-coloured clay rock composed of members of the clay mineral groups, including *montmorillonite*, which commonly have high absorptive properties. Believed to be formed by decomposition under water of finely-divided volcanic ash.

BIOTITE. A flaky, dark-coloured member of the *mica* group of minerals, a hydrous silicate of chiefly iron and magnesium. Common in most igneous rocks, and in many metamorphic gneisses and schists.

BRECCIA. Compacted sedimentary rock composed of angular pieces of pre-existing rocks.

CAIRNGORM. Crystalline quartz, tinted yellow or brown by traces of iron and other impurities.

CALCITE. The common mineral form of calcium carbonate, recognised by distinctive crystal form (six-sided prisms) and especially by habit of cleavage, producing rhomb-shaped fragments with peculiar refractive properties. As a rock component, its presence is usually detected by effervescence in contact with cold, dilute hydrochloric acid.

'CEMENTS'. Local jargon in Westland for cemented quartz-pebble conglomerate beds, in the past crushed in stampers near Reefton to release detrital gold particles.

CHALCEDONY. A cryptocrystalline (q.v.) form of quartz, with a smooth, waxy appearance.

CHALCOPYRITE. The most common sulphide ore of copper, brassy like pyrite, but softer and lacking its distinctive crystal shape (q.v.).

CHERT. Rock composed of one of the cryptocrystalline forms of quartz; light-grey to black when unweathered, found as nodules or in bedded layers, usually with limestone, but sometimes also with geosynclinal volcanic sediments.

CHLORITE. A flaky, greenish magnesium-iron-aluminium hydrous silicate mineral (or mixture of minerals); a common alteration product from pyroxene and biotite in early stages of weathering, and a typical product of low-grade metamorphism. Thus: 'chlorite schist'.

CHROMITE. An oxide of iron and chromium, a black, shiny mineral occurring either as small, perfect crystals, or massive. Found widely in serpentinites and ultramafic rocks.

CHRYSOTILE. A fibrous form of *serpentine* (q.v.); the chief natural source of industrial asbestos.

CLASTIC. A term applied to rock or sediment components to indicate that they are themselves fragments of some pre-existing rock. An individual constituent of a sediment or a sedimentary rock produced by physical disintegration of the parent rock is called a *clast*.

CONGLOMERATE. A compacted sedimentary rock composed of mainly rounded fragments of pre-existing rock.

CRYPTOCRYSTALLINE. Applied to a mineral which is so finely crystalline as to appear non-crystalline to the naked eye, or even under the microscope (e.g. chalcedony is a cryptocrystalline variety of quartz).

DACITE. A variety of *andesite* containing some quartz, usually in the matrix.

DIORITE. A coarse-textured igneous rock composed of plagioclase and hornblende; sometimes biotite or augite. Classed as the plutonic equivalent of andesite.

DOLERITE. As applied in New Zealand, it means a rock of basaltic composition in which the texture of the matrix is coarse enough for individual plagioclase crystal needles to be visible to the eye; usually occurring as sills or dykes.

DOLOMITE. A mineral resembling calcite, but composed of calcium-magnesium carbonate. (Compared with calcite: it effervesces only with *warm* dilute hydrochloric acid.) The name also applied to limestone rock composed partly or wholly of dolomite.

DUNITE. A variety of *peridotite* (q.v.) composed almost entirely of olivine, with pyroxene and chromite as accessory minerals, not always visible.

EPIDOTE. One of a group of complex, hydrous silicates of calcium and aluminium which occurs in many types of metamorphic rocks, in the body of the rock as crystals and in fine veinlets. Often a distinctive olive-green colour.

FELDSPAR. A very important family of aluminium silicate minerals, divided into two main groups, one of which contains potassium and the other varying amounts of sodium and calcium. Common members of the first group are *orthoclase* and *microcline*; of the other, the *plagioclase* feldspars, a series ranging from sodium-rich *albite* at one end to calcium-rich *anorthite* at the other; intermediate varieties include *oligoclase, andesine* and *labradorite*. Feldspars are extremely abundant, occurring in nearly all kinds of igneous rocks and metamorphic rocks. Hard and glassy like quartz, they differ by showing pronounced cleavage. Usually light-coloured.

FELDSPATHOIDS. Related to feldspars, but contain less silica.

GABBRO. Coarse-grained, dark grey or black igneous rock composed of feldspar and augite, sometimes with olivine. Classed as the plutonic equivalent of basalt.

GALENA. The sulphide ore of lead; occurs as metallic, grey cubes and also in massive forms, in veins with quartz, but also disseminated through marble, etc.

GANGUE. Applied to the non-metalliferous minerals associated with metallic ores in mineral veins. Gangue minerals are not necessarily without value, e.g. fluorspar, used as a flux in smelting.

GARNET. A group of hard, silicate minerals which form very regular, usually twelve- or twenty-four-sided crystals. Though commonly dark red, they can occur in other colours. Garnets are most common in metamorphic rocks, but not unknown in igneous rocks, e.g. the rhyolites at Rakaia Gorge.

GLAUCONITE. A hydrous silicate or potassium and iron, of distinctive dark or yellowish green colour, found as grains and pellets in marine sediments. Sedimentary rocks containing much glauconite are called *greensands*.

GNEISS. Coarsely crystalline, usually foliated (i.e. banded) metamorphic rock, nearly always containing feldspar as well as quartz and dark silicate minerals. If produced by metamorphism of igneous rock: *orthogneiss*; if of sedimentary rock: *paragneiss*. (See also Chapter Five.)

GRANITE. The term is applied descriptively to coarsely crystalline plutonic rocks consisting essentially of quartz, potassium feldspar, sodium-rich plagioclase with mica and often hornblende. May be even-textured or porphyritic; in the latter case the phenocrysts are usually large crystals of orthoclase or microcline. (See also Chapter Six.)

GRANODIORITE. A coarsely crystalline igneous rock containing quartz, but also more plagioclase than orthoclase, together with biotite, hornblende or augite.

GRAPHITE. Carbon as a native mineral, very soft, black or dark grey, occurring usually as shiny scales or thin laminae in metamorphic rocks.

GREENSTONE. In New Zealand, this usually means the mineral *nephrite*, made up of densely matted, minute fibres of *actinolite* (q.v.), associated with metamorphism of ultramafic rocks. Varieties recognised by the Maoris under names 'pounamu', 'tangiwai', etc.

GREYWACKE. As used in New Zealand, the term refers to moderately coarse, dark sandstone, well indurated, composed of detrital grains of feldspar and quartz. (See also Chapter Four.)

GYPSUM. A mineral form of calcium sulphate. Soft, light-coloured crystals with a pearly lustre, not uncommon as nodules in marine strata above the Quartzose Coal Measures, as a product of decomposition of pyrite.

HEMATITE. Iron oxide, containing about 70% iron, and the most important iron ore. Occurs as masses of very shiny, black crystals, as smooth brown nodular masses, or as a red earthy substance.

HORNBLENDE. One member of the *amphibole* group of minerals (q.v.), common among the dark, glassy minerals of many igneous and metamorphic rocks.

HORNFELS. A hard, finely speckled, dense black rock due to contact metamorphism presumably of fine sedimentary rocks.

ILMENITE. An oxide of iron and titanium, the usual ore of titanium. Occurs in fine, tabular crystals, black and metallic in appearance, usually in mafic plutonic rocks. Main dark constituent of black beach sands.

JADE, JADEITE. The usual term overseas for jewel-quality nephrite which in New Zealand we call *greenstone* (q.v.)

JASPILLITE. Commonly used in New Zealand for rock composed of red or brown *jasper*, one of the cryptocrystalline varieties of quartz, found with geosynclinal volcanic sediments.

JOINT. A fracture in rock, more or less a plane surface, perpendicular or oblique to the bedding, along which there has been no appreciable movement. Joints tend to occur in roughly parallel sets; two or more regular joint sets in the same rocks constitute a *joint system*.

KAOLINITE. The fundamental aluminium-silicate mineral that is an essential constituent of ceramic clays. Most of it originates in the decomposition of feldspar.

KERATOPHYRE. A sodium-rich variety of *trachyte* (q.v.) containing albite, usually associated with *spilitic* rocks (q.v.) in geosynclinal marine deposits. The dark minerals usually have been altered to chlorite, epidote and calcite.

KYANITE. An aluminium silicate mineral occurring in blue, blade-like crystals in high-grade gneiss and schist.

LAMPROPHYRE. A group of dark, dense dyke-rock varieties, in which the phenocrysts are entirely of the dark minerals, which can be biotite, hornblende or augite, sometimes olivine.

LIMONITE. A vague group of brown-coloured hydrous iron oxide minerals.

LODE. An old miners' term for a well-defined body of ore between distinct rock walls; it may be made up of several closely spaced veins that can all be worked together.

LOESS. An unstratified, even-textured superficial deposit of silt, usually ascribed to deposition by wind of sediment derived from neighbouring glaciated areas; but it has been used in other ways.

MAGNETITE. An oxide of iron commonly found as regular eight-sided (octahedral) individual crystal grains, but also in granular and massive forms; often present as a minor mineral in igneous rocks. Also called 'magnetic iron ore' as it is attracted by a magnet.

MARBLE. Limestone that has been recrystallised, usually as a result of metamorphism.

MARCASITE. One of the mineral forms of iron sulphide; paler in colour than pyrite, and more brittle. Marcasite crystals tend to decompose in a moist atmosphere.

METAVOLCANICS. A rather loose term for more or less metamorphosed volcanic rocks of any kind.

MICA. A group of minerals that has such perfect cleavage that it can be split down into paper-thin sheets. The mica minerals are aluminosilicates of calcium, iron and other elements, and occur as primary constituents of most igneous and metamorphic rocks. As detrital flakes they make up much of the finer substance of sandstones, siltstones and shales. Large sheets can be obtained only from large mica crystals ('books') found in pegmatites. Common varieties are *muscovite* (white mica) and *biotite* (black mica).

MONTMORILLONITE. One of the *bentonite* group (q.v.) of hydrous aluminosilicate minerals with swelling and absorptive properties.

MUSCOVITE. See *Mica*.

MUDSTONE. Compacted, uniformly fine-grained sedimentary rock. If laminated, it can be called *shale*, but the Americans apply this term to all mudstones.

NEPHELINE. A silicate of aluminium, sodium and potassium belonging to the *feldspathoid* group (q.v.); occurs in sodium-rich varieties of igneous rock (e.g. *phonolite*) as small six-sided prismatic crystals.

NEPHRITE. See *Jade, Greenstone*.

NORITE. A coarse-grained, dark plutonic igneous rock, which differs from *gabbro* (q.v.) only as regards the variety of pyroxene (usually enstatite or hypersthene).

OBSIDIAN. A volcanic glass of the composition of rhyolite.

OLIGOCLASE. A member of the *plagioclase* group of *feldspars* (q.v.) rich in sodium.

OLIVINE. Also known as *peridot* (the gem variety); a magnesium-iron silicate mineral, usually green in colour, glassy, and lacking in cleavage (thus distinguished from pyroxenes). It is a normal component of the mafic and ultramafic igneous rocks, and commonly forms phenocrysts in basalt.

OPHIOLITE. A general term for mafic and ultramafic igneous rocks and their alteration products such as serpentinites, associated with a geosyncline.

ORE. A general word for mineral or rock substance from which a metalliferous mineral can be extracted profitably. Thus, as the term is usually applied, a quartz vein containing galena is not necessarily an 'ore' of lead; there should be some prospect of its being worked profitably.

ORTHOCLASE. A member of the *feldspar* group of minerals (q.v.) rich in potassium; an essential component of igneous rocks of the granite and syenite families, including rhyolite, trachyte.

PEGMATITE. Exceptionally coarse-grained plutonic rock, usually in the form of dykes or other sheet-like bodies within or near granitic or other plutons; in Westland, emplaced in Haast Schist belt. (See Chapter Six; also under *mica*.)

PERIDOTITE. A general term for ultramafic plutonic rocks consisting mainly of olivine with amphibole or pyroxene, but little feldspar.

PHONOLITE. Lava rock similar to *trachyte* (q.v.) but containing the feldspathoid *nepheline* and sodium-rich varieties of the dark minerals.

PHYLLITE. A fine-grained metamorphic rock having a silky sheen on cleavage surfaces; intermediate between slate and schist.

PLAGIOCLASE. See under *Feldspar*.

PORPHYRITE. Originally this meant an intrusive rock having the composition of andesite, with phenocrysts of plagioclase, biotite and hornblende in a finely crystalline matrix. It is commonly used now as a textural term for any igneous rock with *porphyritic texture* (phenocrysts in a finer ground mass).

PORPHYRY. Originally applied to a purple, altered intrusive variety of *trachyte*; sometimes confused with porphyrite, so *trachyte-porphyry* is preferable for intruded trachytic rocks.

PUMICE. A glassy lava with the composition of *rhyolite* that has been frothed by expanding gases before solidifying into cellular, light-weight lumps and pieces. The term can still be used when the composition is other than rhyolitic.

PYRITE, IRON PYRITES. Another iron-sulphide mineral (see also *marcasite*), brassy yellow in colour, hard, and commonly occurring as cuboidal crystals with parallel, linear striae on the faces; or in its own characteristic crystal form, enclosed by twelve five-sided faces (pyritohedron). It is found in many kinds of rock, and in quartz veins with other sulphide minerals, and is an important source of sulphur.

QUARTZ. One of the most abundant minerals in the earth's crust, and the most common form of silica (SiO_2). When crystalline, it occurs in six-sided prisms often found lining cavities in rock. When colourless and free from imperfections, it is called *rock crystal*. Usually it is more or less opaque-white, or tinted various colours by impurities (*amethyst* and *cairngorm* mentioned above). Hard, brittle, and lacking in cleavage (thus distinguished from feldspar).

QUARTZITE. Hard, finely granular rock resulting from recrystallisation of sandstone during metamorphism. The term is also applied to sandstone cemented by the deposition of silica between the grains.

QUARTZ-PORPHYRY. An old term still used for rocks of rhyolitic composition, with quartz and orthoclase phenocrysts set in a finely crystalline matrix.

RHYOLITE. The volcanic rock of the granite family, usually having phenocrysts of quartz and orthoclase in a glassy matrix. (See also *pumice* and *obsidian* above, and *ignimbrite* in Chapter Eight.)

SANDSTONE. A self-explanatory term for compacted sedimentary rock of sandy texture.

SERPENTINE. A hydrous magnesium silicate mineral (or group of minerals), believed to be produced by metamorphism or metasomatism from mafic and ultramafic igneous material (see Chapters Four and Six). It usually occurs in a massive form, in a wide range of colours but commonly green, grey or black. The hardness varies, and it has a distinctive 'greasy' appearance and feel. *Serpentinite* is serpentine rock.

SCHIST. A moderately coarse crystalline metamorphic rock in which the orientation of mica and other flaky minerals, in the process of metamorphism, into sub-parallel directions has given the rock a tendency to split and a 'sheen' on the splitting surface (i.e. 'schistosity'). A similar, finer rock could be a *phyllite* (q.v.).

SHALE. A laminated or splittable *mudstone* (q.v.).

SILLIMANITE. An aluminium silicate mineral occurring as long, needle-shaped crystals in high-grade gneisses and schists.

SILTSTONE. A compacted sedimentary rock intermediate in texture between sandstone and mudstone.

SLATE. An early product of progressive, regional metamorphism (see Chapter Five) in which fine-grained rocks have developed one or more distinct directions of cleavage.

SPHALERITE. The chief sulphide ore of zinc. It occurs as yellow or brown crystals, fairly soft, with good cleavage and rather a resinous appearance.

SPILITE. Submarine basaltic lava rocks in which the minerals have been enriched in sodium. Feldspars are altered to albite and the darker minerals to serpentine, chlorite, calcite, etc.

STIBNITE. An antimony sulphide mineral occurring as grey, metallic prismatic crystals, or in masses with a prismatic structure. Usually occurs with chalcopyrite and other sulphide minerals.

SYENITE. A coarse-grained plutonic igneous rock consisting chiefly of orthoclase or microcline and sodium-rich plagioclase with hornblende and biotite.

TALC. Hydrated magnesium silicate, one of the products of alteration of peridotite and other magnesium-rich rock; it usually occurs as greenish-grey, soft, greasy, fibrous or foliated masses. Talc-rock is also called *soapstone*.

TINGUAITE. This rock is similar in composition to *phonolite* consisting of sodium feldspars, nepheline, and sodium-rich varieties of pyroxene and amphibole; it occurs as dykes and sills.

TRACHYTE. As used nowadays, this term refers to a lava rock composed essentially of potassium feldspars with minor biotite and hornblende, in a dense light-grey groundmass in which commonly the minute feldspar crystals are arranged into lines parallel with the direction of flow of the lava.

TRIDYMITE. A high-temperature form of silica, formed at temperatures above 879°C. It is occasionally found in some trachytes.

VEIN. This usually means a more or less regular and continous sheet-like body of ore material enclosed between defined walls and cutting through the country rock. There is no size restriction, but if only a few millimetres thick, they are likely to be called *veinlets*. No consistent distinction from *lode* (q.v.). Also used for very thin intruded sheets of igneous rock.

WOLLASTONITE. A calcium-silicate mineral produced by high-temperature metamorphism of impure limestone containing silica; therefore found in high-grade regional or contact metamorphic rocks. It is a white, glassy and often fibrous mineral, then resembling *actinolite* (q.v.) but it is soluble in hydrochloric acid.

XENOLITH. A fragment of some pre-existing rock now embodied in an igneous rock. (Greek: 'stranger-stone'.)

INDEX

This list of topics, personal and place names, which should be used in conjunction with Appendix 5 (Glossary), includes only substantial references. A page number in italics indicates the reference may be found in some cases in the caption to a figure as well as in the text near by. *Abbreviations: etc* means numerous other references on other pages; *et seq* indicates further references on succeeding pages. (A) Central Auckland district; (BP) Bay of Plenty; (C) Canterbury; (CO) Central Otago; (EA) eastern Auckland Province; (EC) East Cape-East Coast district of North Island; (ES) eastern Southland; (EW) eastern Wellington district; (HB) Hawkes Bay district; (M) Marlborough; (N) central and eastern Nelson; (NA) Northland; (NC) North Canterbury; (NO) North Otago; (NW) North Westland; (NWN) north-west Nelson; (S) central Southland district; (SA) South Auckland; (SC) South Canterbury; (SN) southern Nelson; (SO) South Otago; (SW) South Westland; (SWA) south-west Auckland; (SWN) south-west Nelson; (T) Taranaki; (W) Wellington district; (WN) west Nelson; (WO) west Otago; (WS) western Southland.

Abel Tasman National Park, 79
'absolute' age, 41
Acheron River (C), 369
Adkin, G. L., 194, 259
'acidic' igneous rocks, 132
actinolite, 72, 120, 405
adjustment, to stream pattern, 301, *303*
advance, glacial, 241
agate, 211, 405
'age' of rock, meaning, 61; of landscapes, 340
agglomerate, 72, *214*
aggrade, aggradation, 31, *38*, 300, *302*
aggregate (concrete), sources, 382
Akaroa Harbour (C), 163, 216; caldera, *202*; volcano, *159*, 216
Albatross Point (SWA), 217
albite, 405
Allan, R. S., 180, 195
alluvial fans, 315, *318*; plains (*see* valley-plain), 314
Alpine Facies, 93, 99, 279
Alpine Fault, 31, 57, 79-80, 115, 271, 273, *286*, 287 *et seq*, *288-90*, 305, 349
Alpine Schist, 126
Amberley Limestone, 182
amethyst, 21, 211, 405
amphibole, 156, 405
Amuri Limestone, 182
Anatoki Eugeosyncline, 69; Formation, 74, 87; River (NWN), Track, 74
andalusite, 148, 405
andesite, 405; lava, 199, 201, *etc*; volcanoes of W. Pacific, 295
andesine, 405
Andrews, P. B., 99
angular unconformity, 109, 394
antecedent gorge, 312, *314*
anticline, 394, *etc*; growing, 357; 'New Zealand', 274
anticlinorium, 80, 274, 395

Antipodes Island, 400
anorthite, 407
anthracite, 183, 370, 405
Aorangi Mine Formation, 76, 80
Aorangi Mountains (EW), 266
Aorere Group, 72, 279, 363; River (NWN), 74; 'Slates', 71, 75; stratigraphic problems, 75 *et seq*
Aparima Series, 108
aquifer, 383
aragonite, 21, 119, 405
Arahura River, Valley (NW), 242, 372
Aranuian Stage, 263
Arapuni Dam, 384
arête, *143*
argillite, 73, 86, 112, 405
Arnott Volcanics, 213
artesian spring, wells, 382-3
Arthur Marble, 78
Arthurs Pass, 262, 353, *354*
Arthurton Group, 105
asbestos (chrysotile), 379, 406
asthenosphere, 20
atmosphere, earth zone, 19; weathering agent, 26
Auckland basalts, 220; Peninsula, structural trends, 219; volcanoes, 220
Auckland Islands, 137, 400; eroded basaltic dome, 400; granite, 137, 400
augite, 405
Aukaotere Ash, 225
Avoca (C) glacial outwash, *238*; Oligocene basaltic debris, *58*
Awamoko Stream (NO), *311*
awaruite, 102, 379, 405
Awatere Fault, 113, 291, 327, 353; River, Valley (M), *303*; terraces, *150*, *261*
Axial Facies, 93-4, *etc*; Fault System, 287, 291
axial plane (of fold), 394; spine of New Zealand, 266; trace (of fault), 394

badlands gullying, *190*
Bainham (NWN), 74
Balloon Formation, 73
Banks Peninsula loess, 261; volcanoes, 22, *159*, 198, 284, 339
bar, gravel, sand, *158*, 331
Barrow, G., 121
Barrytown (NW), 84, 137
basalt, 22, 132, 199 *et seq*, 405, *etc*
base-level, -levelling, 299, *301*
base metals, 376-9
basic igneous rocks, 132
basin structure, 395
batholith, 32, *119*, 133
Baton Formation, 86; Group, 71-2, 85-6, 279; River (WN), 84
Battey, M. H., 269
Bay of Islands (NA), 220
Bay Schist, 76
Beck, A. C., 355
bedding (*see* stratification)
bedrock, 24
Beesons Island Volcanic Formation, 218
Bell Hill (NW), 140
Bell, J. M., 14, 77, 361
Benioff Zone, 350
Ben McLeod mountain (C), 211
Benmore Dam, 104
Benson, W. N., 71, 76, 81, 90, 169, 214
bentonite clay, 213, 381, 406; cause of slumping, *155*, 325
Berlins Quartz Porphyry, 209
Big Grey River (NW), 75
biostratigraphic classification, 107
biotite, 121, 406
Biotite Zone, *123*, 126, 128
Birch Hill moraine, 262
Bishop, D. G., 76, 292
Blackwater [ice] Advances, *238, 326*; gold mine, 82, *364*
block mountains, *311*
'blowhole' (fumarole), 203
Blue Bottom Formation, Group, 186
'blue-metal', 22
Bluff (S), 105, 163; 'granite', 157; Hill (HB), 334
Boatmans Harbour, Oamaru (NO), 213
Boulder Bank, Nelson, 144, *333*, 334
boulder-clay (*see* till)
Bounty Islands, 137, 401
Bourne Conglomerate, 190
Bradshaw, J. D., 99, 292
Bradshaw, Margaret, 86
Bradshaw, Gneiss, 152
braided stream channels, 315, *316*
Bream Head (NA), 218
breccia, 66, 165, 406 *etc*
Brighton (SO), *123*
Broken River (C) coal, 189; dyke, *212*
Brook Street Igneous (Volcanic) Formation, 104, 144
Brown, D. A., 403
Browning Pass (C), *28*
Brückner, E., 239
Brunner Beds, Formation, *169-70*, 176, *370*; Coal Measures, 176; clays in, 381; Mine, 366

Buller Geosyncline, 68, 69, 73, 86, 98, 106, 275, 279; volcanoes, 207
Buller Gorge (SWN), 312; River, Valley, 129
Bulman, O. M. B., 25
Burnham Formation, 252
Burnt Hill (C), 215
Burrows, C. J., 265
Burton, Peggy, 360

Cable Bay (N), 334; Granodiorite, 145
cairngorm, 21, 406
calcite, 21, 119, 406 *etc*
caldera, collapse, 201, 227; erosional, *159, 202*, 215
Cameron Glacier (C), 265; Mountains (WS), 266
Campbell Island, 400; Plateau, 267, 294-5
Campbell, J. D., 16, 108, 112
Campbell, K. S. W., 403
Canterbury Plain, 191, 284 *etc*
Cape Farewell (NWN), *172*; Foulwind (SWN), *141*, 163; Providence (WS), 80; Runaway (BP), 210; Terawhiti (W), 339, 360; Turakirae (EW), 339, *357*; Wanbrow (NO), *214*
Caples Group, 100, 105, 113; volcanics in, 207
Carnley Harbour, Volcano, 400
'Carriage Drive, The', Rakaia Valley (C), *355*
Carter, R. M., 182
Castlecliffian Stage, 195, 197
Castle Hill (C), *26, 38*, 302; Basin, *282*
Castle Point (EW), *189*
Cave Stream (C), *326*
'cements', auriferous, 173, 364, 406
Cenozoic Era, 194
Central Seismic Region, 350, *351*
chalcedony, 406
chalcopyrite, 377, 406
Chalky Inlet (WS), 80
chaos-breccia, 277
Charleston (SWN), 23; Gneiss, 137, 152; lignite, 183, 369
Chatham Islands, 399; Rise, 295
chert, 73, 89, 112, 406 *etc*
chlorite, 72, 120, 406 *etc*
Chlorite Zone, 121-2, 126, 128; Sub-zones, II, III, IV, *122*, 128-9
Christchurch (C) artesian water supply, 384; brickworks, 381; well records of post-glacial sea levels, 263
chromite, 377, 406
chronostratigraphic classification, 41, 107
chrysotile (asbestos), 379, 406
Circum-Pacific Seismic (Mobile) Belt, 57
cirque, *143, 319-20*
Clarence Fault, *164*, 291; River Valley (M), *164*, 306, 314
Clark, R. H., 233, 357
clast, 66

clastic components, 66; rocks, 406
clay, bentonitic, 213, 381; deposits, 381; 'minerals', 21; refractory, 173, 381
cleavage (rock), *64, 120,* 395
Clent Hills (C), 112
climate-stratigraphic stages, 252
coal, a sedimentary rock, 22; anthracite, 183, 369, *370*; bituminous, 183; calorific value, 368; coalfields in N.Z., *369*; early discoveries, 365 *et seq*; geological setting, 368-70; grades, 367; lignite, 183, 367-8; 'measures', 173; outcrop, *370*; proximate analysis, 368; Quartzose Coal M., N.Z. major source, 174; search for, 366-7; ultimate analysis, 368
Coalgate (C), 381
coastal erosion, *184*; terraces, *245,* 339
coastline, forms types, 337-8
coasts, prograded, 331, 337; rectified, 331, *332*; retrograded, 331; uplifted, 338-9
Cobb Intrusive Group, 72, 74, 279; River, Valley (NWN), 65, 74
Cobden Limestone, Formation, 183, 188
cold-climate (non-glacial) deposits, 260
Coleman, R. G., *101,* 157
Collingwood (NWN), 25, 75, 173, 363, 374
Colville Ridge, 294-5
'complex carbonate rock', 75, 79
composite volcanic cone, *201, 220,* 339
concretions, spheroidal, *55,* 179
concretionary beds, 180
conduit, of volcano, *203*
conformable sequence, conformity (*opp. of* unconformity, *q.v.*)
conglomerate, 66, 189 *et seq*, 406
Constant Gneiss, 137, 152
constructive processes, 25
contact aureole, *119*
continental accretion, 67; drift, 67; platform, 31
Conway River (M), 179
Cook Strait, 270, 280, *338 etc*
Cookson Volcanic Formation, 214
Coombs, D. S., 68, 101, 108, 124-5, 280
Cooper, R. A., 68, 78, 82, 84, 275
Copland Pass (C/SW), 292
copper deposits, 377
coquina, 189
core, of earth, 20, 267
Coromandel (EA), early gold search, 363; granitic rocks, 136; mid-Tertiary volcanics, 218; Peninsula, 218; Range, 266
correlation, *24,* 69
Cotton, C. A., 96, 164, 261, 298 *etc*
Couper, R. A., 181
covering strata, 164, 177, 303
Cox, S. H., 75, 77, 242
Cranwell, Lucy (Mrs Watson Smith), 248
crater, 201; rings, 221; sink- 201; 'The' (Kakanui Range (NO)), 340
craton, cratonisation, 31, 87
Crawford, J. C., 360

crest, of fold, 394
'Cretaceo-Tertiary', 182
Crook, K. A. W., 403
Crooked River (NW), *289*
cross-folding, 268
crust, of earth, 20; 'discontinuities' in, 267; thickness, 267
crustal mobility, belts of, 348; and earthquakes and volcanoes, 348; plates, 294; warping, 342
cryptocrystalline, 406
crystal fractionation, of magma, 202
cuesta ridges, *304*
Curio Bay, Waikawa (ES), 109

dacite, 133, 199, *219,* 406
Daly, R. A., 240
Darran Diorite, 126, 279
Davis, W. M., 164, 298
Deborah Volcanic Formation, 213
degrading, degradation, of land surface, 298
deltaic deposits, *263*
Denniston (WN), 366
depressed areas, 282
destructive processes, 25
detrital (*see* clastic)
Devil River Formation, Volcanics, 72, 73
Devil's Boot, Rockville (NWN), 25
'Devonian Series', 'Formation', 85
dextral motion, on strike-slip fault, 291, *292*
diagenesis, 118
Diamond Harbour Lava, 216
diastrophism, 31
Dieffenbach, E., 370, 399
differentiation, of magma, 202
diorite, 144, 406 *etc*
dip, of strata, 394 *etc*; of fault-plane, 392
dip-slope, *304, 307*
Disappointment Island, 400
disconformity, *183,* 394 *etc*
discordant pluton, *133*
dissection, 302, *308*
Dobson (NW), 371, 373
Docherty, W., 81
dolerite, 209, 406
dome, 373, 395
Douglas Formation, 76, 78
downthrow, of fault, 392
drag, on fault-plane, *135*; -folding, *168*
'Drift Formation', 196
Drury (A), early coal workings, 366; Fault, 305
Dunedin Volcano, 214-5, 284
dunes, sandhills, 334-5
dunite, 100-1, 104, 157, 407
Dun Mountain (N), 101, *156*; Ophiolite Belt, 101, 157

Dunollie Formation, *170*
Dunstan Range (CO), 171
D'Urville Island, 104; Series, 108
Dusky Sound (WS), 145
dyke, *205, 207, 212 etc*
dynamo-thermal metamorphism, 119

Early Geosynclinal Cycle, 68
earthquakes, 344 *et seq*; aftershocks, 353;
    causes, 346-8; changes of level
    accompanying, 353, *354*; deep,
    intermediate, shallow, 345; hazard in
    N.Z., 349-50, *351*; historic, N.Z., 351 *et
    seq*; Maori traditions, 354; prehistoric,
    354; and volcanoes, 348; swarms, 353;
    terminology of, 344 *et seq*
'earthquake fault-line' myth, 349
'Earthquakes, The', Duntroon (NO), *182*
Eastern Belt, of Paleozoic sediments, 68; of
    New Zealand Geosyncline sediments, 93
economic geology, defined, 358
Edwards River (M), *47*
Eglinton Valley (WS), 148
Egmont Volcano (T), 200-1, 219, *220, 224*
Eiby, G. A., 346, 349
elastic rebound theory, *344, 345*
Ellis Formation, 84, 86
endemic fossils, 86
Enderby Island, 400
engineering geology, 384 *et seq*
englacial debris, *321*
environmental geology, 387 *et seq*
Enys Formation, *183, 190*
epeirogenic (crustal) movements, 182, 184
epicentre, *344, 345*
epidote, 72, 122, 407
erosion, geological concept of, 26
eruptive behaviour, 200
eugeosyncline, 69, 208
eustatic, 240
evert, eversion, 102, 114, *277*
Evison, F. F., 349
exhumed fossil landscape, 171

fabric, 149
facies, 86; -junction, 94, 280
'false gold', 363
fan, alluvial, debris, shingle, 314-5
Farewell Spit (NWN), 334
fault, 19, *47, 135,* 284 *etc*; -geometry,
    392-3; -lines on map, meaning of, 284 *et
    seq*; scarp, 305, *309*; strike-slip, 285 *et seq,*
    393; terminology, 392-3; transcurrent,
    wrench (*see also* strike-slip fault, Alpine
    Fault), 273; trough, *311*
fault-angle depression, *310*

fault-line scarp, *138, 309, 311*; valleys, *290,*
    303, *309-10*
Fearnsides, W. G., 25
feldspar, 153, 407 *etc*
feldspathoid, 215, 407
felsic igneous rocks, 132
'felt intensity', earthquake scale of, 344
Findlay Creek (NW), 242
Finlay, H. J., 181
Finlayson, A. M., 401
Fiord Granite, 137; Intrusives, 148
Fiordland copper deposits, 377; Gneiss,
    149, 279; granites, 137, 140, 148; seismic
    region, 350, *351*
fissile, fissility, *120*
fissure-vein, *365*
Flandrian Transgression, 263
Fleming, C. A., 92, 115, 195, 245, 253
Fletcher Creek coal, 183
Flint, R. F., 236
Flora Formation, 78; Track (WN), 77
Flowers Formation, 106
fluorspar (fluorite), 119
fluvioglacial deposits, forms, 318
flysch deposits, 96
focal depth, focus (earthquake), *344,* 345
folding, *134,* 394-5, *etc*; complex, *178*
foliae, foliation (of metamorphic rock), *54,*
    89, 120-1, 123
foreland, 32, 67, 95, 98, 274, 279
fore-set deposits, of delta, *263*
formation, 24, 76, 106
fossil ammonites, 108-9, 112; barnacles,
    189; belemnites, 112; brachiopods, 84-5,
    86, 108, 112, *182 etc*; bryozoans, 213;
    conodonts, 79, 98; corals, 71, 84-6, 104,
    108, *etc*; crustaceans, 184; foraminifera,
    396 *etc*; forest, 109; graptolites, 71, 79,
    82; gullies, 260; land surfaces, 171, *311*;
    molluscs, 85 *etc*; plants, 112 *etc*; pollen,
    248; reptiles, 112; ripple-marks, *47*;
    sea-urchins, 213; soils, 171; trilobites, 71,
    79, 85; vertebrates, 112, 184; whales,
    185; worms, 112, 185
fossils, distorted, 396; preparation for
    study, 396; use of, 37
Foveaux Strait, 266, 270, 337
Fox Glacier (SW), *236-7, 330*
fracture cleavage, 395
Franklin district (SA) volcanoes, 220, 339
Franz Josef Glacier (SW), *38, 54, 262, 265*
Frog Rock (NC), 25
frost action, *28, 134*
fumarole, 203, 222, 233, 386
Fyfe, H. E., 129

gabbro, 129, 132, 136, 157, 216, 407, *etc*
Gabriels Gully (CO), 173, 364
Gage, M., 195, 245, 273
galena, 378, 407

gangue, 407
garnet, 121, 407 *etc*; in rhyolite lava, Rakaia Gorge, *211*
Garnet-Oligoclase Zone, 126, 128
Garnet Zone schist exposure, *54*
Garvey Creek coalfields, 184
gash veins, 64
geanticline (*see* N.Z. Geanticline), 274
geochemistry, 19
geode, 211
geological column, 41; cycle, *25*, 30, 297; maps, 402; publications, 402-4; surveys, 359-60; time-scale, 37-8; 'thermometres', 121
geomorphology, 19, 296 *et seq*
geomorphic 'cycle', 'stages', 299
geophysical exploration, 371, 385, 390-1
geophysics, 19
geosynclinal phase, 32
geosyncline, 32, 89, 182 *etc* (*see* Buller Geosyncline, N.Z. Geosyn.)
geothermal gradient, 119, 126, 203, 280, 386; energy, 203, 386, *389*
Geraldine (SC), 219
'geyserite', 204
geysers, 203, 386
glacial chronology (Pleistocene), 246, 255; controversy, 235, 242; lake deposits, *134, 236, 256*; 'Lake Speight', *321*; 'Period' in N.Z., 194, 242 *etc*; sculpture of scenery, *51*; smoothing of rock, *38*; terminology, 241, 'Theory', 235-6
glaciation, defined, 241; of N.Z., extent, 236, 243, 318; multiple, 239, 244; North Island, 259; sub-antarctic islands, 400
'glacierisation', 235
glaciers, transporting power, *237*
glacifluvial deposits, forms, 318
glacio-eustatic, 240, 246
glauconite, *182*, 185, 407
Glenhope (N), 126
gneiss, 23, 407, *etc*
Godley River (C), 262
gold, 362-5; 'cements', 173, 364; concentrated on Late Cret. Peneplain, 173, 364; dredge, *366*; exploration, 364-5; 'false', 363; occurrence and origin, 363-4; -quartz veins, *364*
Golden Bay Group, 72, 76, 279
golden sands, source, *54*
Goldlight Formation, *169, 170*, 171
'Gondwanaland', 67, 106
Gore Piedmont Gravel Formation, 190
gorges, 306, 312 *etc*
gouge, *365*
Gouland Downs, (NWN), 171
graben, 305, *311*
graded bedding, *89*, 96; profile, 300
Grange, L. I., 224, 229
granite, 23, 407 *etc*; ages of N.Z., 161, 279; dual origin of, 23, 133 *et seq*; formed in orogen, 122; -gneiss, gneissic granite, 148; layered, 148; outcrop, *54*; sub-antarctic islands, 400

granitisation, 136, 156
'granitisers', 133
granodiorite, 129, 133, 144, 407
graphite, 78, 407
gravity, geological agent, 29
Great Barrier Island, 218, 266, 353, 377
Greenland Group, 61, *64*, 65, 81 *et seq*; 137, 152, 275, 279, 363; 'Series', 75
greensands, 181, *182* (*see* glauconite)
'Green schists', 122
greenstone, 101, 407
Gregg, D. R., 228-9
Greville Formation, *43*, 104
Grey-Inangahua Depression, 188, *277*
Greymouth coalfield, 183, 368, *369*
Grey River, Valley (NW), 354; gorge, 312; -Trough, 280
greywacke, 22, 32, 88 *et seq*; 407 *etc*
Gridiron Formation, 211
Grindley, G. W., 68, 71, 76-7, 124-5, 148, 275, 278, 287
Group (stratigr.), 76, 106
Grove Road (Havelock-Picton) (M), *128*, 246
growing anticlines, 357
Gunn, B. M., 293
guyot, 204
gypsum, 407

Haast, J., 71, 77, 81, 84, 216, 242, 359
Haast Schist Belt, Group, Zone, *54*, 94, 101, 113, 122, *125*, 128, *172*, 216, 276, 280
Hailes Knob Quartzite Formation, 84
Hall, T. S., 81
Hampden (NO), 179
hanging valleys, *320*
Harper Basalt, 215; Pass, *310*; River (C), *27*
Harper Hills Volcanic Formation, 215
Harris, W. F., 248
Hatuma (HB), 189
Hauhungaroa Range (SA), 226
Hauhungatahi volcano, 226
Haupiri Disturbance, 87, 162; Group, 65, 71-2, 157 *etc*; volcanics, 207
Hauraki mining area, 363, 378; Graben, Trough, 283, 305
Hautawan Stage, Substage, 195, 242
Havelock River (C), *320*
Hawera Series, 193, 196, 252; stages in, 245, 263
Hawkes Crag Breccia, 165, *166*; radioactive minerals in, 379
Hay, R. F., 400
Hayward, B. W., 218
Heaphy Track (NWN), 76, 171
Hector, J., 71, 75, 77, 84, 89, 103, 108, 242, 359
Heinz, W. F., 153, 364
hematite, 21, 407

Hen Island (NA), 218
Henderson, J., 83, 85, 129, 152, 244, 271
Henley Breccia, 165, 243; 'Moraine', 243
herringbone ridges, 302, *309*
Hicks Bay (EC), 210, 266
Hikurangi Trench, 350; Volcano, *219*
Hilt's Law, 183
Hochstetter, F., 70, 77, 84, 88, 101, 108,
    242, 270, 359
Hohonu (NW), 140
Hokianga (NA), 217, 372
Hokitika (NW), 259, 262
Hokonui Facies, 93, 97, 99, 108-9, 279;
    Hills, Range (WS), 266, 282; System, 110
Holocene, defined, 196
Homer, L., 16
Hooker Glacier (C), *143*
Hope Fault (*see* Taramakau-Hope F.), 310;
    Saddle (N), 190; Valley (NC), 318
horn, *319, 320*
hornblende, 120, 122, 149, 408 *etc*
hornfels, 84, *119*, 137, 148, 408
Horohoro Bluffs (SA), *222*
Horowhenua (W) artesian systems, 384
horst, 281, *311*
hot springs, 203, *386-7*
Huiarau Range (EC), 266
Hume, B. M., 83, 152
Humphreys Gully (NW), 242
Hundalee Fault, 186
Hunter Fault, 303
Huntly (SA), early coal mines, 366;
    refractory clay, 381
Hunua Hills (SA), 303
Hurunui River (NC) gorge, 312-3
Hutton, C. O., 121-2, 124
Hutton, F. W., 81, 242, 360-1
Hutton, J., 33, 342
hydrothermal alteration (*see*
    metasomatism)

Ice Age (Pleistocene), 194, 234
ice, weathering agent, 27
ice-margin fans, 315; lake sediments, *134,
    237, 238, 256*
igneous rocks, 22; deriv., 199
ignimbrite, 200, 221 *et seq*, 222; plateau,
    224, 284; sheets, 339
ilmenite, 376, 408
Inangahua River, Valley (SN), 85
induration, 65, 89
interglacial intervals, 239, 241 *etc*;
    shorelines, *131*; Stages, 245, 249, 252
interstadial intervals, *241 etc*
intrusion, igneous, 22 *etc*
inversion, of strata, *89*
iron ores, 374 *et seq*; in Quartzose Coal
    Measures, 173, 376; metallurgical
    problems, 374

island arcs, 68, 91
Island Sandstone Formation, *170,* 183
isoclinal folds, 71, 395
isoseismal lines, maps, *344*
isostatic adjustments, 31

Jackson Bay (SW), 337
jade, jadeite, 101, 408
jasper, jaspillite, 112, 208, 408 *etc*
Jenkins, D. G. and T. B., 98
joints, jointing, 19, *28, 47, 54*, 89, 408
joint set, system, *47*
Jones Creek, Ross (SW), 191, *238*, 242
Joyce Stream (C) interglacial sequence,
    249, 250

Kaeo (NA), 218
Kaiangaroa Plain, ignimbrite plateau (SA),
    224
Kaiata Mudstone Formation, 183, *184,
    332*, 373
Kaikoura Coast (M), 338; Orogeny, 186,
    247 *et seq*; climax, 189, 242, 280;
    expressed in present relief, 270; onset,
    277; reviewed after sixty years, 192-3;
    uplift commences, 186; Ranges, *164,
    270, 282, 343*
Kaimai Range (EA), 218
Kaimanawa Mountains, Range, 129, 226,
    266, 280
Kaipara Harbour (NA), 217
Kaitaia (NA), 349
Kaitangata (SO), 165, 366
Kaiteriteri (N), *54*, 79, 140-1
Kaitoke (W), 190
Kaiwharawhara (W), 303
Kakahu (SC), 98, 100, 103; clays, 381
Kakanui Range (NO), 266; River, Valley,
    179, *212*, 214, *300*
Kakapo Granite, 148
Kakaramea Mountain (SA), 226
kame, kame terrace, 318, *322, 324*
Kamo (NA), 176, 219
kaolinite, 381, 408
Kapiti Island (W), 129, 291
Kapuni oil wells, 371-4
Karamea Granite, 76, 140, *141*, 144, *162*,
    275, 279; River, Valley (WN), 74, 77
Karioi volcano, 220, 284
Karori Stream (W), 339
Katiki Beach (NO), 179
Kawakawa coalfield (NA), 366
Kawatiri (SN), 129
Kawau Island (A), 377
Kawerau geothermal power, 387
Kawhia (SWA), 94, 217; Syncline, 279

Keble, R. A., 76
Kekerangu Fault, 291
keratophyre, 104, 208, 408
Kerikeri Basalt, 220
Kermadec Islands, 295, 401; Trench 98,
    295, 350
Kermadec-Tonga Ridge, 294-5
Ketetahi spring, *225*
kettle-hole, -lake, *324*, 328-9, *330*
Keyes, I. W., 52, 398
King, L. C., 300
Kingma, J. T., 80, 99, 274, 287
Kirkliston Range (SC), 322
Knox, G. A., 61, 403
Kongahu Fault, 338; Point (WN), 137
Kotuku (NW), 371
Kowai Bush (C), 249, *250*; Gravel
    Formation, 190, 302; River (C), *135*
Kumara (NW), *237, 255,* 259
Kuriwao Group, 106
kyanite, 121, 408
Kyeburn Formation, 165

labradorite (*see* feldspar), 407
lahar, 201, *224*, 243
Laird, M. G., 68, 82, 140, 291
lakes, deposits in, *170,* 250, *256*; origins of,
    327 *et seq*
Lake Coleridge (C), *328*; Forsyth (C), *331*;
    Ianthe (SW), 255; Kanieri (NW), *263*;
    Mapourika (SW), 255; Ohau (SC), *253*;
    Pearson (C), *158*; Pukaki (SC), 318;
    Rotoiti (N), 163, 287; Rotomahana (SA),
    204, 231; Tarawera (SA), *232*; Taupo
    (SA), 327; Tekapo (SC), 318; Unknown
    (WO), *329*; Wahapo (SW), 255; Wakatipu
    (WO), *51*
lamprophyre, 209, 408
Landis, C. A., 68, 124-5, 280
Lankeys Creek, Reefton (SN), 85, 364
lapies, *26*
Late Cretaceous-Early Tertiary
    transgression, 177, 178
Late Cretaceous Peneplain, 97, 169 *et seq,*
    170, 276, *311,* 364
'Late Tertiary Peneplain', 214
lateral corrasion, *316*
lateral secretion hypothesis, 160
laterite, 376
Lauder, W. R., 94, 100, 144
lava, 24, 199; cones, 200, *221,* 339; dome,
    201, 339; interbedded, *205*; pillows, *100,*
    204, 208, 210, *213*; plains, 200, 221, 339
Law of Superposition, 71
Lawrence, D. B., 264
layered granite, 148
lead deposits, 378
Lee River (N), *43*; Group, 104, 106
Lensen, G. L., 343
Leonard, L., 16

Leslie Formation, 78; River (WN) 77
Levin (W), 357
Lewis, K. B., 357
lignite, 183
Lillie, A. R., 69, 293
limestone, 22, 184, *et seq,* 380 *etc*
limonite, 374, *375,* 408
lineation, *120,* 123
lithification, lithified sediments, 118
lithology, 74
lithosphere, 19
lithostratigraphic classification, 41, 107
lit-par-lit intrusion, 149
'Little Ice Age', 263
Livingstone (NO), *260*; Mountains (WS), 266
Loburn (C), 191, 302
Lockett Conglomerate, 74, 87, 162
lode, 408
loess, 196, 219, *260,* 261, 323, 408 *etc*; use
    as brickearth, 381
Logan Point Flow, 215
'Londonderry Stone', *237*
Longford Coal Measures, 189, 369;
    Formation, *58*
Longwood Group, 100
Lord Howe Island, 45; Rise, 295
Lowry Peaks Range (NC), 266
Lyell (SN), 140
Lyell, C., 33, 194
Lyttelton Harbour (C), 216; tunnel surveys,
    216, 359; Volcano, 24, *159,* 216, 284

Mackays Bluff 'Syenite', 144
Mackenzie Basin, 255; Plains, 318
Macpherson, E. O., 91, 273
Macquarie Island, 294
mafic igneous rocks, 132, 156 *etc*
magma, 99; -chamber, 202, *203*; origin,
    201
'magmatists', 133
magnetite, 376, 408
magnitude-scale, seismic, 345
Main Seismic Region, 350, *351*
Maitai Group, 104, 106; System, 85, 89,
    90, 108
Makara Valley (W), *308*
Malte Brun Range (C), *111*
Malvern Hills (C), 212, 369
Mamaku Plateau (SA), 224
Manapouri (WS), 255
Manawatu Gorge, 190, 312, *313*
Manawatu-Horowhenua coast, 337
Mangaorapa Stream (HB), *187*
manganese ore, 377
Mangatu River (EC), slump and earthflow,
    325
Mangles Valley (SN), 371
Maniototo depression (CO), 171
mantle, zone of earth, 20, 101, 201, 267
Manuherikia depression (CO), 171

Manukau Breccia, 218; Harbour, 217-8; Lowland, 305

Maori Bottom Gravel Formation, 190, 302

Marakopa Valley (SWA), 217

Maramarua opencast coal workings, *367*

marble, 23, 122, 149, 408; terrain, *79*

Marble Hill (SN), 79

marcasite, 176, *184*, 408

marginal facies, 93, 94 *etc*

Marlborough-East Coast Shear Belt, 291

Marlborough Schist, 113, 126; Sounds, sunkland, 282, *337*

Marsden, E., 362, 391

Marshall, P., 14, 57, 89, 179, 223-4, 243, 271, 283

Marshall Line, 57

Marton (W), 357

Maruia Springs Junction (SN), 74, 79; Valley, 128, 343

Marwick, J., 90, 180, 362

Mason, B. H., 111, *122*, 292

mass movement of waste, 29, *38*; role in landscape modelling, 301, 325; slump and earthflow, *155*, 325; under frigid climate, 260

Matakaoa Volcanic Formation, 210, 214; Point (EC), 210

Mataketake Range (SW), 153

Matakitaki River, Valley (SN), 287, 369

Mataura (S), 369

Mataura Island Group, 105

'maturely dissected' landscape ('maturity'), 299, *300*, 302

Maui oil wells, 372

Maungataroto (NA), 219

Mawheraiti (SN), 190

McDonald Limestone, 213

McIntyre, D. B., 181

McKay, A., 77, 80-1, 85, 242, 244, 361, 363

meander-loops, 314, *316*

median ridge (of N.Z. Geosyncline), 91, 100, 103, 114

Median Tectonic Line, 69, 126, 280

Mercalli scale (seismic), 344

metamorphic rock, 22-3; grade, 120; rank, 120; segregation, *123*; zones, 120, *122*

metamorphism, 117 *et seq*; contact, 119, *133*, 141; dynamothermal, 119; progressive, 120, 126; regional, 119, 122, 128; thermal, 119, 140; timing, 127

metasomatic alteration (metasomatism), 118, 136, 160, 219; replacement, 160

metavolcanics, 126, 408

mica, 120, 153, 408 *etc*

microcline (*see* feldspar), 407

microfossils, from sea-bottom drilling, 239

micropaleontology, 181, 371

Milburn (SO), 185

Mildenhall D. C., 181, 248, 252

Milford Sound (WS), 337, 352

mineralisation, mineral veins, 160, 218, 363

miogeosyncline, 69, 208

Moar, N.T., 248

Mobile Belts, of earth's crust, 31, 57

Moeraki 'Boulders', 179; River (SW), 153; Peninsula (NO), 214

Mohaka River (HB), 314

Mohorovičić Discontinuity, 267

monoclinal fold, 395

montmorillonite (*see* bentonite), 408

Moonlight Creek (NW), 137

moraine, 318, *321*; ablation, *143*, 318; end or terminal, 318, *321*; lateral, 143, 264, 318; medial, *321*

Morgan, P. G., 80, 244, 271, 361

Morgan Coal Measures, Formation; volcanics in, 212

Motueka Valley (N), 140

Motunau (NC), *155*

Moturoa oil wells (T), 370 *et seq*

Motutapu Island (A), 52

Mount Arthur (WN), 77; Aspiring (WO), *319*; Cobb (WN), *70*; Dasher (NO), 340; Davy (NW), 169; Egmont (T), 200 *etc*; Evans (NW), *264*; John (C), 255; McLean (SW), 140; Maunganui (BP), 334; Misery (C), 211; Potts (C), 112; Radiant (WN), 140, 377; Somers (C), 381; William (WN), 209

Mount Arthur Group, 72, 77; Marble, 78; 'Series', 75, 77

'Mount Torlesse Formation', 111

Moutere Depression, 129; Gravel Formation, 190; Hills, 190-1; stream pattern, 302, *309*

muscovite, 120, 152, 408 *etc*

mudstone, 22, *55*, 408

Murchison Basin, 280

Murchison, thick Miocene strata, *58*, 188, 280

Murihiku Group, 93, 112

Muriwai (A), 204, *205*

Mutch, A. R., 92, 400

Napier (HB), 225

Naseby (CO), 165

Nathan, S., 83, 140, 209, 339

National Park railway station (SA), 243

Nelson, 103 *et seq etc*; 'Mineral Belt', 377; Syncline, 104, 279

Nelson Creek (NW), 242, *256*

nepheline, 215, 409

nephrite, 102

New Caledonia, 295

New Plymouth (T) petroleum, 370 *et seq*; ironsands, 374-5

Newport, J. N., 366

New Zealand Alps, structure of, 292-3; Anticline, 270-1; Geanticline, 87, 106, 145, 274; Geological Survey established, 360; Geosyncline, 56, 68, 87, 90, 92 *et seq*, 106, 113, *115*, 144, 208, 276, 279 *etc*; Orogen, 274 *et seq*; Platform, *48*, 57, 294

Ngapara (NO), *300*
Ngauruhoe volcano, *63*, 200, 226, *228*, 229, *231*, 339
Ngongotaha volcano, 200, 339
Nine Mile Beach (NW), *184*
Ninety Mile Beach (NA), 335
non-glacial cold-climate deposits, 259, *260*
Norfolk Island, 295, 401; Ridge, 294
norite, 157, 409 *etc*
North Cape (NA), 335
'Notocene', 'Notocenozoic', 197
'Notopleistocene', 197
nuée ardente, 222
Nukumaruan Stage, 195, 242
nunatak, 317

Oamaru (NO), 204, *213-4*, 261 *etc*
obsidian, 200, 409
Ohika Formation, 166, 210
Ohiro Bay (W), *93*
Ohura (T), 189, *369*
Okarito (SW), lagoon, gravel bar, *158*, 258, 333
Okataina eruptive centre, 226
'old age' of landscape, 299, 340
oldermass, 164
Old Man Gravel Formation, Group, 190-1, *195, 242, 252, 302*; Range (CO), 260, 322
oligoclase (*see* feldspar), 409
olistostrome, 277
olivine, 101, 214, 409
Omoeroa Bluff (SW), 258
Omotumotu Formation, 183
Onawe Peninsula Akaroa (C), 163, 216
Onekaka (NWN), limonitic iron ore, 374, 375; Schist, *78*
Ongley, M., 90
ophiolite, 101, 206, 409
Opihi River gorge (SC), *138*
Opunake (T), 224
Orakei Basin (A), 221
orbicular granite, 140
ore, 358, 409
Orepuki (WS), 105
organisms, as transporting agents, 29
orogen, 274; New Zealand, 274 *et seq*
orogenic phase, 32
orogeny, 31 (*see also* Tuhua O., Rangitata O., Kaikoura O,)
Orongorongo River (EW), *315*
orthoclase (*see* feldspar), 409 *etc*
orthogneiss, 149
Otaki River, Valley (W), 259
Otago Peninsula, 339; Schist, 113, 126
Otarama [ice] Advance, *249*, 250; moraine, *250*
Otekaike Limestone Formation, *182*
Otira Glaciation, *249*, 250, 253, *318*
Otiran Glacial Stage, 252
Otuhepe, Whakatane (BP), 218

Oturi Interglacial, *245, 332*; coastal benches, *252*
Oturian Stage, 245, 250
outcrop, 24
outwash, glacial, *151, 238, 251*; plain, 318, *321*; profiles, *258*
overburden, 24
overturning (inversion) of strata, 89; detection of, *89*
Owen Valley (SN), 79, 369
Owhariu Fault, 291
Owhiro Bay, *see* Ohiro Bay
Oxford (C), 215

paired metamorphic belts, 125-6
Pakawau (NWN), 79, 165; Group, *172*
paleogeography, 49, 66
paleogeographic maps, *52, 53*
paleontology, 19
Paleozoic Complex, Nelson, 69 *et seq*
palynology, 248
Panmure Basin crater (A), 221
'papa', *24, 55*, 186
Papahaua Block (WN), 291
Paparoa Coal Measures, 168, *169*; Group, *170*, 368; Granite, 137, 144, 152, 163, 210, 279; Range, 83, 188, 266, 277, 280, 282, 354; Tectonic Zone, 277, 291; Trench, Trough, 169, *182*, 188, *277*, 280
paragneiss, 149
Parahaki Volcanic Formation, 219
Parapara iron ore, 374; Peak (NWN), 106
parasitic eruptions, *203*
Park, J., 81, 136, 243, 361
Patriarch Formation, 76, 78
patterned ground, 322
peat, Chatham Islands, 399; parent coal substance, 368
Peel Formation, 78
pegmatite, 140, 152, 163
Pelorus Group, 100, 104; volcanics, 207
Pencarrow Head (W), 339
Penck, A., 239
peneplain, 33, 299 *et seq* (*see* Late Cretaceous P.)
Pepin Island (N), 94, 144
peridotite, 101, 409 *etc*
periglacial, 260
permafrost, 260, 322
petroleum, geological requirements for, 373-4, *373*; in N.Z., 370 *et seq*
petrology, petrography, 19
phenocryst, 137, 409
Phillip Island, 401
phonolite, 215, 409 *etc*
phyllite, 74, 409
piedmont glacier, *254, 255*
Pihanga volcano, 226
Pioneer Ridge (SW), *95*

Piopio (SA), 101
Pikikiruna Range (NWN), 84, 266, *310*;
    Schist, *78*, 125
pillow lava (*see* lava pillows)
Pink and White Terraces, 204, 231
'Pinnacles, The', Harper River (C), *27*
pinnate stream pattern, 302
Pirongia volcano (SWA), 198, 220, 284, 339
pit crater, 201
plagioclase (*see* feldspar), 409
plate boundaries, crustal, *294*
plate tectonics theory, 103; and
    earthquakes, 348
Playfair's Law, 297-8
Pleistocene-Holocene boundary, 196
Pleistocene Period defined, 195
Plinian eruptions, 201, 210
Pliocene-Pleistocene boundary, 194-6, 242
plug-dome (tholoid), 223
plunging fold, 395
pluton, 32, 114, *119*, 148, 206 *etc*;
    concordant, discordant, *133*
plutonic rocks, 132
Pohangina (W), 357
Pohutu geyser, 204
Point Elizabeth (NW), *245*; Formation, 183
Pomona Granite, 148
Porangahau (HB), *167*, *181*; bentonite, 381
porcellanite, 214
Porika Glaciation, 252
porphyrite, 209, 409
porphyritic texture, 137
porphyry, 379, 409
Porter River (C), *183*
Porters Pass (C), 260; Fault, *355*
Port Hills (C), 24
Port Nicholson depression (W), 283, 338
'Post-Hokonui Orogeny', 192
Pounamu Formation, 101
Poulter River (C), *151*, *321*, 328, *354*
Precambrian rocks, 73, 81 *et seq*
Preservation Inlet (WS), 137, 145;
    Formation, Group, 80-1, 363
Productus Creek Group, 105
prograded coast, progradation, *158*, 330,
    331, 337
propylitisation, 218
'proto-New Zealand', 113, 169, 178
proximate analysis (of coal), 368
pseudomorph, 118
Puhipuhi Syncline, *134*
Pukerua Fault, 291
Puketeraki Range (C), 266
Puketoi Range (HB/EW), 302
Pullar, W. A., 225
pumice, 204, *224*, 409 *etc*
Punakaiki (Pancake Rocks) (NW), 188
Pupu Conglomerate, 106
Puysegur Point (WS), *172*
pyrite, 176, *184*, 409 *etc*
pyroclastic components, rocks, 66, 199
pyroxene, 156 (*see* augite, 405)
pyroxenite, 136, 157

quartz, 20-1, 119, 409 *etc*; -conglomerates,
    171 *et seq*; -diorite, 133; -sand deposits,
    380; -porphyry, 209, 409
quartzite, *70*, 122, 409 *etc*
quartzo-feldspathic schist, 122
Quartzose Coal Measures, *172*, 173 *et seq*,
    183, 369, *370 etc*
Quaternary Era, 194 *etc*

radial stream pattern, *307*
radioactive minerals, 165, 379
radiocarbon dating, 240, 248
radio-isotope, 'radiometric' dating, 37, 61,
    162, 240
radius of earth, 44
Rahu Saddle (SN), 317
Rakeahua Granite, 148, 279
Rakaia Gorge (C), *30*, *211*; rhyolite, 211;
    River, Valley, 215, 259, *316*
Ranft, T., 85
Rangitata Orogeny, 113 *et seq*, 127, 166 *et*
    seq, 192; climax, 276; grain, 268, *269*;
    structural features from, 279-80, 295;
    volcanism, 209
Rangitikei (W), 261
Rangitoto Island (A), *58*, 200, 221, 338-9
Raoul Island, 401
Rapahoe (NW), *332*, 373
Raumukara district (EC), 169 *etc*
Recent, defined, 196
rectification, of coast, 331, *332*
recumbent fold, *282*, 395
red colouration, 21; of breccias, 166;
    weathering, 246
Red Crater, Tongariro volcano, 227
Red Rocks Point (W), *100*, 208
re-deposited beds, 96
Reed, J. J., 124, 129, 136
Reefton (SN), 81, 85 *etc*; gold-quartz veins,
    365; Formation, Group, 'Series', 79, 85 *et*
    seq
regolith, *24*
re-grading, of land profile, *301*
Reid, Gabriel, 363
Reid, H. F., 347
retreat, of glaciers, 241
retrograded coast, retrogradation, 331
rhyolite, 410 *etc*; lava, 132, 199, 200; with
    garnets, 211
rhyolitic volcanoes, forms, *222*
Richmond (N), 108; Range, 280
Rimutaka Range (W), 266, 306
ring-plain, *224*
river-plain, 314-15
Roaring Lion Formation, 76
rock-creep, 120; -crystal (*see* quartz 410);
    -disintegration, 26; -falls (seismic), *354*;
    -flowage, 120-1; geological concept of,
    24; nature of, 21; -stratigraphic
    classification, 106-7

Rock and Pillar Range (CO), 171
Ross Glaciation, 191, 195, *238*, 242, *243*, 252
Ross volcano (Auckland Island), 400
Rotoroa Diorite, 141; Gneiss, 126; Igneous Complex, 126, *129*, 157, 279
Rotorua-Taupo Volcanic Zone, 219, 226, 233, 283; heat outflow, 386
Rough Ridge (CO), 171
Ruahine Range, 52, 266, 282, *302*
Ruapehu volcano, 201, *225*, 226, *227*; glaciation, 259

Saddle Hill, Dunedin, 366
sag-ponds, 327
Saint Arnaud Range (SN), 266
sandhills, dunes, 334-5
Sandhills Creek Formation, 74
sandspits, bars, 331
sandstone, 22, 410 *etc*
Sara, W. A., 404
schist, 32, 410 *etc*
schistosity, 121, 410
Schofield, J. C., 99, 218
scoria, 58, 200; cones, *221*
scree, *28*, 260
sea-floor spreading hypothesis, 57, 274, *294*, 295
sea-mounts, 204
sea-level oscillations, Pleistocene, 49, 240, *263*
Searle, E. J., 221, 403
Seaview Formation, 80
'secular' (absolute) age of rock, 41
Seddon (M), *261*
sediment, origin of, 27; transport, 29
sedimentary basins, 280; ecology, 181; rocks, 22
sedimentology, 19
seiche, 229
seismicity, 31
seismic regions, 350, *351*
seismology, 344 *et seq*
Selwyn River (C), 112
Separation Point Granite, *54*, 140, 144, 162, 279
septaria, septarian concretions, 179
'sequential forms' (geomorph.), 299
Series, 76, 106-7
serpentine, 100, 410 *etc*
serpentinite, 101, 104, 410 *etc*
Seventeen Mile Bluff (NW), *28*
Shag Point (NO) early coal workings, 365
shale, 410 *etc*
Shelley, D., 82, 140, 152
Shirley, J., 84
Shoal Bay crater (A), 221
shorelines, types, 330 *et seq*; uplifted, *131*
Sibbalds Island moraine, 262
sigmoidal cleavage, *64*

silicic igneous rocks, 132
silica, industrial, sources, 173; terraces (*see* sinter, 'geyserite'), 204
sill, 72, *205, 207, 212 etc*
sillimanite, 121, 148, 410
siltstone, 410 *etc*
sinistral motion, on strike-slip fault, 291, *292*
sinter, 204
Skey, W. F., 360
Skippers-Skelmorlie Fault, 280
slate, 23, 410 *etc*
slump, 325
Smith, W., 358
Snares Island, 137, 400
soil, 24; -creep, 325
Solander Island, 219; Trough, 295
'solid flowage', 78, 121
solifluction, *146*, 260
Southland Plain, 284; Syncline, 105, 279
Speden, I. G., 98, 108, 169
Speight, R., 112, 216, 244
Spenser Range (SN), 266
sphalerite, 378, 410
spherulitic lava, 219
spilite, 104, 410
spit, gravel, *158*, 331
sponge spicules, 262
Spooner Range (N), 190; stream pattern, 302
spotted slate, 84, *119*
stack, *245*, *332*
stadial, 241
'stage' (geomorph.), 299; glacial, 241; stratigraphic, 106
Stephens Formation, 104
Stevens, G. R., 261, 291, 305, 403
Stewart Island, 137, 140, 148 *etc*
stibnite, 378, 410
Stillwater Creek (WN), 188
Stockton (WN) coal outcrop, *370*
stone polygons, rings, stripes, 322
Stony Creek, Reefton (SN), 85
strata (*sgl.* stratum), 19
stratification, stratified rocks, 19, 23
stratigraphy, 19
striae, 236
strike, *22*, 392, 394
structural divisions, of N.Z., 278 *et seq*; framework, 266 *et seq*; geology, 19; grain, 268; landforms, 268, *304*; plan of S.W. Pacific, 294; terraces, 323, *324*; trends, 266 *et seq*
structures, development of ideas of N.Z., 270 *et seq*; influence in landscape, 268, 301 *etc*; major volcanic -s, 268, 283; N.Z. Alps, 292 *et seq*; strike-slip faults, 285 *et seq* (*see* Alpine Fault)
subduction zone, 295
Suess, E., 270
Suggate, R. P., 85, 111, 113, 115, 127, 196, 244-5, 246, 253, 263, 343
sulphur, in coal beds, 176; fumarolic, volcanic, 233

summit-height accordance, 306, *312*
Sumner (C), 24
superglacial debris, *321*
superposed gorge, *138*, 313, *314*
syenite, 137, 144, 216, 410
syncline, *183*, 394
synform, 292
syntaxis, 271
System, 106-7

tachylite, 213
Tahunanui (N), 22
Taieri (SO), 377; Ridge, 266
Takaka (NWN), 366 *etc*; marble, 79
Takapau Plain (HB), 284
Takitimu Group, 105
talc (soapstone), 379, 410 *etc*
Tangihua Volcanic Formation, 210, 218, 340
Tank Gully (C), 112
Tapuwaeroa Formation, *167*
Tarakohe (NWN), *55*, 185
Taramakau-Hope Fault (Hope F.), 113, 287, 291, *310*
Taramakau River, Valley, 128, 305, *310*
Taranaki Basin, 281
Tararua Range (W), 52, 259, 266, 282, 306
Taratu Coal Measures, 176
Tarawera 'Chasm', Rift, 231, *232*; volcano, 200-1, 229 *et seq*
Tasman Bay, 144; Formation, 73; Geosyncline, 68 *etc*; Glacier, 265, 318; Igneous Complex, 145; Intrusive Group, 144, 279; Metamorphic Belt, 125; River, Valley, 262, 318; Sea, 48 *etc*
Tatas Beach (NWN), *54*, 141
Tauhara volcano (SA), *224*
Taupo Graben, 282; Lake, 327; pumice eruption, *224*, 225; Volcanic Zone (Rotorua-Taupo Volcanic Zone), 226 *etc*
Tauranga (BP), 218
Taylorville (NW), 262
Te Anau Group, 105; 'System', 71, 105
Te Anau-Manapouri glacial features, 255
Te Aute Limestone, *189*
tectonic axis of N.Z., 294; 'highs', 'lows', 186; phase, 32; relief 'ridges', 281; scarp, *310*; thickening, 123
'tectonics', tectonism, 32
Tekapo (C), 255
Te Kinga (NW), 140
Te Kuiti Group, *185*
Te Maari Crater, Tongariro volcano, 227
Te Maika (SWA), 217
Te Miko Point (NW), *131*
Ten Mile Bluff (NW), *170*
tephra, 199, 248 *etc*; eruptions, 225
tephrachronology, 248-9
Te Puke (SA), 218
Te Punga, M. T., 246, 261, 323, 356-7

Terangi Interglacial, Terangian Stage, 245, 250; coastal terraces, *258*
terraces, 323-5; coastal, *245*, 246, 248, *258*, 338-9; climatic, *151*, *325-6*; Pink and White, 204, 231; river, *261*, *303*; sinter, 204; structural, 323; valley-plain, *150*, 323, *324*
Te Waewae Bay (WS), 295
texture, of dissection, 302, *308*
thermal event, 126; metamorphism, 119, 140, 148
tholoid, 223, 228
Thomas Formation, *26*, *183*, 214
Thomson, J. A., 192-3, 195-7, 361
'thrust' fault, *283*
tidal rock platform, 22
till (glacial), 236, *238*, *255*, 318, *321*
tillite, 237
Timaru (SC), 219, 261
time-stratigraphic classification, 41, 107
'time-transgressive' strata, *176*
tinguaite, 209, 410
titanomagnetite, 376
Tokatoka volcanic neck (NA), *217*, 339
tonalite, 136
Tongariro volcano, Volcanic Group, 198, *225*, 226 *et seq*, 283
Torlesse Group, Supergroup, *47*, *93*, *95*, *111*, *112 et seq*, 363; Range (C), 112, 266
Totara Limestone, 213
trachyte, 410 *etc*; lava, 199, 201; tuff, 215
Tramway Sandstone, 104
transport, of weathering products, 26, 29
Trechmann, C. T., 90
trellised stream pattern, *304*
'Trias-Jura', 89, 103
tridymite, 21, 410
trough, of fold, 394
Tuapeka Group, 113
Tuatapere (WS), 372
tuff, 199, 204 *etc*; rings, 221
Tuhua Granite, 87, 137 *et seq*; Intrusive Group, 144; Orogeny, 86, 103, 106, 109, 127, 268, 275, 279
Tui Mine, Te Aroha (EA), *378*, 379
turbidite facies, 96
turbidity currents, 96
Turner, F. J., 121, 124, 126

Ulrich, G. F., 361
ultimate analysis, of coal, 368
ultramafic rocks, 100, 157
unconformity, 97, 109, *172*, 173, 177, 186, *189*; angular, 109, 186, 394
undermass, 164, 177, 303
Uniformitarian Principle, 299, 342
uplift rates, 342
upthrow, of fault, 392

valley-plain, 314; -train, *321*
'varve' silts, *257*
vein, 363, 410, *etc.*
vesicularity, of lava; vesicles, 200, *206*
Victoria Range (SN), *138, 162*
View Hill (C), 213; Fault, Mokihinui (WN), *283*
viscosity, of lava, 200
vitric tuff, 72
'volatiles', volatile components of magma, 160, 202
volcanic 'ash', 199; -erosional landforms, 340; history of N.Z., 198 *at seq*; igneous rocks, 133; neck, 205; plug, 205, 215; skeleton, 339; tuff, 199
volcano, defined, 198-9; as landform, 339; geosynclinal, 206, 209; in orogenic phase, 206
volcanogenic components, of sedimentary rocks, 66
volcanology, 19

Waiareka Volcanic Formation, 213
Waiau Basin (WS), 281; Syncline, 295; River, Valley (WS), 188
Waiho Loop Moraine, *262*; River, Valley (SW), *54, 372 etc*
Waikato Coal Measures, 176; Heads, 108; River, Valley (SA), 224, 329, 384
Waimakariri canyon, *47*; Glacier (Pleist.), 244, *250, 321*; Valley, 259 *etc*
Waimamaku (NA), 372
Waimangu geyser, 232
Waimate (SC), 303
Waimaunga Glaciation, Waimaungan Glacial Stage, 250, 253
Waimea [ice] Advance, Glaciation, 250, 253; Fault, 287
Waimean Glacial Stage, 250
Waingaro Schist, 74
Waiohine River (EW), 343
Waiotapu geothermal field, 387
Waipahi Group, 105
Waipaoa Valley (EC), 325
Waipapa Group, 218
Waipara River (NC), concretions, *55*, 179; terraces, *307*
Waipawa River (HB), *302*
Waipiata Basalt, 215
Waipipi ironsands, *377*
Waipoua basalt plateau (NA), 217
Waipuna Bay (NO), 215
Wairakau Andesite, 218
Wairakei geothermal power, 386-7, *389*
Wairarapa Faults (East, West), 291, 353
Wairau Fault, 113, 287, 291; Valley, offset terraces, 343
Wairoa Fault, Hunua Hills (SA), 303, 305; River (N), *47*, 110; (NA), *217*
Waitahu River (SN), 82, 85 *etc*

Waitakere Range (A), 340
Waitaki River, Valley (SC-NO), 255-6, 318, 384 *etc*
Waitangi Hill, Gisborne, early oil well, 371, River (NA), 328; (SW) *38*
Waitati Phonolite, 215
Waitemata Group, 218
Waitewhena (T), 189
Waiua Formation, *47*, 104
Waiuku (A), 376
Waiuta Group, 82
Wakaepa 'Series', 112
Wakamarama Range (NWN), 74, 266; Schist, 74
Wakapuaka (N), 94, 144
Wakatipu Metamorphic Belt, 125, 128
Walker Quartzite Formation, 106
Wanganui district, 189 *etc*; Series, 196, 252 *etc*
Wangapeka (NWN), 71, Formation, 78-9; River, Valley, 77, 86; Track, 74, 76
Wardle, P., 264
Warren, G., 108, *254*
waste mantle (*see* regolith)
Waterhouse, J. B., 65, 90, 103, 107, 124, 160
water supply, *382 et seq*; artesian, *382, 383*; Christchurch, 384
water-table, *28, 29*, 383
Watters, W. A., 16, 400
wave-cut platform, *245, 332*; uplifted, 338
weathering, 20 *et seq*
Webb Formation, 76
Weber Formation, *187*
Weka Pass Limestone, 182, *307*
Wellington Fault, 305, *343*; periglacial features, 323
Wellman, H. W., 80, 92-3, 153, 166, 177, 271, 273, 278, 287-8, 306, 343, 357
western belt, of Paleozoic strata, 68; of N.Z. Geosyncline rocks, 93
Whakarewarewa, Rotorua (SA), 204
Whakatane (BP), 218
Whangaehu Valley (W), 225
Whangai Formation, *181, 187*
Whangamarino swamp (SA), 305
Whangarei (NA), 219
Whangaroa (NA), 104, 218
Wharanui (M), 213
Whareama Stream (EW), *316*
Whatipu (A), 218
Whitcombe Valley (SW), *264*
Whitecliffs (C), 112, 211
White Creek, Oxford (NC), 213; Island (BP), 233, 357
White Creek Fault (SN), 303, 353, *354*
Whitianga Group, 219
Wilkies Bluff Limestone, 189
Wilkinson Glacier (SW), *264*
Willett, R. W., 271, 273, 287, 359
Williams, G. J., 362
Willis, I., 85
Wilmot Pass (WS), 148

Wilson River (ws), 81
Wollastonite, 79, 410
Wood, B. L., 80, 109, 287
Wooded Peak Limestone, 104
Woodstock [ice] advance, *249*

'Younger Rock Series', 179
'youth' (geomorph.), 299, *300*, 340

'zero' age, of landscape, 341
zinc ores, 379
zone, biostratigraphic, 193

xenolith, 136, 411